cyberarts

Cyberarts, International Compendium Prix Ars Electronica - .net, Interactive Art, Computer Animation, Computer Music - Edition 97 - **Publisher:** Dr. Hannes Leopoldseder, Dr. Christine Schöpf - **Editor:** Christian Schrenk, ORF, Landesstudio Oberösterreich, Europaplatz 3, 4010 Linz - **Translation:** Aileen Derieg, Helmut Einfalt - **Cover-Design, Layout:** Arthouse, Hansi Schorn - **Frontispiece:** Project Taos, "Web Hopper/sensorium" - **Coordination/Copy Editing:** Ingrid Fischer-Schreiber - **English Proof-Reading:** Aileen Derieg - **Offset Reproduction, Assembly:** Typeshop, Linz - **Copyright** 1997 by Österreichischer Rundfunk (ORF), Landesstudio Oberösterreich - **Photo Credits:** P. 7: Hansi Schorn, Linz; p. 17: Erich Wimmer, Fotostudio Stasny, Linz; p. 117/1,2: ORF, p. 131: Tatsuya Yusa; p. 135: Josephine Jasperse; p. 149/3: Markku Tuuli; p. 165/2: SA White; p. 175/2: Jean-Pierre Joly; p. 181/2: SA White; p. 201,.103/2: Friedemann Simon, Hamburg, p. 205: ORF;p. 207: ORF; p. 208: Véronique Hug; p. 212: Nola Lopez; p. 216: André Hoekzema;p. 218: Third Eye, Austin, TX; p. 219: Brett Robertson, Victoria University. The portrait and group photos on pages 11, 15, 21, 27, 33, 35, 43, 49, 55/56, 57, 59, 62, 63, 65, 67, 105/106, 107, 108/109, 110, 152/153, 155, 156, 157, 195, 196/197, 198, 199, 214-228 are by Erich Wimmer, Fotostudio Stasny, Linz.
 Printing: A. Holzhausens Nfg., A-1070 Wien. Printed on acid-free and chlorine-free bleached paper.
Prix Ars Electronica 97 - International Competition for Computer Arts - **Organizer:** Österreichischer Rundfunk (ORF), Landesstudio Oberösterreich - **Idea:** Dr. Hannes Leopoldseder - **Conception:** Dr. Christine Schöpf - **Financing/Copyright:** Dkfm. Augner - **Liaison Office:** Prix Ars Electronica, ORF, Europaplatz 3, A-4010 Linz, Telefon: 0043/732/6900-267, Telefax: 0043/732/6900-270, Telex: (02)1616, E-mail: info@prixars.orf.at. © 1997 Österreichischer Rundfunk (ORF), Landesstudio Oberösterreich. **SPIN:** 10631578. **ISBN** 3-211-82998-9 Springer-Verlag Wien New York

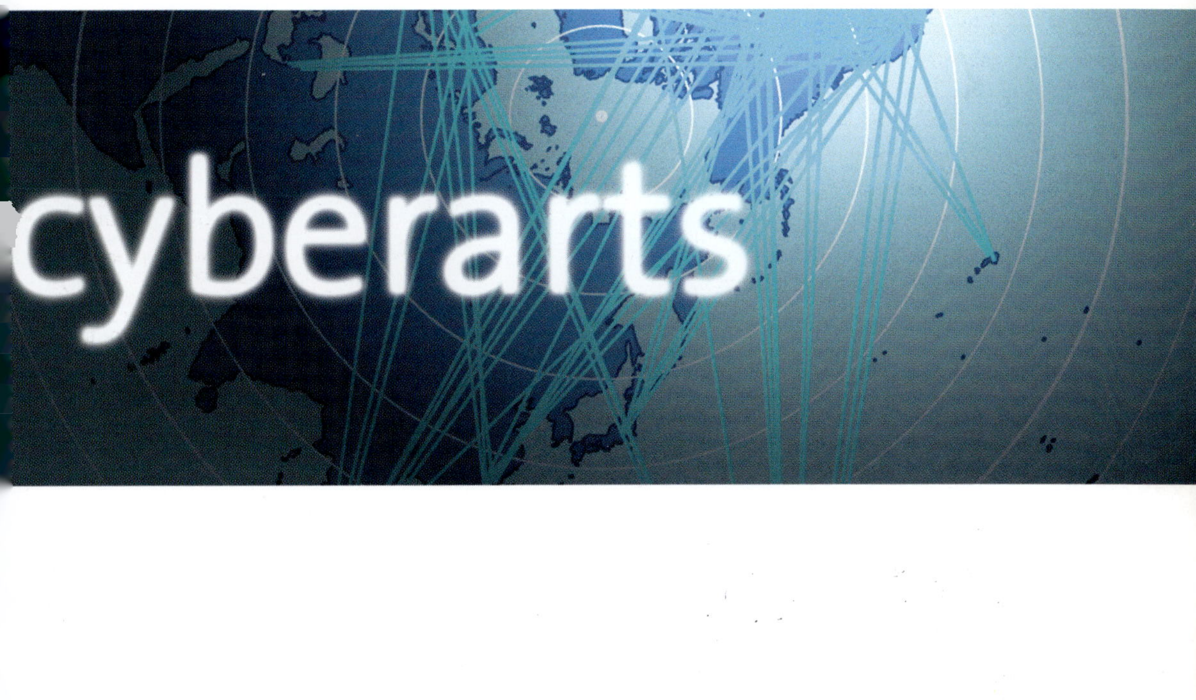

SpringerWienNewYork

Cyberarts

.net

Interactive Art

Computer Animation

Computer Music

CYBERART: ART OF THE FUTURE – THE FUTURE OF ART

HANNES LEOPOLDSEDER

"I want people to recognize the crisis as the center of modern art," stated Catherine David, director of documenta X in Kassel, formulating her intention for this world exhibition of contemporary art. "I am seeking to redefine the place of art."

Where is the new place of art? "Have the fine arts come to an end," as Paul Virilio said, "is there nothing left?"

The answers to this question of seeking a redefinition of art at the threshold of the millennium are many. One of the reasons for this is the way that a new communication society has set off in new directions. Communication has attained a new quality. New technologies, bandwidth, convergence, new services, mobility, networks – all of these are building blocks of the communication society.

Information technology and telecommunications are about to establish themselves as the largest industry in the world. Globalization and media convergence are among the characteristics of the upheaval. The language of the dawning 21st century is digital, the new universal language of the communication society.

After a ten-year period of development since 1987, the Prix Ars Electronica as a competition for digital media is on the verge of its second decade, on the verge of the

„Ich will, daß man als Zentrum der modernen Kunst die Krise erkennt", formuliert Catherine David als Ausstellungsmacherin der documenta X in Kassel ihre Absicht für diese Weltausstellung zeitgenössischer Kunst. Davids Zielsetzung: „Ich suche, den Ort der Kunst neu zu bestimmen".

Wo ist der neue Ort der Kunst? „Ist die bildende Kunst am Ende", wie Paul Virilio meint, „bleibt nichts von ihr übrig?"

Die Antworten auf diese Fragestellung mit der Suche nach einer Neubestimmung von Kunst an der Schwelle der Jahrtausendwende sind vielfältig. Eine der Ursachen liegt im Aufbruch einer neuen Kommunikationsgesellschaft. Die Kommunikation erhält eine neue Qualität. Neue Technologien, Breitbandigkeit, Konvergenz, neue Dienste, Mobilität, Vernetzung – all das sind Bausteine der Kommunikationsgesellschaft.

Informationstechnologie und Telekommunikation sind dabei, sich als größte Industrie der Welt zu etablieren. Globalisierung und Konvergenz der Medien zählen zu den bekannten Kennzeichen dieses Umbruchs. Die Sprache des anbrechenden 21. Jahrhunderts ist digital, die neue universelle Sprache der Kommunikationsgesellschaft.

Der Prix Ars Electronica als Wettbewerb für digitale Medien steht nach einer zehnjährigen Entwicklung seit 1987 an der Schwelle seiner zweiten Dekade. An der Schwelle zur Milleniumsdekade von 1997 bis zum Jahr 2006.

Setzte die Idee des Prix Ars Electronica 1987 auf den Computer als universelles Medium, allerdings aufgefächert in unterschiedliche Bereiche wie Computeranimation, Computergraphik, Computer-

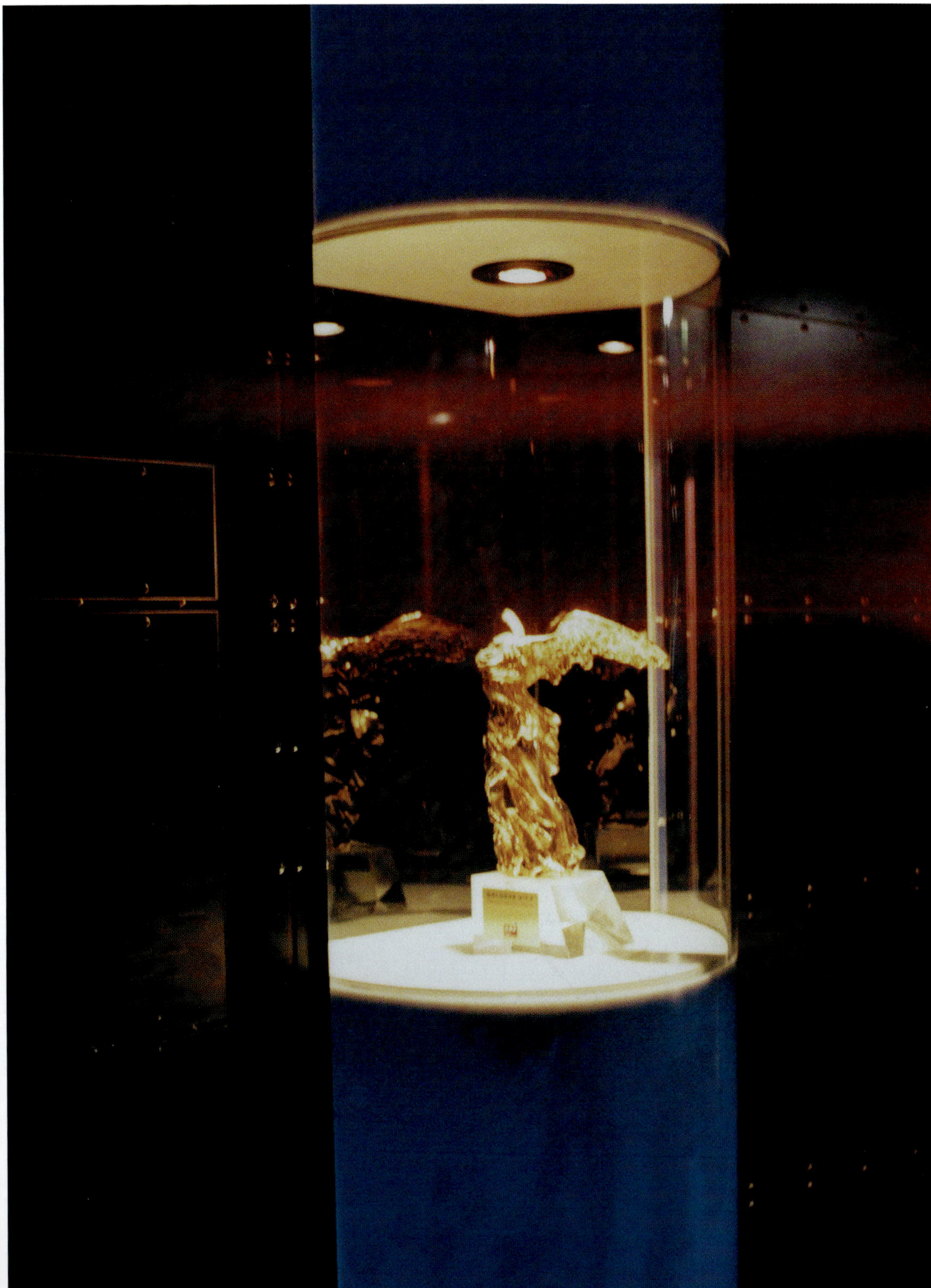

decade of the millennium from 1997 to the year 2006.

In 1987, the idea of the Prix Ars Electronica was based on the computer as a universal medium which was still distributed among different fields such as computer animation, computer graphics, computer music, and subsequently interactive art and .net. Now at the beginning of the second decade, however, we are seeing the beginning of the process of integration and the convergence of originally separate areas. If the crisis of art is addressed today, then it especially involves traditional video art, media art and computer art, because these disciplines in particular are directly affected by the development in technology.

Since 1987, the Prix Ars Electronica has consciously and purposely used the term "computer art", in order to tie into the historical tradition of this discipline. Its history and aesthetic has been elucidated particularly by Herbert W. Franke, one of the co-founders of the Ars Electronica, in numerous pioneering publications.

Walter Benjamin attempted a redefinition of art in 1936 in his revolutionary essay "The Work of Art in the Age of Mechanical Reproduction" in light of the emerging medium of film ("The shooting of a film, especially of a sound film, affords a spectacle unimaginable anywhere at any time before this"), yet the attempt of a similar redefinition today on the verge of the digital age presents itself in its full radicality.

The Prix Ars Electronica has always understood itself to be a laboratory and testing ground, in which the outlines of a new digital art may be identified.

The example of the Prix Ars Electronica 97 shows that digital and online are increasingly becoming the characteristic features of this development. Traditional media and computer art is on the way to becoming cyber art.

For this reason, the Prix Ars Electronica 97, now in the first year of the second

musik, in der Folge Interaktive Kunst und Netz, so steht am Beginn der zweiten Dekade der Prozeß der Integration und der Konvergenz ursprünglich getrennter Bereiche. Wenn heute die Krise der Kunst angesprochen wird, sind traditionelle Viedeokunst, Medienkunst und Computerkunst in besonderem Maße miteingeschlossen, denn gerade diese Disziplinen sind von der technologischen Entwicklung unmittelbar betroffen.

Seit 1987 hat sich der Prix Ars Electronica bewußt zum Begriff „Computerkunst" bekannt, um an die historische Tradition dieser Disziplin anzuknüpfen, deren Historie und Ästhetik vor allem von Herbert W. Franke, einem der Mitbegründer von Ars Electronica, in zahlreichen Publikationen pionierhaft dargestellt wurde.

Wenn Walter Benjamin 1936 in seinem bahnbrechenden Aufsatz „Das Kunstwerk im Zeitalter seiner technischen Reproduzierbarkeit" angesichts des aufkommenden Mediums Film („Eine Film- und besonders eine Tonfilmaufnahme bietet einen Anblick, wie er vorher nie und nirgends denkbar gewesen ist.") eine Neubestimmung der Kunst zu formulieren sucht, so stellt sich heute der Versuch einer ähnlichen Neudefinition an der Schwelle des digitalen Zeitalters in seiner ganzen Radikalität.

Der Prix Ars Electronica versteht sich seit jeher als Labor und Testbeet, in dem sich Umrisse einer neuen digitalen Kunst erkennen lassen.

Das Beispiel des Prix Ars Electronica 97 zeigt, daß digital und online mehr und mehr zu den charakteristischen Kennzeichen dieser Entwicklung werden. Traditionelle Medien- und Computerkunst ist auf dem Weg zur Cyberkunst.

Aus diesem Grund stellt der Prix Ars Electronica 1997 bewußt im ersten Jahr der zweiten Dekade dieses Wettbewerbes den Begriff Cyberkunst auf den Prüf-

decade of this competition, has purposely put the term cyber art to the test of the discourse. There is hardly another word in the nineties that has had a career comparable to that of the word "cyber". Even though it has also been seriously overused, it nevertheless stands a good chance of being transformed from a prefix to one of the key terms of the 21st century. Beginning with the Greek word "kybernan" ("steer") through Norbert Wiener's "Cybernetics" (1948) – which thus gave its name to the scientific field of cybernetics – to William Gibson's science fiction novel "Neuromancer" (1984), "cyber" has unremittingly developed into the universally used prefix and signal word for everything having to do with online and digital. Cyberlaws, cyber-architecture, cybermedicine, cybercash – all of these fields have one thing in common: the universal digital language, the new language of digital communication, the language of a new cyberculture.

"Where is the new place of art?" If we attempt to respond to Catherine David's question with at least a single answer: one new place of art will be cyberspace. The Prix Ars Electronica intends to provide a forum for forging into this new place of art.

If the Prix Ars Electronica has been able to fulfill its function as a laboratory in the field of digital media since 1987, then this should be especially true now for the second decade of development.

In the concurrent field of contemporary art and media festivals, Ars Electronica has become a relevant contact point, with its global grassroots network of over 11,000 works from participants in the Prix Ars Electronica from 60 countries around the world. Each year the achievements are judged by a qualified international jury, and the results are presented comprehensively at the Ars Electronica Festival – in the media and also on the Net, at exhibitions and events. Awards of Distinction, Honorary Mentions and Golden Nicas have been

stand des Diskurses. Kaum ein anderes Wort hat in den neunziger Jahren eine vergleichbare Karriere hinter sich wie das Wort „Cyber", das aber nicht nur eine Überstrapazierung hinter sich hat, sondern insbesondere auch die Chance hat, von einem Präfix zu einem der Schlüsselbegriffe des 21. Jahrhunderts zu werden. Ausgehend vom griechischen „kybernan" („steuern") über Norbert Wieners „Cybernetics" (1948) – der damit zum Namensgeber für die Wissenschaftsdisziplin der Kybernetik wurde – bis zu William Gibsons Science- Fiction-Roman „Neuromancer" (1984) entwickelt sich „Cyber" unaufhaltsam zum universell eingesetzten Vorwort und Signalwort für das gesamte Umfeld von online und digital. Ob Cyberrechte, Cyberarchitektur, Cybermedizin, Cybercash, Cyberpunk – alle Bereiche haben eines gemeinsam: die universelle digitale Sprache, die neue Sprache einer digitalen Kommunikation, die Sprache der neuen Cyberkultur.

„Wo ist der neue Ort der Kunst?" Wenn wir auf Catherine Davids Frage zumindest eine einzige Antwort versuchen: Ein neuer Ort der Kunst wird der Cyberspace sein. Für das Vordringen in diesen neuen Ort der Kunst will der Prix Ars Electronica ein Forum anbieten.

Wenn der Prix Ars Electronica von 1987 an seine Laborfunktion im digitalen Medienbereich erfüllt hat, soll dies in besonderer Weise nun für die zweite Dekade der Entwicklung gelten.

Im konkurrenten Umfeld von zeitgenössischen Kunst- und Medienfestivals ist Ars Electronica durch das globale Graswurzelnetz von über 11.000 Arbeiten der Teilnehmerinnen und Teilnehmer am Prix Ars Electronica aus weltweit 60 Nationen zu einem relevanten Ansprechpartner geworden. Die jeweiligen Leistungen werden von einer qualifizierten internationalen Jury beurteilt und die Ergebnisse innerhalb des Festivals Ars Electronica umfassend präsentiert – in den Medien, auch im Netz, bei Ausstellungen und Veranstaltungen. An 500 Künstler und Wissenschafter wurden Auszeichnungen, Anerkennungen und Goldene Nicas verliehen, an 120 KünstlerInnen wurden Geldpreise vergeben. Das bedeutet, daß sich innerhalb eines Jahrzehnts eine Community von

awarded to about 500 artists and scientists, roughly 120 artists have received money prizes. That means that within the space of one decade, a community of over 600 digital media artists who have been distinguished in Linz has been gathered. They form the spearhead of the digital transformation and belong to the "digerati", who, to use a term first coined by the "Times" reporter John Markoff in 1992, belong to the driving creative and intellectual personalities of a networked world.

The Golden Nica of the Prix Ars Electronica intends to continue and enhance the signal function that this "media art Oscar" has attained during its first decade. The boundary lines between the individual fields, but also between art and commerce, are changing and flowing more and more into a "digital river", to quote Richard Dawkins.

The Prix Ars Electronica has become the heart of the Ars Electronica Festival, which has been organized since 1979 by the city of Linz (by Brucknerhaus from 1979 to 1995 and by the Ars Electronica Center since 1996) and by the Austrian Broadcasting Corporation. The Prix Ars Electronica is organized independently by the Austrian Broadcasting Corporation and forms an integrative part of the Ars Electronica Festival.

The funding of the entire Prix Ars Electronica project is achieved in cooperation with the Province of Upper Austria and the city of Linz, but also with leading businesses such as Siemens Nixdorf, VOEST-ALPINE STAHL, the Gerhard Andlinger Company, as well as a number of other companies and institutions.

Since the institution of the Prix Ars Electronica, a total of 13 million dollars in prize money has gone out to computer and media artists.

Where is the new place of art? The next decade of the Prix Ars Electronica, the millennium decade, will attempt at least one answer. Media and computer art is

über 600 digitalen Medienkünstlern als Ausgezeichnete in Linz zusammengefunden haben, die die Speerspitze des digitalen Wandels bilden, die zu jenen „digerati" zählen, die, um den erstmals von „Times"-Reporter John Markoff 1992 verwendeten Begriff zu zitieren, zu den treibenden kreativen und intellektuellen Persönlichkeiten einer vernetzten Welt zählen.

Die Goldene Nica des Prix Ars Electronica will die Signalfunktion, die dieser „Oscar der Medienkunst" innerhalb des ersten Jahrzehnts erhalten hat, weiterführen und ausbauen. Die Grenzen zwischen den einzelnen Bereichen, aber auch zwischen Kunst und Kommerz verändern sich und münden mehr und mehr in einen „digitalen Fluß", um mit Richard Dawkins zu sprechen.

Der Prix Ars Electronica hat sich zu einem Herzstück des Festival Ars Electronica entwickelt, das von 1979 an von der Stadt Linz (vom Brucknerhaus von 1979 bis 1995 und ab 1996 vom Ars Electronica Center) und dem Österreichischen Rundfunk veranstaltet wird. Der Prix Ars Electronica wird eigenständig vom Österreichischen Rundfunk organisiert und bildet einen integrativen Teil des Festival Ars Electronica.

Die Finanzierung des gesamten Projektes Prix Ars Electronica erfolgt in Zusammenarbeit mit dem Land Oberösterreich, der Stadt Linz, aber auch mit renommierten Wirtschaftsunternehmen wie Siemens Nixdorf, VOEST-ALPINE STAHL, der Gerhard Andlinger Company sowie einer Reihe weiterer Firmen und Institutionen.

Seit Bestehen des Prix Ars Electronica sind insgesamt 13 Mio. Dollar Preisgeld an Computer- und Medienkünstler gegangen.

Wo ist der neue Ort der Kunst? Das nächste Jahrzehnt des Prix Ars Electronica, das Millenniumsjahrzehnt, wird zumindest eine Antwort darauf versu-

being transformed into cyber art. At the turn of the millennium, computer art, media art and cyber art will most likely be considered „mainstream" in contemporary art.

Nevertheless, the decade that lies before us is still very much a terra incognita. Although the landscape is coming into view, the contours are not yet visible. „The Digital Revolution is whipping through our lives like a Bengali typhoon," wrote Louis Rossetto in the first edition of WIRED. If we continue this comparison, the typhoon will not only whip through the fields of economics, but also the constructs of ideas and art, yet at the same time, it will also be clearing the ground for a new cultural era. The participants of the Prix Ars Electronica will take part in shaping this process, not as onlookers but as agents.

chen. Medien- und Computerkunst wandeln sich zur Cyberkunst, an der Jahrtausendwende zählen voraussichtlich Computer-, Medien- und Cyberkunst zum „Mainstream" im zeitgenössischen Kunstgeschehen.

Trotz allem: Das vor uns liegende Jahrzehnt ist nach wie vor eine terra incognita, deren Landschaft zwar sichtbar wird, aber noch nicht die einzelnen Konturen. „The Digital Revolution is whipping through our lives like a Bengali typhoon", schrieb Louis Rossetto in der ersten Ausgabe von WIRED.

Wenn wir diesen Vergleich gelten lassen, wird der Taifun nicht nur Wirtschaftsbereiche, sondern auch Gedanken- und Kunstkonstrukte einreißen, gleichzeitig aber den Boden für eine neue kulturelle Ära freimachen. Die Teilnehmerinnen und Teilnehmer des Prix Ars Electronica werden nicht als Zuschauer, sondern als Akteure diesen Prozeß mitgestalten.

HANNES LEOPOLDSEDER

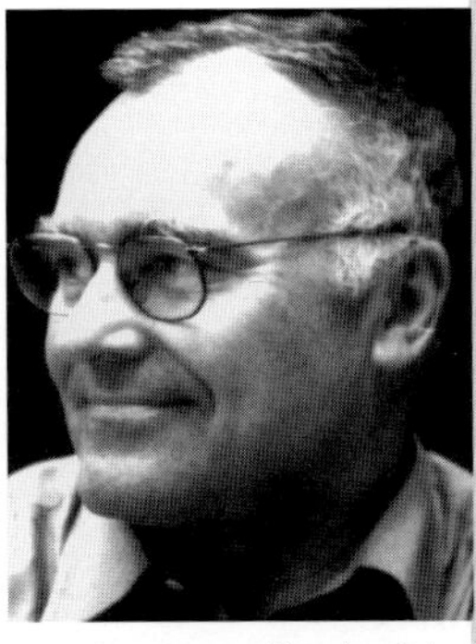

Hannes Leopoldseder (A), Managing Director of the Upper Austrian regional studio of the Austrian Broadcasting Corporation. Co-founder of the Ars Electronica Festival, and the Ars Electronica Center.

Hannes Leopoldseder (A) ist Landesintendant des ORF, Landesstudio Oberösterreich, Mitbegründer der Ars Electronica und des Ars Electronica Center.

Prix Ars Electronica 97

CHRISTINE SCHÖPF

At the transition from the industrial age to the digital era, a multitude of unanswered questions arise concerning all areas of life – our communal, social system and global economic constellations, as well as education, communication and cultural positions.

Since 1987 the Prix Ars Electronica as a competition for computer animation, music, interactive art and .net has become an international node in this discourse. It is a research and experimental laboratory for art as a decisive factor in the shaping of the digital information society.

Within the field of contemporary art, the Prix Ars Electronica has become a focal point for cyber artists – those creative beings that are leaving the bonds of tradition behind and crossing over the boundaries between disciplines.

The mass medium of the Internet opens up new forms of artistic creativity to the cyber artists; they are the ones, in the midst of research and development dominated by industry, who are setting out in different directions with regards to programs and interfaces, shaping interactivity as the key cultural technology in complex systems. They are the creators of new worlds of image and sound in art, science, research and entertainment, dissolving the traditional boundaries in making imagination and

Am Übergang vom Industriezeitalter zur digitalen Ära bricht eine Vielzahl unbeantworteter Fragen auf, die alle Lebensbereiche betreffen – unser gesellschaftlich-soziales System und globale wirtschaftliche Konstellationen ebenso wie Bildung, Kommunikation und kulturelle Positionen.

Der Prix Ars Electronica ist als Wettbewerb für Computeranimation, Musik, Interaktive Kunst und .net in diesem Diskurs seit 1987 zu einem internationalen Knotenpunkt geworden. Er ist ein Forschungs- und Versuchslabor für Kunst als entscheidendem Gestaltungsfaktor in der digitalen Informationsgesellschaft.

Innerhalb des zeitgenössischen Kunstgeschehens ist der Prix Ars Electronica zum Fokus für Cyberkünstler geworden – jener Kreateure, die die Fesseln der Traditionen verlassen und die Grenzen zwischen den Disziplinen überschreiten.

Cyberartisten eröffnen das Massenmedium Internet neuen Formen künstlerisch-kreativer Gestaltung, sie sind jene, die in einer industriedominierten Forschung und Entwicklung alternative Wege in bezug auf Programme und Interfaces beschreiten und Interaktivität als Schlüsselkulturtechnik in komplexen Systemen gestalten, sie sind die Kreateure neuer Bild- und Klangwelten in Kunst, Wissenschaft, Forschung und Unterhaltung, die die hergebrachten Grenzen der Be-Greifbarmachung von Imagination und Wissen auflösen.

Mit 877 Einreichungen aus 34 Ländern weltweit ist der vom Österreichischen Rundfunk, Landesstudio Oberösterreich, zum elften Mal veranstaltete Prix Ars Electronica 97 in seinen Ergebnissen einmal mehr Seismograph für den aktuellen Stand digitaler Mediengestaltung im Aufbruch zur Cyberkunst.

Vier internationale Fachjuries haben

knowledge comprehensible.

With 877 entries from 34 countries around the world, the results of the Prix Ars Electronica 97, which is organized for the eleventh time by the Upper Austrian Regional Studio of the Austrian Broadcasting Corporation, once again act as a seismograph for the current state of digital media design on the way to cyber art.

Now in 1997, four juries of international experts have awarded a total of eleven money prizes (4 Golden Nicas and 7 Awards of Distinction) and 47 Honorary Mentions. With total prize money amounting to ATS 1.25 million (about US $ 105,000), the Prix Ars Electronica is the best endowed competition in the field of digital media.

In 1997 the four main prizes, the Golden Nicas, are equally divided between the USA and Japan; the seven Awards of Distinction go to the USA, Great Britain and Germany.

The winners of the Prix Ars Electronica 97

The Golden Nica, with ATS 100.000 (US $ 8,300), in the competition category .net goes to Project Taos, an interdisciplinary group of media designers and scientists from Japan, for "sensorium".

"sensorium" is a Web site enabling users to experience the world as a living organism. Data on communication on the Internet, earthquake activity around the world, netsound and more is constantly updated and turns cyberspace into a "breathing" sensorium. The two Awards of Distinction (ATS 50.000 or US $ 4,170 each) go to Rolf Schmidt (USA) for "MMF – Make Money Fast" and to a British team working with Rycharde Hawkes for "TechnoSphere". "Make Money Fast" is a humorous attack on the irritating chain letters and pyramid games circulating on the Internet. Anyone engaging in such annoying activities is publicly exposed. "TechnoSphere" is the home of virtual pets and a place for communication among their owners.

1997 insgesamt elf Geldpreise (4 Goldene Nicas und 7 Auszeichnungen) vergeben und 47 Anerkennungen ausgesprochen. Mit einem Preisgeld von ATS 1,25 Mio ist der Prix Ars Electronica der höchstdotierte Wettbewerb im digitalen Medienbereich.

1997 gehen die 4 Hauptpreise, die Goldenen Nicas, zu gleichen Teilen in die USA und nach Japan, die 7 Auszeichnungen in die USA, nach Großbritannien und nach Deutschland.

Die Gewinner des Prix Ars Electronica 97

Die mit öS 100.000 dotierte Goldene Nica in der Wettbewerbssparte .net erhält Project Taos, eine interdisziplinäre Gruppe von Mediendesignern und Wissenschaftern aus Japan, für „sensorium".

„sensorium" ist eine Web-Site, die den Benutzer die Welt als lebenden Organismus erfahren läßt. Daten zur Kommunikation im Internet, weltweite Erdbebenaktivitäten, Netsound u. a. werden permanent aktualisiert und lassen den Cyberspace zum „atmenden" Sensorium werden. Die beiden Auszeichnungen (je öS 50.000) erhalten Rolf Schmidt (USA) für „MMF – Make Money Fast" und ein britisches Team um Rycharde Hawkes für „TechnoSphere".

„Make Money Fast" nimmt die Unsitte der Kettenbriefe und Pyramidenspiele im Internet aufs Korn. Wer solche anzettelt, wird an den Pranger gestellt. „TechnoSphere" ist die Heimat virtueller Haustiere und Kommunikationsort für deren Besitzer.

Die mit öS 150.000 dotierte Goldene Nica für Computermusik erhält der Amerikaner Matt Heckert für „Munich Samba". „Munich Samba" ist Maschinenmusik. Eine gigantische Maschinerie von Metallobjekten ist das Instrumentarium, auf und mit dem der Computer zum virtuosen Instrumentalisten wird.

Die beiden mit je öS 50.000 dotierten Auszeichnungen erhalten die Amerikanerin Maryanne Amacher für „The Levi-Montalcini Variations" und der Brite Jonty Harrison für „Unsound Objects".

Die „Levi-Montalcini Variations" sind in den Raum projizierte, dreidimensionale Klangflächen,

The Golden Nica for computer music, with ATS 150.000 (US $ 12,500), goes to the American Matt Heckert for "Munich Samba". "Munich Samba" is machine music. Gigantic machinery made of metal objects is the instrumentarium on and with which the computer becomes a virtuoso instrumentalist.

The two Awards of Distinction, with ATS 50.000 (US $ 4,170) each, go to Maryanne Amacher from the USA for "The Levi-Montalcini Variations" and Jonty Harrison from Great Britain for "Unsound Objects". "The Levi-Montalcini Variations" are three-dimensional sound surfaces projected in space, which turn listening into a body experience – a work with which the experimentalist Amacher has once again opened up new dimensions in sound experience. "Unsound Objects" is cinema for the mind. Harrison has sampled everyday noises and combined them with wit and craftsmanship to form a fascinating, associative journey of discovery through his sonic world.

The Golden Nica for interactive art, with ATS 200.000 (US $ 17,000), goes to Toshio Iwai and Ryuichi Sakamoto of Japan for "Music plays Images x Images play Music". This is an interactive multimedia performance with two computer grand pianos, which are played by Ryuichi Sakamoto and Internet users and controll Iwai's graphic installation in real time.

The two Awards of Distinction, with ATS 50.000 (US $ 4,170) each, go to the American Paul Garrin and the Canadian David Rokeby for "Border Patrol" and the duo Joachim Sauter / Dirk Lüsebrink of ART+COM in Berlin for "The Invisible Shape of Things Past". "Border Patrol" is an installation, in which visual sensors control robot cameras. When a visitor steps into the center of the cross hairs of a target sight, a "shot" is fired. "The Invisible Shape

die das Hören zur Körpererfahrung werden lassen – ein Werk, mit dem die Experimentatorin Amacher einmal mehr neue Dimensionen der Klangerfahrung öffnet. „Unsound Objects" ist Kino für den Kopf. Harrison sampelt dafür Alltagsgeräusche und fügt sie mit Witz und Kunstfertigkeit zu einer spannenden, assoziativen Entdekkungsreise durch seine Klangwelt zusammen.

Die mit öS 200.000 dotierte Goldene Nica für Interaktive Kunst erhalten Toshio Iwai und Ryuichi Sakamoto aus Japan für „Music plays Images x Images play Music", eine interaktive Multimedia-Performance mit zwei Computerflügeln, gespielt von Ryuichi Sakamoto und Internet-Benützern, über die Iwais Graphik-Installation in Echtzeit gesteuert wird.

Die beiden Auszeichnungen zu je öS 50.000 erhalten der Amerikaner Paul Garrin und der Kanadier David Rokeby für „Border Patrol" und das Duo Joachim Sauter/Dirk Lüsebrink von ART+COM aus Berlin für „The Invisible Shape of Things Past".

„Border Patrol" ist eine Installation, in der visuelle Sensoren Roboterkameras steuern. Gerät der Besucher ins Zentrum des Fadenkreuzes, fällt ein Schuß. „The Invisible Shape of Things Past" ist ein weiterer Schritt im „Cyber City Projekt" von ART+COM. In ein virtuelles 3D-Stadtmode[ll] werden historische Filme als Objekte integriert und ermöglichen dem Benutzer so die Erforschung eines virtuellen Raumes auch in der zeitlichen Dimension.

Die mit öS 300.000 dotierte Goldene Nica für Computeranimation erhält Scott Squires, ILM, USA, für die Special Effects [in] „Dragonheart", eine mit öS 100.000 dotierte Auszeichnung geht an Chris Wedge für die Computeranimation „Funky Towel" aus „Joe's Apartment".

Aus den insgesamt sieben Einreichun

of Things Past" is a further step in the "Cyber City Project" by ART+COM. Historical films are integrated as objects into a virtual 3D model city, enabling the visitor to explore virtual space in a time dimension as well.

The Golden Nica for computer animation, with ATS 300.000 (US $ 25,000), goes to Scott Squires, ILM, USA for the special effects in "Dragonheart". One Award of Distinction with ATS 100.000 (US $ 8,300) goes to Chris Wedge for the computer animation "Funky Towel" from "Joe's Apartment".

Of the seven entries from Industrial Light & Magic ("Mission Impossible", "101 Dalmatians", "Jumanji", "Star Trek", "Mars Attacks!", "Twister" and "Dragonheart"), the jury selected "Dragonheart" as an outstanding example of character animation, modeling, rendering, motion capture, etc. All of this makes Draco, the last of his kind, a believable leading actor in the story.

"Funky Towel", a sequence from "Joe's Apartment", is an excellent example in the tradition of character animation, with which Chris Wedge has successfully created a perfect, humorous genre satire of the film musicals of the 50's (Esther Williams films).

Sponsors of the Prix Ars Electronica 97

The realization of the Prix Ars Electronica is only made possible through the financial support from sponsors and benefactors.

The Prix Ars Electronica 97 was made possible through the support of Siemens Nixdorf, VOEST-ALPINE STAHL, the city of Linz, the province of Upper Austria and the Gerhard Andlinger Company. For additional support, the Prix Ars Electronica is also grateful to Lufthansa, the Ramada Hotel Linz, Mailfast, the Casinos Austria AG, Silicon Graphics Austria, the EDP-Center of the Linzer Johannes Kepler University.

CHRISTINE SCHÖPF

gen von Industrial Light & Magic („Mission Impossible", „101 Dalmatiner", „Jumanji", „Star Trek", „Mars Attacks!", „Twister" und „Dragonheart") wählte die Jury „Dragonheart" als ein herausragendes Beispiel für Character Animation, Modelling, Rendering, Motion Capture etc. All dies macht Draco, diesen Letzten seiner Gattung, zum glaubhaften Hauptdarsteller der Story.

„Funky Towel", eine Sequenz aus „Joe's Apartment", ist ein herausragendes Beispiel in der Tradition der Character Animation, mittels derer Chris Wedge eine perfekte, humorvolle Genrepersiflage des Filmmusicals der 50er (der Filme Esther Williams) gelungen ist.

Sponsoren des Prix Ars Electronica 97

Die Durchführung des Prix Ars Electronica kann nur durch das finanzielle Engagement von Sponsoren und Förderern realisiert werden.

Der Prix Ars Electronica 97 wird von Siemens Nixdorf gestiftet und von der VOEST-ALPINE STAHL, der Stadt Linz, dem Land Oberösterreich und der Gerhard Andlinger Company gefördert. Für weitere Unterstützung dankt der Prix Ars Electronica Lufthansa, dem Ramada Hotel Linz, Mailfast, der Casinos Austria AG, Silicon Graphics Österreich und dem EDV-Zentrum der Linzer Johannes Kepler Universität.

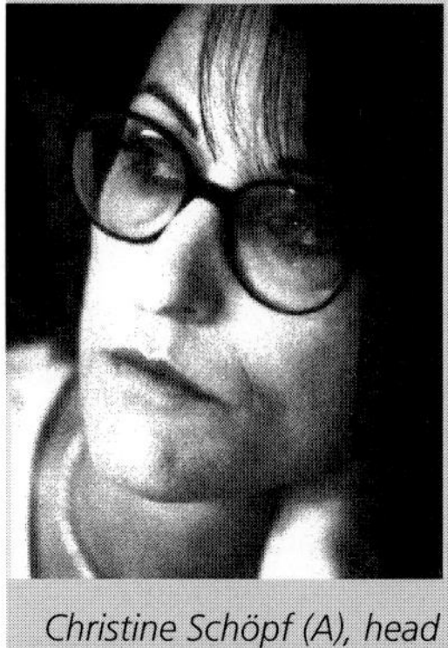

Christine Schöpf (A), head of the department arts and science at the Upper Austrian regional ORF studio, emphasis Ars Electronica.

Christine Schöpf (A), Leiterin der Abteilung Kultur/Wissenschaft im ORF-Landesstudio Oberösterreich, Schwerpunkt Ars Electronica.

The Aesthetics of the Internet – Context as a Medium

Die Ästhetik des Internet – Kontext als Medium

Joichi Ito

The Internet connects computers, people, sensors, vehicles, telephones, and just about anything together in a global network which is fast and cheap.

This interconnectedness is the context. Context represents the way and the timing in which nodes are connected together. If content were the noun part of information, then context would be the verb part. New forms of media and communications tend to mimic their predecessors. Carl Malamud gives the example of early television where television shows often consisted of a radio announcer and a microphone on the screen. The Internet has often been called a method of online publishing or online broadcasting. Magazine publishers tell me that Internet advertising on a computer screen doesn't compare to an excellent full page ad in a magazine. Television producers often compare gritty Internet video to the power of an excellent television commercial. The Internet as a medium is not suited

Das Internet verbindet Computer, Menschen, Sensoren, Fahrzeuge, Telephone und so ziemlich alles andere in einem globalen Netzwerk, das schnell und billig ist.

Dieses Verknüpft-Sein ist der Kontext. Der Kontext repräsentiert die Art und Weise, das Timing, in dem Knoten miteinander verbunden werden. Wenn der Inhal sozusagen den substantivischen Anteil der Information beschreibt, dann spielt der Kontext die Rolle des Verbs.

Neue Medien- und Kommunikationsformen neigen dazu, zunächst einmal ihre Vorgänger zu imitieren. Carl Malamud hat als Beispiel die frühen Fernsehshows erwähnt, in denen nichts anderes zu sehen war als ein Radiosprecher samt seinem Mikrophon – aber eben auf dem Bildschirm. Auch das Internet wurde oft als ei Mittel zum Online-Publizieren oder -Senden bezeichnet. Zeitschriftenverleger berichten mir, daß eine Computerbildschirm-Anzeige im Internet nicht mit einer gut gestalteten Anzeigenseite in einer Zeitschrift konkurrieren kann, und die Fernse produzenten vergleichen oft das pixelkörnige Internet-Videobild mit der polierten Oberfläche eines exzellenten TV-Spots. Aber das Internet als Medium ist nicht geeignet dafür, große Mengen derselben Information an viele Leute gleichzeitig zu verteilen. Derzeit verbindet das Internet alle Benutzer miteinander – bei begrenzte Bandbreite und niedrigen Kosten. Das Internet liefert Kontext, und daraus sollten wir die Zukunft des Internet formen.

Die Halbwertszeit von Information wird heutzutage immer kürzer: fünfzehn Minuten alte Aktienkurse der Börse gibt e umsonst, während die aktuellen Daten

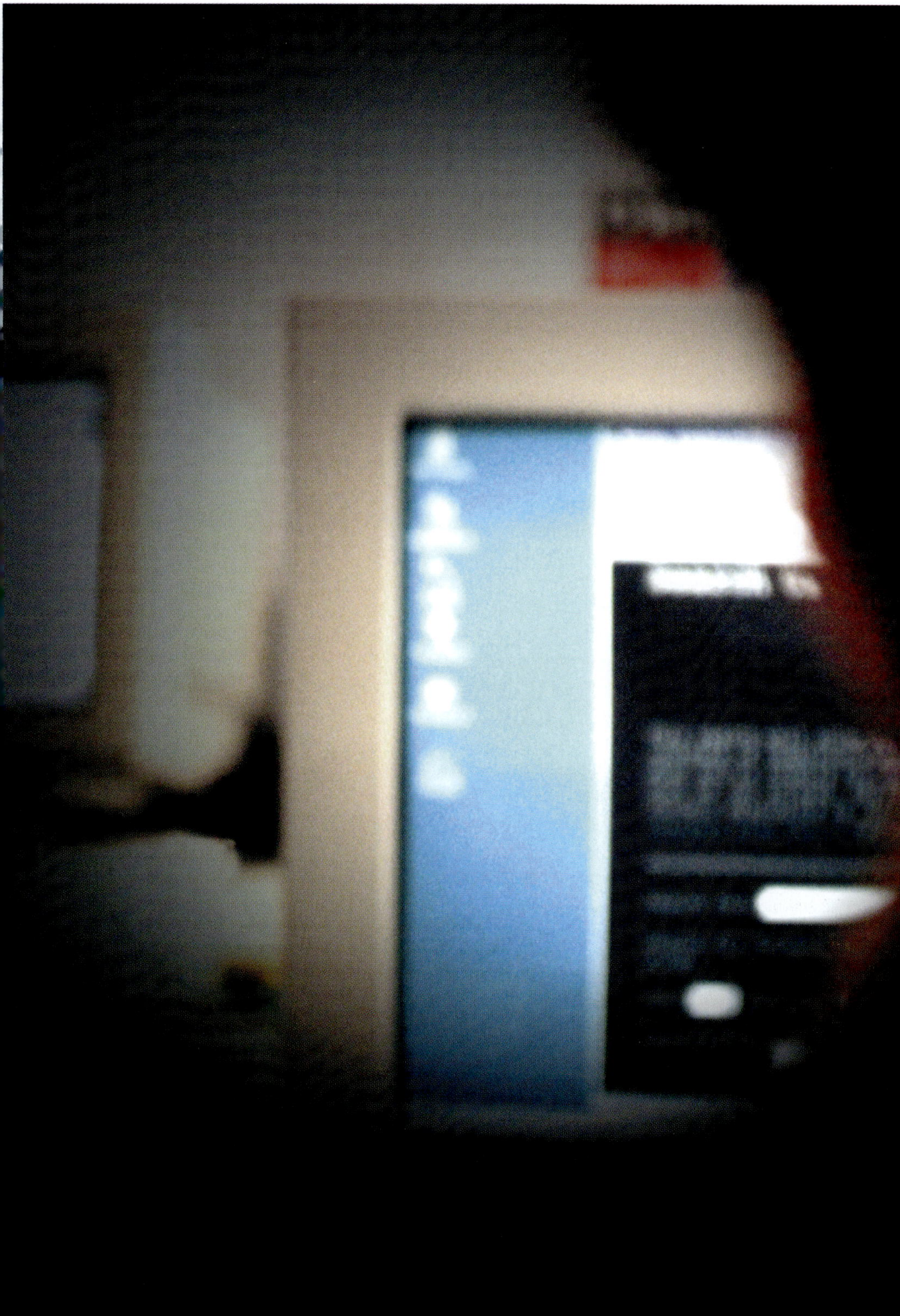

for the delivery of high volumes of the same information to many people. Currently, the Internet connects everyone together at rather low bandwidth at low cost. The Internet delivers context, and it is of this that we should be building the future of the Internet.

Much of the information in today's world and on the Internet expires very quickly. Fifteen minute old stock quotes become free, instant stock quotes cost money. Yesterday's newspapers are free on the Internet, but today's (or tomorrow's) can cost you money. It is a relationship with the newspaper and its reporters that is more important than the database of old articles. Your Netscape browser will expire in weeks. Stealing an old Netscape diskette at a computer shop makes very little sense. Rather than downloading lots of software, on the Internet people remember where to find what they need, or better yet, whom to ask or where to search. It is information about information about information ... Just as our monetary system has become very abstract, our currencies represent something that really has no physical reality, most information on the Internet is about context, rather than content. Instead of the hard data of yesteryear that could be bound in a book, stacked in a warehouse and distributed by trucks, the information on the Internet is about being connected LIVE and about being in the right place at the right time.

Communities on the Net consist of a group of people connected to each other in the form of discussions, games, or some other form of two-way connectedness. People invest time and energy into these communities, and these communities evolve into a complex aggregate of relationships between people mediated by a technology and a context. It becomes a kind of place. These communities are influenced by the underlying technology, but grow far beyond the technology itself. The technology is a kind of genetic basis on which a new

Geld kosten. Die gestrigen Zeitungen sind im Internet gratis zu haben, aber die heuti gen (oder morgigen) meist nur gegen Gebühr. Und die Beziehung zur Zeitung und ihren Redakteuren ist viel wichtiger al: die Datenbank alter Artikel. Ihr Netscape-Browser wird in ein paar Wochen veraltet sein, und eine alte Netscape-Diskette im Computershop zu klauen, macht nur wenig Sinn. Anstatt Berge von Software herunterzuladen, merken sich die Leute ir Internet, wo sie finden können, was sie brauchen, oder noch besser: Sie wissen, wen sie fragen müssen oder wo sie suche sollen. Es geht um Information über Information über Information ... So wie unser Währungssystem immer abstrakter geworden ist und unser Geld etwas darstellt, was eigentlich keine physische Realität mehr besitzt, so handelt auch der Großteil der Information im Internet vom Kontext, nicht vom Inhalt. Statt harter Fakten von gestern oder vorgestern, die auch in ein Buch gebunden und per Lastwagen verteilt werden könnten, geht es bei Information im Internet darum, live verbunden und zur richtigen Zeit am richt gen Ort zu sein.

Die Gemeinschaften im Netz bestehe aus Leuten, die miteinander verbunden sind, sei es in Diskussionen, durch Spiele oder beliebige andere Formen von Zweiweg-Kommunikation. Menschen inve stieren Zeit und Energie in diese Gemeinschaften, und diese Gemeinschaften entwickeln sich zu einem komplexen Aggregat von Beziehungen zwischen Menscher in dem Technologie und Kontext die Vermittler sind. Das Internet wird zu einer Ar „Ort", die Gemeinschaften werden zwar durch die ihnen zugrundeliegende Techno logie beeinflußt, aber sie wachsen weit über sie hinaus. Die Technologie ist so etwas wie die genetische Basis, auf der ne Organismen wachsen können, und erhält Input aus ihrer Umgebung durch die Mitwirkenden.

Auch Kunstwerke, Schriften und

organism can grow, receiving input from its environment through its participants.

Artwork, writing and other forms of content which are often nearly static in the slow moving physical world can also become living things in the fluid, high-speed context of the Internet. An interesting idea or design can quickly become a popular item to be sampled, edited and redistributed. The artist can watch their work, or their child, quickly develop into something quite different from what it was originally intended to be. The original artist is the parent, but unlike a child raised in complete isolation, work on the Internet is educated and formed, for better or for worse, into a product of its environment and society. Putting work on the Internet is more like giving birth than creating a static object.

Communities, multi-user games systems, markets, search engines and router configurations are all context oriented. The aesthetic of context is the design of such context-oriented systems which are outstanding in their nature. A good context-oriented system causes the network of living connections to converge, interact and grow. It adds value to the network and attracts users and connections.

The Internet is a self-organizing adaptive system. As John Casti from the Santa Fe Institute pointed out in his talk at the Ars Electronica Memesis symposium last year, one can understand completely the process in which a complex adaptive system works, but it is impossible to predict what it does. The Internet self-organizes itself in the very interesting area between total chaos and order. Eric Hughes has called it a working anarchy. When order is forced onto the Internet such as rigid protocols or singular ubiquitous operating systems, that layer becomes very brittle and as one learns in catastrophe theory, a shock to the system can cause a huge amount of damage. One virus or bug in the system could take the whole system down. The more inefficient

andere Formen von Inhalt, die in der sich langsam bewegenden physikalischen Welt häufig fast statisch sind, werden im schnellen, liquiden Kontext des Internet zu lebendigen Organismen. Eine interessante Idee oder ein Entwurf kann binnen kurzer Zeit zu einem populären Element werden, das gesampelt, editiert und neu verteilt wird. Die Künstler können ihre Arbeit – oder ihr Kind – sich in etwas völlig anderes entwickeln sehen, als sie eigentlich geplant hatten. Der ursprüngliche Künstler ist ein Elternteil, aber anders als ein in völliger Isolation erzogenes Kind wird das Werk im Internet zu einem Produkt seiner Umwelt geformt und erzogen – zum Besseren oder Schlechteren. Eine Arbeit ins Internet zu stellen hat mehr mit Geburt und Ins-Leben-Entlassen zu tun denn mit der Schöpfung eines statischen Objekts.

Gemeinschaften, MUDs, Märkte, Suchmaschinen und Router-Konfigurationen sind alle kontext-orientiert. Die Ästhetik des Kontexts ist das Design solcher kontext-orientierter Systeme, die alle ihrer Art nach einmalig sind. Ein gutes kontext-orientiertes System läßt ein Netzwerk aus lebendigen Verknüpfungen konvergieren, interagieren, wachsen, es erhöht den Wert des Netzwerks und zieht Benutzer und Verbindungen an.

Das Internet ist ein sich selbst organisierendes adaptives System. Wie John Casti vom Santa Fe Institute in seinem Vortrag beim Ars Electronica Memesis-Symposion vergangenes Jahr gesagt hat, kann man zwar den Prozeß, innerhalb dessen ein komplexes adaptives System funktioniert, vollständig verstehen, aber trotzdem nicht voraussagen, was es tun wird. Das Internet organisiert sich im sehr interessanten Zwischenbereich zwischen totalem Chaos und völliger Ordnung, Eric Hughes hat es eine „funktionierende Anarchie" genannt. Wenn dem Internet Ordnung aufgezwungen wird – sei es durch ein rigides Protokoll oder ein einziges ubiquitäres Betriebssystem –, dann wird diese tragende Schicht sehr brüchig, und wie man aus der Katastrophentheorie weiß, kann eine einzige Erschütterung des Systems riesigen Schaden anrichten. Ein Virus oder Fehler im System könnte das Netz zum völligen Zusammenbruch bringen. Die wesentlich uneffizientere und diversifizierte Struktur des gegenwärtigen Mem- bzw.

and diverse nature of the current memetic/software pool allows the risk to be distributed. Many small earthquakes can help prevent a catastrophic earthquake. It is the inefficiency and the small errors that can help the Internet adapt and grow without imploding or exploding.

Ordered efficient systems are also very susceptible to fluctuation amplification. With feedback going in the wrong direction, small fluctuations in economy, politics, traffic or opinion can be amplified by the super-efficient network and explode or crash. Nature uses feedback systems that dampen such fluctuations in an elegant way to contain the energy and balance the systems. This non-linear balance is becoming exceedingly more important than making the system faster or more efficient. This balance can also be explained as the aesthetic of the context.

Nearly complete chaos can also be found on the Internet in the sheer number of disorganized pieces of content and people. Total chaos can also be made much more useful by adding just enough context to help group the content and people into useful communities and networks.

Therefore, I would conclude that both complete order and complete chaos offer very little information, value or energy. Systems that help order chaos or disorder order are useful. In addition, the way in which these systems cause this non-chaos/non-ordered system to manifest should retain or create as much energy as possible while keeping a feedback system that prevents it from amplifying into destruction or dampening into nothing. This requires a group of rules or memes that attracts energy in the form of people, content, traffic, money, etc. and organizes this content in a way that grows and adds value. It is almost a kind of memetic engineering.

The memetic engineer/Internet artist is interested in coming up with an idea, software protocol or image that grows and evolves on the Internet. It is more about

Software-Pools erlaubt es, das Risiko zu verteilen. Viele kleine Erdbeben verhindern das eine große, katastrophale. Und gerade die Ineffizienz und die kleinen Fehler können dem Internet helfen, sich anzupassen und zu wachsen, ohne zu implodieren oder zu explodieren.

Geordnete effiziente Netzwerke sind auch sehr empfindlich gegenüber einer Fluktuationsverstärkung. Wenn ein Feedback in die falsche Richtung läuft, können kleinste Fluktuationen in Wirtschaft, Politik, Verkehr oder in der öffentlichen Meinung vom super-effizienten Netzwerk so verstärkt werden, daß es explodiert oder zusammenbricht. Die Natur verwendet Feedback-Systeme, die solche Fluktuationen auf elegante Weise dämpfen, um Energie zu sparen und das System in der Balance zu halten. Ein derartiges nicht-lineares Gleichgewicht ist viel wichtiger als eine Steigerung der Leistung oder Geschwindigkeit des Systems. Und dieses Gleichgewicht kann auch als die Ästhetik des Kontexts definiert werden.

Im Netz existiert ein fast vollkommenes Chaos – schon allein in der Anzahl nicht-organisierter Elemente von Inhalten und Personen. Dieses totale Chaos kann dann von viel größerem Nutzen sein, wenn gerade soviel Kontext hinzugefügt wird, daß diese Inhalte und Leute in nützliche Gemeinschaften und Netzwerke strukturiert werden können. Ich schließe daraus, daß sowohl totale Ordnung wie auch komplettes Chaos ziemlich wenig Information, Wert oder Energie zu liefern in der Lage sind. Systeme, die dazu beitragen, Chaos zu ordnen oder Ordnung zu entordnen, sind nützlich. Darüber hinaus sollte die Art und Weise, in der diese Nicht-Chaos/Nicht-Ordnung-Systeme zur Entfaltung gebracht werden, so wenig Energie wie möglich verbrauchen bzw. so viel Energie wie möglich schaffen und gleichzeitig ein Feedback-System aufrechterhalten, das eine Expansion bis zum totalen Kollaps ebenso wie eine Aufweichung bis

creating life than about creating a non-living piece of art. The memetic engineer seeks to have the particular meme copied and replicated where traditional artists are protective of their work. It is the use, familiarity and reproduction that makes a meme powerful and proves its aesthetic quality. The Internet artist and the meme both work in the medium of context rather than content.

zum Nichts verhindert. Doch dies erfordert ein Set von Regeln oder Memen, das Energie in Form von Menschen, Inhalt, Verkehr, Geld usw. anzieht und diesen Inhalt so organisiert, daß er wachsen und seinen Wert steigern kann – mit anderen Worten: Es bedarf einer Art „memetischen Engineering".

Der memetische Ingenieur (oder Internet-Künstler) ist daran interessiert, eine Idee, ein Software-Protokoll oder ein Bild zu erarbeiten, das wächst und sich im Internet entwickelt. Es geht ihm mehr um die Schöpfung von Leben als um die Schaffung eines nicht-lebenden Kunstwerks. Der memetische Ingenieur will, daß das jeweilige Mem kopiert und repliziert wird, während der traditionelle Künstler sein Werk zu schützen sucht. Ein Mem wird nur durch Verwendung, durch Vertrautheit und Reproduktion stark, nur so beweist es seine ästhetischen Qualitäten. Die Internet-Künstler und das Mem arbeiten beide eher in einem Medium des Kontexts denn des Inhalts.

JOICHI ITO

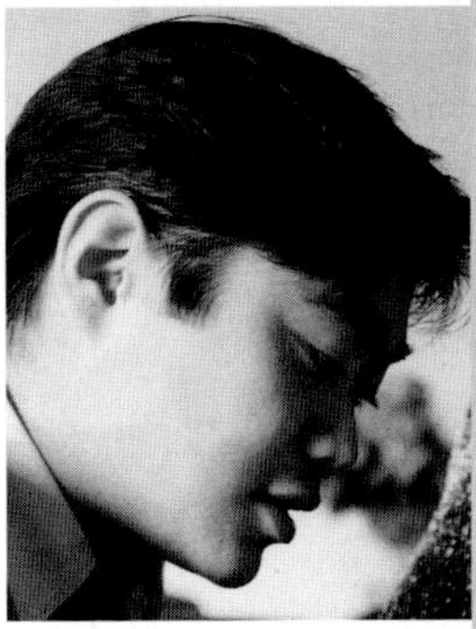

Joichi Ito (J) is developer and producer in the areas of virtual reality and multimedia. He was member of the .net-jury.

Joichi Ito (J) ist Entwickler und Produzent im Bereich Virtual Reality und Multimedia. Er war Mitglied der .net-Jury.

Signed By Artist

MACHIKO KUSAHARA

Die Verschmelzung von Computer und Telekommunikationstechnologie hat neue Möglichkeiten des Kunstschaffens eröffnet.

The merging of computers and telecommunication technologies has opened up new possibilities for art-making.

This has not only provided new ways for the artists to present their imagination by means of still and moving images; it has also made it possible for artists and other lovers of art to share what has been imagined. Digital data can be copied and emitted without loss of quality. Interactive art requires audience participation at its maximum degree. The ideals of Dada or Fluxus have finally found an appropriate technical infrastructure.

This possibility is important enough to change the traditional concept of making and appreciating art. Copyright on the network is an example of a traditional issue facing the necessity of a radical transformation. Similarly, the concept of originality is undergoing a transition with the advent of digital technology. Interactive art, often offering a system designed by an artist with which visitors are asked to create imagery or sound, explicitly contradicts the traditional concept of art that a piece should be the result of the artist's original creative activity.

Sie hat nicht nur Künstlern neue Wege geebnet, ihre Vorstellungskraft in bewegten und unbewegten Bildern auszudrükken, sie erlaubt es ihnen auch, ihre Imagination mit anderen Künstlern und Kunstliebhabern zu teilen. Digitale Daten können kopiert und ohne Qualitätsverlust übertragen werden. Interaktive Kunst verlangt die größtmögliche Mitwirkung des Publikums, und die Ideen von Dada und Fluxus haben endlich eine adäquate technische Infrastruktur gefunden.

Diese neuen Möglichkeiten sind bedeutsam genug, um auch das traditionelle Konzept des Kunstschaffens und der Kunstwahrnehmung zu verändern. Die Frage de. Copyright im Netz ist ein Beispiel für jene traditionellen Anliegen, die einer radikalen Umformulierung bedürfen. Und auch das Konzept der Originalität erlebt seit dem Auftreten der digitalen Technologie eine tiefgreifende Veränderung. Interaktive Kunst, die häufig als ein vom Künstler konzipiertes System auftritt, das von den Besuchern verlangt, Bild- oder Klangwelten zu erzeugen, steht in eklatantem Widerspruch zum traditionellen Begriff des Kunstwerks, der das Werk als Resultat der originären kreativen Arbeit des Künstlers definiert.

Aber halt … von welchen Traditionen sprechen wir eigentlich? Als Japanerin muß ich fragen: Sind das denn unsere Traditionen?

Das japanische Verständnis von Originalität wurde von der westlichen Kultur of in Frage gestellt, sobald Copyright-Fragen

But, wait ... whose traditions are we talking about? As a Japanese I need to ask: are these our traditions?

The Japanese idea of originality has often been put on trial by Western culture when the copyright issue has been raised. However, we have a different tradition, a different way of thinking about originality or copyright. The oldest official Japanese book of songs (edited by the Emperor's Court) appeared in the ninth century. It includes many poems created all over Japan during that century by anonymous ordinary people. There are love songs from different provinces, poems by soldiers missing their families, etc.

This book already demonstrates that in our traditional poetry, citing from other people's poems is considered a virtue, not a theft. Knowing old literature and being capable of using words from another person's work was considered a sign of intelligence. Of course, the audience/readers had to share the same knowledge to be able to understand the connection and to get the hidden meaning. It is part of the Japanese culture to "borrow" ideas or landscapes from others in order to create a richer poem or a more impressive garden. The other side of this tradition is that the Japanese haven't paid much attention to preserving things. Old Japanese prints, for example, were things to be consumed rather than to be kept.

It is interesting to observe that young Japanese artists now feel so much at ease breaking the rules (or myths) of originality by means of digital technology. For them the net is a tool for integrating different imaginations, or even different egos. Open your mind and enjoy art with other people, they insist. Presenting the products of one's imagination is one way of making art; exchanging ideas and collaborating with others is another. Here the idea of making art exists within the idea of communication. Finishing a piece is not the main goal.

involviert waren. Wir haben nun einmal eine andere Tradition, eine unterschiedliche Auffassung von Ursprung, Herkunft und Copyright. Das älteste offizielle japanische Liederbuch (herausgegeben vom kaiserlichen Hof) stammt aus dem 9. Jahrhundert. Es enthält viele anonyme Gedichte, die in jenem Jahrhundert in ganz Japan entstanden waren – Liebeslieder aus verschiedenen Provinzen, Gedichte von Soldaten, die sich nach ihrer Familie sehnen, und so weiter.

Bereits dieses Buch zeigt, daß es in unserer traditionellen Dichtung keineswegs als Diebstahl, sondern vielmehr als Tugend gewertet wird, wenn man aus fremden Werken zitiert. Alte Literatur zu kennen und in der Lage zu sein, die Worte anderer zu verwenden, wurde und wird als Zeichen von Intelligenz angesehen. Natürlich mußten die Zuhörer/Leser über denselben Hintergrund verfügen, um den Zusammenhang zu verstehen und den versteckten Sinn zu begreifen. Es ist Teil der japanischen Kultur, Ideen oder Landschaften von anderen zu „entlehnen", um selbst ein reicheres Gedicht oder einen noch beeindruckenderen Garten zu schaffen. Die andere Seite dieser Einstellung ist natürlich, daß die Japaner wenig Wert auf die Bewahrung alter Dinge gelegt haben – die frühen Drucke zum Beispiel waren zum Konsum gedacht, nicht zur Aufbewahrung.

Es ist interessant zu beobachten, daß junge japanische Künstler sich heute dabei sehr wohl fühlen, wenn sie die Regeln (oder Mythen?) der Originalität mit Hilfe der digitalen Technologien brechen. Für sie ist das Netz ein Werkzeug zur Integration unterschiedlicher Vorstellungswelten, ja, sogar verschiedener Egos. „Öffne deinen Geist und erfreue dich zusammen mit anderen an der Kunst", sagen sie. Die Präsentation der Produkte der eigenen Phantasie ist eine Möglichkeit, Kunst zu machen, der Austausch und die Zusammenarbeit mit anderen ist eine andere. Und hier lebt die Vorstellung vom Kunst-Machen innerhalb der Idee von der Kommunikation. Ein Werk zu vollenden ist dabei nicht das Hauptziel.

Und diese Praxis hängt mit der langen Geschichte der japanischen Kultur zusammen: von den „Kollektiven Gedichten", die schon vor Jahrhunderten von etlichen Künstlern gemeinsam geschaffen wurden, bis hin zu den gemeinschaftli-

This practice is related to the long history of Japanese culture: from the collective "linked poems" created together by several poets hundreds of years ago to the collaborative wall paintings in today's elementary schools, familiar all over our country. As digital technology has matured and become commonplace, artists have begun to introduce Japanese traditions into new forms of art-making.

The most well-known of such projects is "Renga" ("linked image"), which has been continued by artists Toshihiro Anzai and Rieko Nakamura since 1992. The title "Renga" is a play on words, referring to the Japanese word for traditional linked poems (ren = linked and ga = verse); yet the word for "image" also has the same sound, "ga". As artists working with the computer and also as heavy users of computer networks, Anzai and Nakamura conceived of a new method of collaboration on the network (see p.130 of this book). Anzai sent one of his digital paintings to Nakamura via e-mail and asked her to modify it. She did as requested and sent the modified image back to Anzai. As they continued, the series of images became a mixture of their styles leading beyond what they would have been able to create independently.

The idea sounds simple, but in fact it fully utilizes the features of digital technology. In real life one would hesitate to "modify" someone else's original painting. But here, on the net, there is no original, no copy. Images can go through changes just like a landscape that changes from season to season.

Since then, Anzai and Nakamura have investigated different possibilities of Renga among themselves and together with other artists. Participating in a Renga session is a totally new experience for anyone. One finds out that modifying the image created by someone else (or more typically as the result of a preceding collaborative process) does not mean deleting or neglecting a

chen Wandgemälden, die in allen Volksschulen unseres Landes zu finden sind. In dem Maße, in dem die digitale Technologie gereift und allgemein zugänglich geworden ist, haben die Künstler japanische Traditionen in neue Wege des Kunstschaffens hineingetragen.

Das bekannteste Projekt dieser Art ist „Renga" („Verknüpftes Bild"), das seit 1992 von Toshihiro Anzai und Rieko Nakamura betrieben wird. Der Titel „Renga" ist ein Wortspiel und bezieht sich auf den traditionellen japanischen Namen für „verknüpfte Poesie" (ren = verknüpft, und ga = Vers), gleichzeitig aber wird auch der Begriff für Bild als „ga" ausgesprochen. Als Künstler, die mit dem Computer arbeiten, und als intensive Netzwerk-User haben Anzai und Nakamura eine neue Methode der Zusammenarbeit im Netz entwickelt. Anzai schickte Nakamura eines seiner digitalen Gemälde über E-Mail und forderte sie auf, es zu verändern. Sie folgte seinem Wunsch und schickte das Bild zurück an Anzai. Im Zuge dieses Austauschs entstand eine Serie von Bildern, die in ihrer Ausdruckskraft weit über das hinausgingen, was jeder von beiden für sich alleine hätte schaffen können.

Der Gedanke scheint simpel, aber tatsächlich werden so alle Fähigkeiten der digitalen Technologie ausgenützt. Im „realen" Leben hätte man wahrscheinlich Hemmungen, ein Originalgemälde eines anderen Künstlers zu verändern. Aber hier, im Netz, gibt es kein Original, keine Kopie. Bilder können sich wandeln, wie sich Landschaften je nach Wetter und Jahreszeit verändern.

Seit den ersten Versuchen haben Anzai and Nakamura die verschiedenen Möglichkeiten von Renga gemeinsam und auch mit anderen Künstlern ausgelotet. Die Teilnahme an einer solchen Renga-Sitzung ist für jeden eine völlig neuartige Erfahrung. Man erkennt, daß die Veränderung eines Bildes, das von jemand anderem geschaffen wurde (oder das typischerweise

part of it; the act of modification requires an understanding of the nature of the image; much imagination is also needed to open a new landscape within the existing image. Producing a series of images is not the goal of Renga. It is more important to be in it, enjoy the process, and get something out of the experience. What one has taken for granted might change.

In 1996, a Renga session took place in Beijing in collaboration with a celebrated Chinese calligraphy artist. The Chinese artist sent his image to Japan with his signature – the signature is usually considered an important part of calligraphy. Anzai and Nakamura modified the image eventually erasing his signature and sent it back. A modified image came back from China, again with his signature. But after the second exchange of images, the calligraphy artist gave up signing the image he had painted. He understood that his imagination had become a part of the collective image. It was a new experience for him, and he enjoyed it.

"Mega Diary" was another interesting project. It was created by Kazuhiko Hachiya. It took place on a telecommunication network sponsored by NTT/ICC. One hundred participants "opened" their diaries on the network for a hundred days, to virtually share their lives with others. Participants (mostly students) came to know each others' lives in detail over the course of three months. The idea was to mix one's daily life with others' lives and to psychologically remove the boundaries between one another. It was an exciting experience for the participants. After a hundred days, they came to feel like a big family. On the other hand, writing a diary to be read by those whom one has never met in real life makes a participant confront the gap between his/her real life and the self-image, and its reflection onto the network. The collective diary that came out from the project is interesting to read because of its diversity and the interaction

bereits das Ergebnis eines vorhergegangenen Gemeinschaftsprozesses ist), nicht bedeutet, einen Teil davon zu zerstören oder zu vernachlässigen. Die Veränderung des Bildes erfordert Verständnis für und Kenntnis der Natur des Bildes, viel Phantasie wird benötigt, um eine neue Landschaft im bereits existierenden Bild eröffnen zu können. Die Produktion einer Serie von Bildern ist nicht das Ziel von Renga, wichtig ist die Teilnahme am Projekt, die Freude am Prozeß, die gewonnene Erfahrung. Was man für garantiert angesehen hatte, könnte sich verändern.

1996 fand eine Renga-Sitzung in Beijing statt, in Zusammenarbeit mit einem renommierten chinesischen Kalligraphiekünstler. Der Chinese schickte sein Bild nach Japan, säuberlich unterschrieben, denn die Signatur ist traditionellerweise ein wichtiges Element in der Kalligraphie. Anzai und Nakamura modifizierten das Bild, wobei auch die Signatur entfiel, und schickten es zurück. Ein neuerlich verändertes Bild kam zurück – wieder mit Signatur. Nach der zweiten Veränderung des Bildes gab der Künstler es auf, das Werk zu signieren, das er gemalt hatte. Er hatte eingesehen, daß seine Imagination ein Teil des kollektiven Bildes geworden war – das war eine neue Erfahrung für ihn, und er genoß sie.

„Mega Diary" war ein anderes interessantes Projekt. Es wurde von Kazuhiko Hachiya geschaffen und fand auf einem von NTT/ICC gesponsorten Telekommunikationsnetzwerk statt. Einhundert Mitwirkende „öffneten" diesem Netzwerk ihre Tagebücher für hundert Tage, sie teilten sozusagen ihr Leben mit anderen. Die Teilnehmer – zumeist Studenten – lernten in diesen drei Monaten das Leben der anderen in allen Details kennen. Der Grundgedanke des Projekts war es, das eigene tägliche Leben mit jenem von anderen zu vermischen und die Grenzen zwischen den Menschen auf psychologischer Ebene aufzuheben. Es war für alle Teilnehmer eine aufregende Erfahrung, nach den hundert Tagen fühlte man sich wie in einer großen Familie. Ein Tagebuch zu führen, das von Leuten gelesen werden kann, die man noch nie im Leben getroffen hat, führt einem auch jenen Widerspruch vor Augen, der zwischen der eigenen Lebenswirklichkeit und dem jeweiligen Selbstbild besteht, und läßt ihn

between the participants (the text is used in Hachiya's new work "Seeing is Believing" in an interesting way, see p. 126 of this book), but the resulting text itself may not be the essential part of the project. The more important part of the work is the experience the participants went through. They discovered something about themselves, about communication, about media, by thinking about the nature of privacy on the net. Through their experiences others come to think about the issue the project dealt with.

But then, what is the role of an artist? If images and texts were created by the participants, is an artist simply a designer of a system (or a framework) to guide participants?

I would say, what an artist has created was a field, not simply a system. A magnetic field is not visible, but it exists as a space that has a possibility, a power to influence a piece of steel when brought in it. You can see it only through the way it works. In the film "The Field of Dreams" what was built in the middle of the corn field was not just a baseball field, but a space with an invisible power to make a dream come true. Contemporary artist Walter DeMaria did not produce lightning by himself in his work "Lightning Field". He turned a no-man's-land land into a field of expectation and incredible experience with an original idea that no one had ever thought of. The metal rods DeMaria built in the desert are not the same as the lightning rods one sees on top of buildings in big towns. Lightning rods on rooftops are neutral – they are there for their regular function. An artist might use the same material for a different purpose, with one's own aesthetics and style, to show something people do not see otherwise. The signature of the artist is not on each rod, or on each bolt of lightning. It exists on the project itself and in the experience of each person. Similarly, artists can build a field on the net where people are invited

im Netz transparent werden. Das kollektive Tagebuch, das sich aus diesem Projekt ergeben hat, ist interessant zu lesen – wegen seiner Unterschiedlichkeit ebenso wie wegen der Interaktion zwischen den Mitwirkenden (der Text wurde in Hachiyas neuem Werk "Seeing is Believing" auf interessante Weise verarbeitet; siehe S. XXX dieses Buches) –, aber der so entstandene Text ist vielleicht gar nicht das Hauptergebnis des Projekts. Der wichtigere Teil ist die Erfahrung, die die Mitwirkenden dabei gemacht haben. Sie haben etwas über sich selbst erfahren, über Kommunikation, über Medien, über Privatheit und Intimsphäre im Netz. Durch ihre Erfahrung werden auch andere dazu angeregt, über das Anliegen des Projektes nachzudenken.

Aber was ist dabei eigentlich die Rolle des Künstlers? Wenn Bilder und Texte von den Mitwirkenden geschaffen werden, ist der Künstler dann nur jemand, der ein System (oder einen Rahmen) entwirft und die Regeln für die Mitwirkenden vorgibt?

Ich meine, was ein Künstler hier geschaffen hat, ist ein Feld, nicht bloß ein System. Ein Magnetfeld ist unsichtbar und doch existiert es als Raum voller Möglichkeiten, der ein Stahlstück beeinflussen kann, wenn es in seinen Bereich gelangt. Man erkennt es nur an der Art, wie es wirkt. Im Film „The Field of Dreams" war das, was in einem Kornfeld aufgebaut wurde, nicht bloß ein simples Baseball-Feld, sondern ein Raum mit der unsichtbaren Macht, einen Traum Wirklichkeit werden zu lassen. Der zeitgenössische Künstler Walter DeMaria hat in seinem „Lightning Field" nicht selbst Blitze erzeugt. Er verwandelte ein Niemandsland mithilfe einer originellen Idee, die vorher noch nie jemand gehabt hatte, in ein Feld voller Erwartung und unglaublicher Erfahrungen. Die Metallstäbe, die DeMaria in der Wüste aufgestellt hat, können nicht mit den gewöhnlichen Blitzableitern auf Gebäuden verglichen werden. Blitzableiter auf Dächern sind neutral, sie sind ausschließlich

to come and play. If the field proves to be effective and it clearly shows the trace of the artist in the air, it can be a piece of art.

In old Japan, organizing a renga party for a group of people was an important role of celebrated haiku masters. It meant turning a normal space into a magical field of collaborative creativity. Guided by the master, personal visions meet others' imagination to be finally integrated into a panorama. This kind of tradition has returned now, this time on the net.

There might be another question. Are such collaborative approaches related to the totalitarian, collective tendencies so typical of the Japanese society? Possibly, because they arise from a common shared cultural background. However, young artists use the traditional concepts with playfulness, aiming to break down the boundaries between people and to establish better modes of communication. Every country has its traditions. In the era of global telecommunications we should look for positive ways of respecting and enjoying these different cultural backgrounds.

MACHIKO KUSAHARA

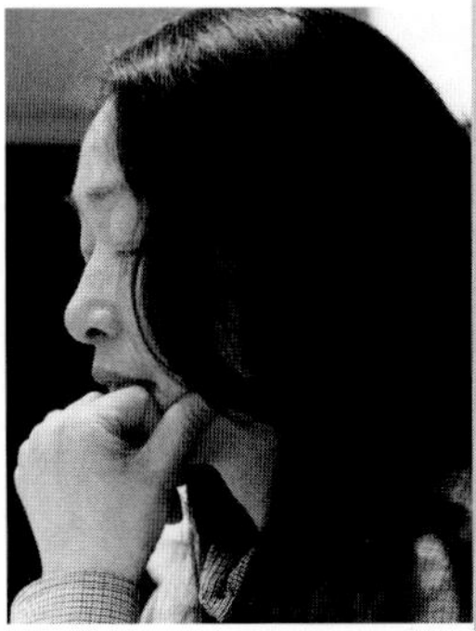

Machiko Kusahara (J) is Associate Professor of Media Art at the Tokyo Institute of Polytechnics. She was a member of the interactive art jury.

ihrer Funktion wegen da. Ein Künstler könnte dasselbe Material für einen völlig anderen Zweck verwenden, mit seiner eigenen Ästhetik und seinem eigenen Stil, um etwas zu zeigen, was die Menschen sonst nicht sehen würden. Die Signatur des Künstlers ist nicht auf jedem einzelnen Stab oder auf jedem Blitz angebracht, sie existiert auf dem Projekt selbst und in der Erfahrung eines jeden Teilnehmers. Und auf die gleiche Weise können Künstler ein Feld im Netz aufbauen und Leute einladen, dort zu spielen. Wenn das Feld sich als effizient erweist und klar die Handschrift des Künstlers erkennen läßt, dann ist es ein Kunstwerk.

Im alten Japan stellte die Organisation eines Renga-Festes für eine Gruppe von Leuten eine wichtige Aufgabe für die verehrten Haiku-Meister dar. Das bedeutete, einen normalen Raum in ein Magnetfeld kollektiver Kreativität zu verwandeln. Unter Anleitung des Meisters trafen persönliche Visionen auf die Phantasie der anderen, um schließlich in ein Panorama integriert zu werden. Und diese Tradition lebt wieder auf, diesmal im Netz.

Eine andere Frage drängt sich auf: Hängen solch kollaborative Ansätze mit jenen totalitären, kollektiven Tendenzen zusammen, die für die japanische Gesellschaft so typisch sind? Wahrscheinlich ja, weil sie aus einem gemeinsamen kulturellen Hintergrund erwachsen. Aber die jungen Künstler verwenden diese traditionellen Konzepte auf spielerische Weise, zielen auf die Durchbrechung der Barrieren zwischen den Menschen ab und darauf, bessere Kommunikationsmodi zu etablieren. Jedes Land hat seine Traditionen. Im Zeitalter der globalen Telekommunikation sollten wir nach positiven Wegen suchen, diese unterschiedlichen kulturellen Hintergründe zu erkennen und uns an ihnen zu erfreuen.

Machiko Kusahara (J) ist Associate Professor für Medienkunst am Tokyo Institute of Polytechnics. Sie war Mitglied der Jury für Interaktive Kunst.

Interactive Art
Maybe it's a bad idea

Vielleicht ist das gar keine so gute Idee

Seit gut zwanzig Jahren hören wir nun schon von „Interaktiver Kunst", besonders im Zusammenhang mit dem Zugang von Künstlern zu Computern.

Michael Naimark

We've been hearing about "interactive art" for a good twenty years now, particularly in the context of computers accessible to artists.

A great many very smart and talented people have worked very hard exploring its potential. The visionaries have waxed eloquent about it, the artists have promised us bold new artforms, and the entrepreneurs have promised us vast new industries.

And what have we seen? Well certainly many interesting things, some of which are pretty good. But as good as the first two decades of photography? Of paints in tubes? Of sound recording? Of cinema?

I don't think so.

So I really must wonder: maybe interactive art is a bad idea. At least in the ways we've been thinking, practicing, and (especially!) promising it.

When I ponder very carefully how much interactive art I've experienced that has truly moved me, I can recall four works.

Only four.

And two of the four were made by Czechs (go figure), so let's start there.

Of all the interactive artforms promised, the "interactive movie" has undoubdtedly been the most flaunted. The concept is

Eine große Menge sehr gescheiter und talentierter Leute hat schwer gearbeitet, um ihr Potential zu erkunden, die Visionäre haben große Reden darüber geschwungen, die Künstler haben uns mutige neue Kunstformen versprochen, und die Unternehmer haben riesige neue Industriezweige angekündigt.

Und was haben wir zu sehen bekommen? Nun, sicherlich viele interessante Sachen, und einige davon waren ziemlich gut. Aber waren sie so gut wie die in den ersten zwei Jahrzehnten der Photographie? Der Ölmalerei? Der Schallaufzeichnung? Des Kinos?

Ich glaube nicht.

Und so frage ich mich, ob die Interaktive Kunst vielleicht doch keine so gute Idee ist, zumindest in jener Form, in der wir sie bisher gedacht, praktiziert und vor allem versprochen haben.

Wenn ich ernsthaft darüber nachdenke, wieviel Interaktive Kunst ich erlebt habe, die mich wirklich bewegt hat, fallen mir vier Werke ein.

Nur vier.

Und zwei von den vieren stammt von Tschechen – man stelle sich das vor! –, also wollen wir auch dort beginnen.

Von all den versprochenen Formen Interaktiver Kunst war zweifellos das „Interaktive Kino" die am marktschreierischsten angepriesene. Sein Grundkonzept ist, daß das Publikum den Handlungsablauf oder Stand- und Blickpunkte oder auch diverse Eigenschaften des Werkes – den Anteil von Sex und Gewalt etwa – steuern kann. Zahlreiche groß angekündigte Versuche wurden auf Kinosaal-Ebene realisiert – und sind bei Kritik und Publikum durchgefallen. Berge von bescheideneren interaktiven Video- und Trickfilmen wur-

that the audience can control the plot, or point of view, or degree of various properties (such as sex and violence). Several well-publicized attempts have been realized on a theatrical level and failed, both critically and commercially. Scores of more modest video and animated interactive movies have been produced, but nothing as intense and memorable as the two interactive movies made by the Czechs.

The first was the "Kino-Automat" film in the Czech pavilion at Expo '67 in Montreal, and was promoted as the world's first interactive movie. It was film-based, and all members of the audience had a red and a green button in front of them (the results of voting were displayed around the screen). The movie itself was a dark comedy about a man who believes he was responsible for his apartment building burning down, and is structured as a series of flashbacks leading up to the fire. After each scene the film would stop and a live performer would walk onto the stage and ask the audience to vote. Immediately, as if by magic, the voted scene was played.

How did they do it? Deceit, of sorts. The branching structure wasn't tree-like, doubling the number of scenes needed at each choice, but rather always remained only two. They did this by carefully crafting a story such that no matter which of the two options were chosen, it would end up back at the same next choice. The vote was executed by the projectionist switching one lens cap between the two synchronized projectors. The artfulness, ultimately, was not in the interaction but in the illusion of interaction. The film's director, Raduz

den produziert, aber nichts, was so intensiv und in jeder Hinsicht bemerkenswert gewesen wäre, wie die zwei interaktiven Filme der Tschechen.

Der erste war der Film „Kino-Automat". Er wurde bei der Expo '67 in Montreal als der erste interaktive Film der Welt propagiert. Er basierte auf traditioneller Filmtechnik, und jeder Zuseher im Publikum hatte einen roten und einen grünen Knopf vor sich (die Abstimmungsergebnisse wurden auf der Leinwand angezeigt). Die Handlung war eine düstere Komödie über einen Mann, der glaubte, seinen Wohnungsbrand selbst verursacht zu haben, und war als Folge von Rückblicken bis zum eigentlichen Brand strukturiert. Nach jeder Szene blieb der Film stehen, ein Schauspieler betrat die Bühne und ließ das Publikum abstimmen. Und wie durch ein Wunder wurde daraufhin die vom Publikum gewünschte Fortsetzung gespielt.

Wie das gemacht wurde? Nun ja, durch eine Art Täuschungsmanöver. Die Verzweigungsstruktur war nicht wirklich baumförmig, es wurde nicht an jedem Knotenpunkt die Anzahl der erforderlichen Szenen verdoppelt – es blieb immer bei zwei Handlungsverläufen. Dies wurde durch sorgfältige Planung der Geschichte erreicht, die – ganz egal, welche der beiden Möglichkeiten jeweils ausgewählt worden war – am nächsten Verzweigungspunkt stets wieder genau zwei Optionen bot. Und die gewünschte Fortsetzung wurde gespielt, indem der Kinooperateur einfach einen der beiden synchron laufenden Projektoren abblendete. Die Kunst lag also letztlich nicht in der Interaktion, sondern in der Illusion einer Interaktion. Der Regisseur des Films, Raduz Cincera, machte daraus eine Satire auf die Demokratie, in der jeder wählt, ohne daß es einen Unterschied macht.

Das zweite tschechische Stück, „Cinelabyrinth", wurde als das „erste labyrinthische kine-

Cincera, made it as a satire of democracy, where everyone votes but it doesn't make any difference.

The second Czech piece was called "Cinelabyrinth," promoted as "the world's first labyrinth-style Cinematic System," made 23 years later for the Gas Pavillion at Expo '90 in Osaka (and directed, amazingly, also by Raduz Cincera). Everyone began in a single large theater sitting on the floor. At the front on each side of the screen was a door. After the first scene was played out, the audience was asked to chose one of two options by walking through one of the two doors. The story was an ecology yarn about kids trying to save a grand old tree from greedy developers. After seeing several scenes and making several choices, enough to totally lose one's sense of direction inside the pavillion, one watches the final scene, where the children successfully save the tree.

At the moment when the kids shout "we did it!" the screen in front raises up to reveal a full-size replica of the tree used in the film. Simultaneously, three other screens on the other three sides of the tree rise up, revealing four theaters with everyone who began in the first room, now all facing each other. The gag was that no matter which options were chosen, the kids successfully saved the tree. It was as manipulative as the first Czech piece made 23 years earlier, but this time they made it transparent. The effect of realizing everyone "won" and now were all in the same room together was really very powerful.

Another of the four most personally moving interactive art experiences is Alan

matographische System der Welt" angekündigt und entstand 23 Jahre später für den Gas Pavillion der Expo '90 in Osaka. Erstaunlicherweise war der Regisseur wiederum Raduz Cincera. Zu Beginn saßen alle Besucher in einem großen Vorführraum vor einer Leinwand, neben der sich links und rechts jeweils eine Türe befand. Nachdem die erste Szene abgespielt war, wurden die Besucher gebeten, den Raum durch die eine oder die andere Türe zu verlassen. Die Handlung war ökologisch angehaucht und erzählte vom Bemühen von Kindern, einen riesigen alten Baum vc der Gier von Stadtentwicklern zu bewahren. Nach etlichen Szenen und ebensovielen Entscheidungen – genug jedenfalls, ur die Orientierung im Inneren des Pavillons völlig zu verlieren – betrachtet man die Schlußszene, in der es den Kindern tatsächlich gelingt, den Baum zu retten.

In dem Augenblick, in dem die Kinde ausrufen „Wir haben es geschafft!", geht die Leinwand hoch und enthüllt einen lebensgroßen Nachbau des im Film verwendeten Baums. Gleichzeitig heben sich auc drei andere Leinwände auf den anderen Seiten des Baumes und öffnen den Blick i drei andere gleichartige Kinosäle, in dene sich all die anderen Besucher befinden, d ursprünglich mit einem selbst im ersten Raum waren – und jetzt blicken sich alle an. Der Gag dabei war, daß unabhängig von der Entscheidung der Besucher die Kinder in jedem Falle den Baum retten würden, und das Stück war nicht wenige manipulativ als das erste 23 Jahre zuvor. Bloß wurde diesmal die Manipulation transparent gemacht, und der Effekt, daß

Lomax's "Global Jukebox" project. Alan is an 82 year old ethnomusicologist who has spent sixty years collecting thousands of recordings of song and dance from every culture on the planet. Moreover, he and his colleagues at Columbia University spent twenty years (1962–82) developing a scheme to codify song and dance, breaking each down to several dozen parameters. These data, when cross-correlated with itself, and when cross-correlated with other ethnographic databases of general culture, strongly suggests that the performing arts around the world are all closely related, and is the most robust and accurate carrier of cultural traits.

For example, based purely on factor analysis, Lomax has shown statistical significance for how the traditional music of Georgian Russia is similar to (and may be a descendent from) that of Central Africa, and how the music of Patagonia is similar to that of the Inuit. And how dances with strong lateral knee movement correlate with cultures with a tradition of pottery (and potter's wheels), and how dances with narrow heel-to-toe movements correlate with societies whose main crop is planted in narrow rows (like rice). And how songstyles with large amounts of rasp in the male voice correlates with societies who raise boys to be independent rather than team players. And ones with a high degree of tonal blend (sounding as one) correlates with societies where women account for most of the food production. The list goes on. It's strong, romantic, arguable stuff.

These data, along with audio and video

alle Besucher „gewonnen" hatten und sich letztlich alle in einem Raum befanden, war wirklich eindrucksvoll.

Eine andere der persönlich berührenden Erfahrungen Interaktiver Kunst ist Alan Lomax' „Global Jukebox"-Projekt. Alan ist ein 82jähriger Ethnomusiker, der 60 Jahre damit verbrachte, Tausende von Tanz- und Gesangsaufnahmen aller Kulturen und Ethnien dieses Planeten zusammenzutragen. Und darüber hinaus haben er und seine Kollegen von der Columbia University 20 Jahre (1962–82) damit zugebracht, ein Schema zur Kodifizierung von Tanz und Gesang zu entwickeln, indem jedes einzelne davon in einige Dutzend Parameter zerlegt wird. Werden die so gewonnenen Daten mit sich selbst und anderen ethnographischen Datenbanken zur allgemeinen Kultur verknüpft, so suggerieren sie, daß die darstellenden Künste der ganzen Welt eng miteinander verwandt und insgesamt die tragfähigsten und exaktesten Träger kultureller Züge seien.

So beweist Lomax rein auf der Basis von Faktorenanalyse, daß es eine statistisch signifikante Ähnlichkeit der Georgischen Musik mit jener Zentralafrikas gibt und wie die Musik Patagoniens jener der Inuit ähnelt. Und wie sehr Tänze mit betonten Seitwärtsbewegungen der Knie mit jenen Kulturen korrelieren, in denen Töpferei und Töpferscheibe eine Tradition haben, während Tänze mit überwiegenden Ferse-Zehen-Abrollbewegungen mit jenen Kulturen korrelieren, deren wichtigste Feldfrüchte in geraden Reihen angebaut werden, wie etwa Reis. Gesangsstile mit krächzenden, räuspernden Lauten in den Männerstimmen korrelieren mit jenen Kulturen, in denen Knaben eher zur Unabhängigkeit erzogen werden denn zur Teamarbeit, während hingegen jene mit hochgradiger tonaler Verschmelzung (also einer Tendenz zur Einstimmigkeit) mit jenen

clips of song and dance styles, make up the "Global Jukebox". You can start with a map of the world, point to a spot, hear songstyles and see danceforms from that region, and begin a query about how it all relates with other cultures. Most everyone who first engages with it begins with their own roots. It becomes a personal investigation, toggling between intellect and emotion – and in the end synthesizing them.

Strictly speaking, the "Global Jukebox" is not interactive art as much as an interactive delivery system about art. Alan's group, the Association for Cultural Equity (relocated at Hunter College), has worked on the current system for more than ten years. It's a massive project and still not complete (particularly in terms of interface). Perhaps the system itself will be considered interactive art when it's finally completed.

The fourth and final favorite interactive art piece didn't use a computer. It was installed at the 1980 New York Avant Garde Festival, on the roof of the massive Cunard Terminal where the festival was being held. It consisted of a long wide plank of wood, a box of pencils (the standard orange kind with little pink erasers on the ends), and several hammers. That was it.

I literally stumbled over it as it was being set up, and kept an eye on it much of the day. The artist set the plank and the hammers down on the floor, took out a knife and sharpened the pencils, and left. Visitors would walk by, look at the assemblage, and, more often than not, kneel down and try to hammer a sharp-

Gesellschaften korreliert, in denen Frauen den Hauptanteil der Lebensmittelproduktion tragen. Diese Liste ließe sich beliebig weiterführen. Es ist starker Tobak, romantisch und diskutabel.

Zusammen mit Audio- und Videoclips von Gesang und Tanz bilden diese Daten die „Global Jukebox". Man kann mit einer Weltkarte beginnen, die Gesangsstile und Tänze der Regionen Stelle um Stelle abhören und ansehen und eine Abfrage starten, mit welchen anderen Kulturen etwas auf welche Weise zusammenhängt. Fast jeder, der sich erstmals damit beschäftigt, beginnt bei seinen eigenen Wurzeln – es wird zu einer persönlichen Forschungsreise, zu einem Wechsel zwischen Intellekt und Emotion und letztlich zu einer Synthese aus beiden.

Im engeren Wortsinne ist die „Global Jukebox" nicht so sehr ein interaktives Kunstwerk wie vielmehr ein interaktives Informationssystem über Kunst. Alans Gruppe, die am Hunter College beheimatete Association for Cultural Equity, arbeitet seit zehn Jahren am derzeitigen System. Es ist ein enormes Projekt und noch lange nicht fertiggestellt (besonders was das Interface betrifft), aber vielleicht wird das System als solches nach seiner Vollendung als interaktives Kunstwerk anzusehen sein.

Das vierte und letzte großartige interaktive Kunstwerk wurde ohne Computer geschaffen. Es wurde 1980 beim New York Avantgarde Festival installiert, auf dem Dach des Cunard Terminal, wo das gesamte Festival stattfand. Es bestand aus einer langen, breiten Holzplanke, einer

ened pencil into the plank. It was difficult (mainly due to those damn erasers!). It was funny. And every action left a mark on the plank, successful or not.

The artist was anonymous (although I recall he said he was Australian). Throughout the day he stopped by regularly to sharpen the pencils, now getting smaller and smaller. Then, at the end of the day, he put the pencils and hammer away in his pack, heaved onto his shoulder the plank (which now had several dozen pencils sticking out of it and hundreds of marks), and left.

Of course I really don't believe interactive art is a bad idea. And twenty years is not a long time for a new genre to fully evolve. I have seen a lot of good stuff. But these four are the only ones that really left strong permanent impressions.

Maybe we've been barking up the wrong trees.

MICHAEL NAIMARK

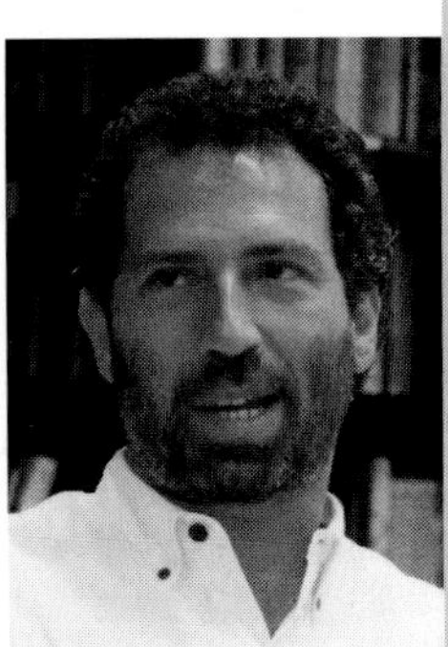

Michael Naimark (USA), co-editor of "Presence" and "Leonardo Electronic Almanac", was a member of the interactive art jury.

Schachtel Bleistifte (von der traditionellen, orange lackierten Sorte mit dem Radiergummi am einen Ende) und aus mehreren Hämmern. Das war alles. Ich bin im wahrsten Sinne des Wortes darüber gestolpert, als es gerade installiert wurde, und habe es den Rest des Tages im Auge behalten. Der Künstler legte Brett und Hämmer auf den Boden, zog ein Messer heraus, spitzte die Bleistifte und ging. Immer wieder kamen Besucher vorbei, betrachteten das Ganze, und meist blieben sie stehen und versuchten, einen gespitzten Bleistift in das Brett zu hämmern. Und das war schwer (nicht zuletzt wegen der verdammten Radiergummis!). Es war lustig. Und jede Aktion, ob erfolgreich oder nicht, hinterließ ihre Spur auf der Planke.

Der Künstler war anonym (ich erinnere mich allerdings, daß er sagte, er sei aus Australien). Den ganzen Tag über kam er immer wieder vorbei, spitzte die immer kürzer werdenden Bleistifte. Dann, am Abend, packte er Bleistifte und Hämmer in seine Tasche, schulterte das Brett, aus dem nun mehrere Dutzend Bleistifte herausragten und das hunderte Spuren aufwies, und ging fort.

Natürlich glaube ich nicht wirklich, daß Interaktive Kunst etwas Schlechtes ist, und 20 Jahre sind auch nicht sehr viel Zeit, in der sich ein neues Genre vollständig entwickeln könnte. Ich habe eine Menge guter Sachen gesehen, aber diese vier sind die einzigen, die einen bleibenden Eindruck hinterlassen haben.

Vielleicht haben wir bloß immer aufs falsche Pferd gesetzt?

Michael Naimark (USA), Mitherausgeber von „Presence" und „Leonardo Electronic Almanac", war Mitglied der Jury für Interaktive Kunst.

Music and the Computer or Music from the Computer?

Musik und Computer oder Musik aus dem Computer?

Robert Normandeau

Wir schreiben das Jahr 1997 - das Jahr vor dem fünfzigsten Jubiläum der „Erfindung" der Musique Concrète[1] durch Pierre Schaeffer.

The year now is 1997 – one year before the fiftieth anniversary of the "invention" of musique concrète[1] by Pierre Schaeffer.

Where are we today nearly half a century after this major revolution in the history of music, and more than one hundred and twenty years after the invention of sound recording? Is there any point left today in categorizing electroacoustic music according to the criteria of – let us say – instrument making? Has the computer – and which computer? – actually changed the way this music is composed and expressed? And which electroacoustic music are we even talking about here? Does this term still make sense? Which reality does it comprise?

Writing on a platform ...

Since the beginnings of time, music has always been an art of interpretation. Up until the twentieth century, if you wanted to hear music, then you either had to play it yourself or else move physically to a designated place, where you could see and hear musicians of flesh and blood. Since the Middle Ages, western composers have used a transmission code (keep in mind that the western musical tradition is the

Wo stehen wir heute, fast ein halbes Jahrhundert nach dieser großen Revolution in der Musikgeschichte und mehr als hundertzwanzig Jahre nach der Erfindung der Tonaufzeichnung? Hat die Kategorisierung der elektroakustischen Musik nach – wenn man so will – Kriterien des Instrumentenbaus heute noch einen Sinn? Hat der Computer – und welcher Computer? – tatsächlich die Kompositions- und die Ausdrucksweise dieser Musik verändert? Und überhaupt: Von welcher elektroakustischen Musik sprechen wir hier eigentlich? Hat dieser Begriff noch Sinn? Welche Realität umfaßt er?

Auf ein Trägermaterial zu schreiben ...

Seit Anbeginn der Zeit war die Musik stets eine Kunst der Interpretation. Bis zum zwanzigsten Jahrhundert mußte man, wollte man Musik hören, sie entweder selbst spielen oder sich an einen dafür vorgesehenen Ort begeben, um dort Musiker aus Fleisch und Blut zu hören und zu sehen. Die westlichen Komponisten haben sich seit dem Mittelalter (man erinnere sich daran, daß die westliche Musiktradition die einzige verschriftlichte ist, und diese Tatsache ist nicht zu unterschätzen!) eines Übertragungscodes bedient, einer instrumentalen Notation auf liniertem Papier, um die Um- und Übersetzung ihrer Gedanken anderen Interpreten zu übertragen – denn um eine Übersetzung handelt es sich, und als Beweis will ich nur die große Interpretationsspannweite anführen, die man

only written musical tradition, a fact that should not be underestimated!), an instrumental notation on lined paper to communicate the realization and translation of their ideas to other musicians (and indeed it is a case of translation, as is evident even in just the scope of possible interpretations of the work of a composer as precise as Beethoven or Mahler), and to ultimately enable an audience to hear the work. Until 1948, music remained essentially an interpretative art like dance or theater. It is the representation of an abstract sequence ranging from the composer and his/her coded signs through the score all the way to the listener, implemented through the instruments used by the musicians.

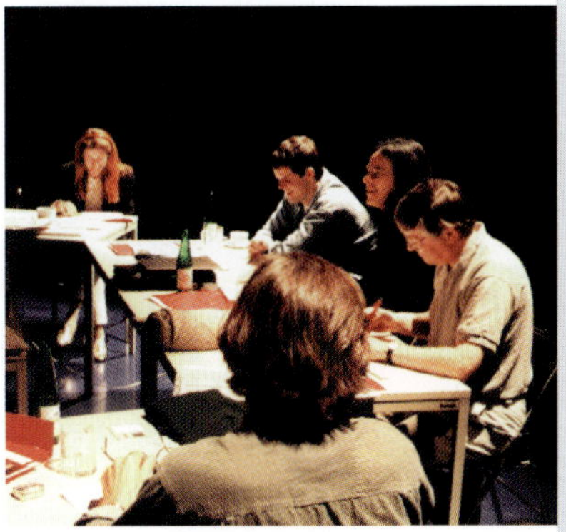

However, after 1876 – when Edison invented the phonograph and Charles Cros the gramophone – a crack appeared in the landscape of music. This crack was caused by the possibility of recording music onto a fixed medium. Although the mediocre quality of these recordings was certainly not likely to make us confuse the copy with the original, it was nevertheless the first time in history that musicians were no longer always required if you wanted to listen to music. And they have never recovered from it since (just think of the thousands of musicians' jobs that have been made redundant at broadcasting institutions over the last fifty years). Yet in those early years, recording was merely an ersatz for the actual performance – and that is what it was to remain until 1948, with only a few exceptions, notably on the part of John Cage, and even there in a thoroughly anecdotal form. To begin with, recording

selbst bei so präzisen Komponisten wie Beethoven oder Mahler findet –, und ihr Werk auf diese Weise letztlich einem Publikum zu Gehör zu bringen. Bis 1948 ist die Musik ihrem Wesen nach eine Interpretationskunst, wie der Tanz oder das Theater. Sie ist die Abbildung einer abstrakten Folge, die sich, ausgehend vom Komponisten und seinen codierten Zeichen, in der Partitur bis hin zum Hörer erstreckt, umgesetzt durch die von den Musikern bedienten Instrumente.

Seit 1876 jedoch – als Edison den Phonographen und Charles Cros das Grammophon erfunden hat – ist eine Bresche in der musikalischen Landschaft entstanden. Diese Bresche wurde durch die Möglichkeit zur Aufzeichnung der Musik auf ein fixes Medium geschlagen. Deren mediokre Qualität war nun keineswegs dazu angetan, uns das Original mit der Kopie verwechseln zu lassen, aber dennoch waren zum ersten Mal in der Geschichte die Interpreten nicht mehr ständig erforderlich, wollte man Musik hören. Und davon haben sie sich niemals mehr erholt (man denke an die Tausenden von Musikerstellen, die seit einem halben Jahrhundert bei den Rundfunkanstalten abgebaut wurden). Aber die Aufzeichnung war dennoch in jenen Jahren – und das sollte bis 1948 so bleiben, mit einigen wenigen Ausnahmen freilich, besonders seitens John Cage, und auch dort noch in durchaus anekdotischer Form – nur ein Ersatz für die tatsächliche Aufführung, sie war nichts weiter als die mehr oder weniger getreue Transkription einer Kunst, die davor schon existiert hatte und die in dem neuen Medium nur ihre natürliche Verlängerung fand. Vor 1948 war die Tonaufzeichnung in keiner Weise als neue Ausdrucksform bestimmend – um meinen Landsmann McLuhan zu paraphrasieren: Das Medium war noch nicht die Botschaft.

Was geschah also 1948? Zum ersten Mal in der Geschichte realisiert ein Komponist (ich weiß nicht, ob Schaeffer den Begriff gutheißen würde, aber da er nun einmal Werke hinterlassen hat, wollen wir ihn ihm posthum zuerkennen) bewußt und nicht zufällig, daß die Tonaufzeichnung jenseits ihrer Fähigkeit, die Realität zu reproduzieren – Schaeffer ist damals Leiter des Versuchsstudios des französischen Rundfunks und verantwortlich für die klangliche Umsetzung von Hörspielen –,

was no more than the more or less authentic transcription of an art that already existed and which found its natural extension in this new medium. Before 1948, sound recording was in no way definitive as a new form of expression – to paraphrase my compatriot McLuhan: the medium was not yet the message.

Then what happened in 1948? For the first time in history, a composer (I do not know if Schaeffer would approve of the term, but since he did leave some works to posterity, we will apply it posthumously) realized consciously and not just by chance that far beyond the ability of sound recordings to reproduce reality – at the time Schaeffer was the director of the experimental studio of the French broadcasting company and responsible for the realization of radio plays – they also made it possible to treat sound as having the same status as a material. He recognized that the sound material was no longer a dependent component of the two elements that had previously determined its nature. It was no longer necessary to play it "for real", it could also be liberated from the real time of the musician (and also, I would like to add, from realistic time because it was no longer subject to the physiological limitations of the human being – the maximum speed of repeated notes as the limit for shortness and the volume of breath as the limit for length). What had never been heard before could be allowed to emerge, the unheard-of. We must not forget that what seems quite ordinary even to the inexperienced listener of today, such as hearing a sound played in reverse, was new and extraordinary in 1948, something one simply could not imagine without having heard it. On the basis of this fact alone, music became an experimental art, in which a work is the result of various attempted approaches, numerous experiments, the result of just as many successes as failures and errors. Like the painter's brush, the recorded gesture returns on the

es auch erlaubt, dem Klang den Status eines Materials zu geben. Er erkennt, daß das Klangmaterial nicht mehr ein abhängiger Bestandteil jener beiden Elemente ist, die seine Natur bisher bestimmt haben. Es ist nicht mehr nötig, es „als echt" zu spielen, man kann es befreien von der realen Zeit des Interpreten (und von der realistischen Zeit, wie ich hinzufügen möchte, weil es auch den physiologischen Beschrän kungen des menschlichen Wesens unterworfen ist – die maximale Geschwindigkei wiederholter Töne als Grenze für die Kürze und das Atemvolumen als Begrenzung in der Länge), man kann daraus auftauchen lassen, was niemals zuvor gehört wurde, das Un-Erhörte. Man darf nicht vergessen, daß das, was selbst für den unerfahrenen Hörer der Gegenwart nichts Besonderes ist, nämlich einen Klang umgekehrt abgespielt zu hören, 1948 neu und außergewöhnlich war, etwas, das man sich gar nicht vorstellen konnte, wenn man es noch nicht gehört hatte. Allein durch diese Tatsache wird die Musik zu einer experimentellen Kunst, in der das Werk das Ergebnis verschiedener Versuche der Annäherung ist, zahlreicher Experimente, ein Ergebnis aus ebensovielen Erfolgen wie Mißerfolgen und Irrtümern. Wie der Pinsel des Malers kehrt der auf Band aufgenommene Gestu zurück, wird durch einen neuen überlager korrigiert, ersetzt, zurückgenommen, neuerlich ersetzt und so fort.

Glücklicherweise trocknen die Töne nicht ein wie Farben! Vor den erstaunten Ohren Schaeffers öffnete sich eine klangliche Unendlichkeit, ein Schwindel erfaßte ihn – zumindest würde er unsereinen ergreifen –, und davon hat er sich nicht wirklich wieder erholt (am Ende seines Lebens hat er diese seine fortgesetzte Arbeit allerdings verleugnet). Und der Großteil seiner Zeitgenossen, die aus der instrumentalen Musik kamen, verstand davon rein gar nichts. Erst seine Nachfolger konnten die von diesem Zauberlehrling freigesetzten Kraftlinien entwirren.

tape to be overlaid, corrected, replaced, erased and newly replaced and so on.

Fortunately, the sounds do not dry out like paint! A sonic eternity revealed itself before Schaeffer's astonished ears, he became dizzy – at least, we certainly would in his place – and he never really recovered from the experience (at the end of his life, he repudiated this continuation of his work). And the majority of his contemporaries, coming from a background of instrumental music, understood absolutely nothing. Only his later successors have become able to untangle the lines of force that this sorcerer's apprentice set free.

And computer music?

What is noticeable in all of Schaeffer's writings and also in the writings of the Groupe des Recherches Musicales in Paris (of which the greatest flaw, in my opinion, is that they are exclusively in French) is that there is practically no mention at all of instrument making. Thus no particular notice was taken of an explicit computer, even in the early years in the GRM studios (first in 1980 in Studio 123, then through the creation of the SYTER system in the mid-80's). In this case, as in all manifestations of musique concrète (the term is too frequently applied only to the origins of electroacoustic music, but the meaning of it still has significance for anyone attempting to define this medium), the means that were utilized were not considered as important as the sonic result thus created and particularly the mechanisms of perceptions involved and the understanding that grew out of this. For this reason, a method of sound classification – typomorphology – and various methods of analysis were created that always revolve around the same basic question: the question of the perception of sound.

In the discourse revolving around that which is "mediating", the media arts (in other words, categories such as video, cinema, computer graphics, etc.), of which

Und die Computermusik?

Was in allen Schriften Schaeffers und auch in jenen der Groupe des Recherches Musicales in Paris (deren größter Irrtum meiner Ansicht nach darin besteht, daß sie ausschließlich französisch sind) auffällt, ist, daß in ihnen praktisch niemals vom Instrumentenbau die Rede ist. So schenkte man auch der Existenz eines speziellen Computers schon in frühen Jahren in den Studios der GRM (zuerst 1980 im Studio 123, dann durch die Schaffung des SYTER-Systems Mitte der 80er Jahre) keine besondere Beachtung. Wie bei allen Ausformungen der Konkreten Musik (ein Begriff, der allzuhäufig nur auf den Ursprung der elektroakustischen Musik angewendet wird, obwohl sein Sinngehalt noch heute für jeden von Bedeutung ist, der versucht, dieses Medium zu definieren) sind auch hier die verwendeten Mittel nicht so wichtig wie das dabei entstehende Klangergebnis und vor allem die Wahrnehmungsmechanismen, die im Spiel sind, sowie das daraus resultierende Verständnis. Deshalb hat man eine Methodologie der Klang-Klassifikation– die Typomorphologie – und verschiedene Analysemethoden geschaffen, die als Besonderheit immer die gleiche Grundfrage stellen: jene nach der Wahrnehmung des Klangs.

Von daher kommt auch der Leitgedanke dieses Aufsatzes: Musik wird gemacht, um gehört zu werden. Wobei man noch hinzufügen könnte: und nicht nur, um konzipiert, gedacht zu werden.

In jenem Diskurs, der sich um die „vermittelnden", die mediatisierenden, also Medien-Künste dreht (um solche Kategorien wie Video, Kino, Computergraphik usw.), deren historisch erster Vertreter eben die elektroakustische Musik war, gibt es eine Anzahl von Begriffen, die meist von Theoretikern der visuellen Künste formuliert wurden und bisweilen zeigen, daß ihnen die fast fünfzigjährige Geschichte der elektroakustischen Musik völlig fremd ist. Nehmen wir nur den Terminus „virtuell", der nach und nach alle Gebiete rund um die „Neuen Technologien" erobert hat: Man sollte sich darüber klar sein, daß er ein Grundkonzept aller Medienkünste aufgreift, ein Konzept, das Pierre Schaeffer gemeinsam mit Jérôme Peignot schon vollständig erfaßt hatte, als er den Terminus „akusmatisch[2]" wieder aufgriff, um eine

electroacoustic music was historically the first representative, a number of terms have been formulated, mostly by theorists of the visual arts. Many of these terms show that the nearly fifty-year history of electroacoustic music is entirely unknown. Take the term "virtual", for example, which has successively impenetrated all the areas associated with the "new technologies": it should be realized that this is goes back to a basic concept of all media arts, a concept that Pierre Schaeffer, together with Jérôme Peignot, grasped entirely when he re-introduced the term "acousmatic[2]" to describe a music, of which the causes could not be identified. And that as early as 1956! Of course the word itself is different, but the concept it covers is basically the same.

It is equally remarkable, that the history of contemporary media art runs counter to that of electroacoustic music. We can indeed see that music has taken a direction liberating it from the interpreting musician and allowing it to be understood as an object ultimately fixed to a supportive platform[3], while visual art is moving in the opposite direction. From the cave paintings in Lascaux to the later works of Picasso, the fixation of all painted works on a stabile background has been an omnipresent factor in human history. This is the reason why artists in the second half of the twentieth century – and the theorists following their lead (and occasionally passing them up) – have been so interested in liberating the visual arts. And thus the concept of virtuality emerged, making the work non-haptic, moveable, untouchable in every sense (although it is an irony of fate that with interactive systems using force feedback mechanisms to enable users to have physical contact, this is starting to go back in the opposite direction again). And this even goes so far that the term "work" itself is beginning to disappear, making way for "processes", which are becoming a warranty for an artistic statement.

Musik zu beschreiben, deren Ursachen man nicht identifizieren kann. Und das bereits 1956! Sicherlich ist das Wort ein anderes, aber das Konzept, das es umfaßt, ist grundsätzlich dasselbe.

Gleichermaßen bemerkenswert ist, daß die Geschichte der gegenwärtigen Medienkunst jener der elektroakustischen Musik entgegenläuft. Man kann tatsächlich feststellen, daß die Musik einen Weg beschreitet, der sie vom Interpreten befreit und ihr erlaubt, als ein auf einer unterstützenden Plattform endgültig fixiertes Objekt[3] verstanden zu werden, während die visuelle Kunst in entgegengesetzter Richtung vorgeht. Von den Höhlenmalereien von Lascaux bis zu den Spätwerken Picassos war die Fixierung aller malerischen Werke auf einem festen Untergrund ein allgegenwärtiger Faktor in der Geschichte der Menschheit. Weshalb sich auch die Künstler dieser zweiten Hälfte des zwanzigsten Jahrhunderts – und die Theoretiker, die in ihrem Kielwasser segeln (oder sie, manchmal, überholen) – sich besonders für ihre Befreiung einsetzen. Und so tauchte das Konzept der Virtualität auf, die das Werk unberührbar, beweglich, un(an)tastbar macht (obwohl man – Ironie des Schicksals – bei interaktiven Systemen mit Rückkoppelungsmechanismen, die dem Benutzer einen physischen Kontakt ermöglichen, gerade wieder den umgekehrten Weg zu beschreiten beginnt). Und man geht sogar so weit, daß der Begriff „Werk" selbst verschwindet zugunsten der Prozesse, die zu Garanten einer künstlerischen Aussage werden.

Prozess–musik?

Wenn wir am Anfang dieses Textes einen kurzen historischen Rückblick gemacht haben, so deswegen, um unser Anliegen im besonderen Kontext der konkreten Musik zu verorten, einer Musik, die essentiell auf der Phänomenologie basiert, also auf der Wahrnehmung. Und das kann man in der folgenden Frage zusammenfas-

Process music?

The brief historical review at the beginning of this article was meant to locate our intention in the particular context of musique concrète, a music based essentially on phenomenology, in other words, on perception. And this may be summarized in the question: what exactly do we hear in what is given to us to listen to?

In Germany, in Cologne, to be precise, parallel to musique concrète as described above, first Herbert Eimert and later Karlheinz Stockhausen in the early fifties defined a music based on an entirely different foundation. In this case, musicality is not formed from perception – a vague term, subjective and therefore deprecated in the structuralist context of the post-war era – but rather from the codification and determination of all sonic parameters and the conscious and precise control of all factors at every stage of the creative process. And thus, process music has emerged since the fifties as one of the basic elements of electroacoustic music. But what is the role of the computer in this context? To put it briefly, on the conceptive level it may be disregarded, because the concepts were essentially defined from the beginning, in Paris as in Cologne, but not only there, if we consider Xenakis or Varese. In this case, the protagonists of the scene immediately comprehended the intellectual impulses introduced here (a remarkable phenomenon, if we consider the blindness of their predecessors[4]), and the computer, which emerged in the early sixties, was no more than additional support for an approach that already existed.

And on the level of realizing the works, in other words, the practical aspect, it had no other function than that of providing the tools which had already been in existence since the invention of sound recording with a final polish – the "virtual" cut, for example, is merely a clone of the actual cut made using magnetic tape, thus merely an extension of the same; this is

sen: Was hören wir genau in dem, was man uns zu hören gibt?

Parallel zur erwähnten Musique Concrète definierten in Deutschland – um genau zu sein, in Köln – zuerst Heibert Eimert und später Karlheinz Stockhausen Anfang der fünfziger Jahre eine Musik, die auf einer völlig anders gearteten Basis aufbaut. Dabei wird das Musikalische nicht von der Perzeption her gestaltet – ein verschwommener Begriff, subjektiv und daher im strukturalistischen Kontext der Nachkriegszeit verpönt –, sondern von der Kodifizierung und Determination aller klanglichen Parameter und der bewußten und genauen Steuerung aller Faktoren auf allen Etappen des kreativen Prozesses. Und so taucht auch seit den fünfziger Jahren eine Prozeßmusik unter den Grundbausteinen der elektroakustischen Musik auf.

Welche Rolle aber spielt der Computer in diesem Zusammenhang? Auf der konzeptiven Ebene ist er – kurz gesagt – zu vernachlässigen, weil diese Konzepte im wesentlichen schon von Anbeginn definiert waren, in Paris wie in Köln, und nicht nur dort – man denke an Xenakis oder an Varese. Die ins Spiel gebrachten Denkanstöße waren von den unterschiedlichen Protagonisten der Szene sofort verstanden worden (ein bemerkenswertes Phänomen, wenn man an die Blindheit ihrer Vorgänger denkt[4]), und der Computer, der Anfang der sechziger Jahre auftauchte, war dann nichts anderes denn Unterstüzung bei einem Ansatz, der schon vorher existiert hatte.

Und auf der Ebene der Realisierung der Werke, also vom praktischen Gesichtspunkt her, hatte er keine andere Funktion, als den bereits seit Erfindung des Tonbandes existierenden Werkzeugen den letzten Schliff zu geben – der „virtuelle" Schnitt beispielsweise ist nur ein Klon des tatsächlichen Schnittes mittels Magnetband, also bloß seine Erweiterung; dasselbe gilt für den Sampler, der nichts anderes als eine Verlängerung des „Phonogène" der GRM darstellt – oder die Funktion hat, einige zusätzliche Werkzeuge einzubringen, die sich gut in die Reihe jener einfügen, die in den Jahrzehnten zuvor benutzt worden waren. Im Falle der Musique Concrète reihen sich diese Werkzeuge in jenes Arsenal ein, das das Unerhörte hörbar gemacht hatte, denn sie ermöglichen es, das Klangmaterial so zu bearbeiten, daß

also true for the sampler, which is nothing other than an extension of the "phonogène" of the GRM – or introducing additional tools which easily complemented those used in previous decades. As far as musique concrète is concerned, these tools were simply integrated into those used to make "the unheard-of" audible, since they made it possible to treat sound material so as to develop the aspects of perception that some had considered unimaginable. As far as process music is concerned, on the other hand, it became possible to refine the concepts and levels of interaction. Yet on the whole, it was simply a matter of continuing the development of elements that had already existed long before the computer.

There also seems to be a fundamental difference between the world of sound and the world of images. And this difference is practically underlined in red in a book that was published in 1996 on the tenth anniversary of the Prix Ars Electronica. As a member of the computer music jury explained it: "To put the matter in its bluntest form, it appears that the more technology is thrown at the problem, the more boring the results. People set out for new timbral horizons, get lost along the way in the writing of the code, the troubleshooting of the systems, and the funding to make the whole thing possible, then fail to notice that the results do not justify the effort. It is interesting to note that the jury for computer animation found an opposite result: in animation at least, the difference in quality between work done with cutting edge versus commonplace technology is immediately apparent to even the untrained observer. (...) In computer music, on the other hand, the merits of the works done with cutting edge versus commonplace technology are certainly opaque to the uninitiated, and often discernible only to those who have invested time and effort in acquiring expertise in the very same technology."[5] What is remarkable here is the observation that the computer as a

einige für undenkbar gehaltene Aspekte der Wahrnehmung herausgearbeitet werden. Im Falle der Prozeßmusik wiederum werden die Konzepte und Interaktionsebenen verfeinert. Aber alles in allem handelt es sich lediglich um eine Weiterentwicklung von Elementen, die bereits lange vor dem Computer existierten.

Es scheint übrigens einen grundlegenden Unterschied zwischen der Welt des Klangs und der Welt der Bilder zu geben. Und dieser wird geradezu mit Rotstift in einem Buch unterstrichen, das 1996 aus Anlaß des zehnjährigen Jubiläums des Prix Ars Electronica herausgekommen ist. Ein Mitglied der Computermusik-Jury erläutert es folgendermaßen: „Um es ganz kraß auszudrücken – je mehr Technologie, desto langweiliger die Ergebnisse. Menschen machen sich auf die Suche nach neuen timbralen Dimensionen, verirren sich beim Codeschreiben, Systemeinrichten und beim Geldauftreiben und übersehen dabei, daß die Resultate den Aufwand gar nicht rechtfertigen. Es ist interessant, daß die Jury für Computeranimation zu einem gegenteiligen Schluß kam: Zumindest in der Animation ist der Qualitätsunterschied zwischen Werken, die mit hochentwickelter Technologie, und solchen, die mit herkömmlicher Technologie produziert wurden, auch für den unbedarften Beobachter sofort erkennbar. [...] In der Computermusik hingegen ist die relative Wertigkeit von Werken, die mit hochentwickelter bzw. herkömmlicher Technologie produziert wurden, dem Uneingeweihten keineswegs einsichtig und wird oft nur von denen erkannt, die sehr viel Zeit und Energie dafür aufgebracht haben, sich ein Fachwissen bezüglich dieser speziellen Technologie anzueignen."[5] Bemerkenswert daran ist die Feststellung, daß der Computer als konzeptives Werkzeug offenbar keineswegs gehalten hat, was man sich von ihm zum Zeitpunkt seines Erscheines versprochen hatte. Und es scheint – immer im Lichte des eben zitierten Textes –, daß die signifi

conceptive tool has apparently not been able to hold what it seemed to promise when it first appeared. And it seems – still in light of the quotation above – that the most significant change that the computer has introduced into electroacoustic music is more to be found in the PC – which a composer may personally own – than in institutional mainframe computers.

This phenomenon may certainly be explained in a variety of ways, but if we think of the seventies, when it was nearly impossible to work comfortably at home and dependency on an institution was the rule, then this development naturally brought a type of part-time work into the studios, because the resources had to be shared with a number of other users, who also needed time. Since this is still the case in institutions today, it might explain why the works created there are sometimes quite disappointing. It is not so much a question of money – what does one machine cost in comparison with another? – but rather a question of time: how much time may a composer use to have control over his/her instruments? How much time may be devoted to the development of a work? It is obvious that from this perspective, the works created in private studios that have to make do with more modest means – which, on the side, also requires more imagination – stand a better chance of being more interesting. From a historical perspective, the PC is the most important development since the invention of sound recording, but not because it is a computer, but rather because it allows the composer complete command of the instrument without dependence on the dictates of an institution.

I am not sure how my colleagues from the computer animation jury would react to this observation, but it is obvious that the situation of independent computer filmmaker is much less enviable than that of the independent composer.

kantesten Veränderungen, die der Computer in der elektroakustischen Musik mit sich gebracht hat, noch eher beim PC zu suchen sind – den der Komponist selbst besitzen kann – als bei den institutionellen Großrechnern.

Dieses Phänomen kann man zweifellos auf verschiedene Art erklären, aber wenn man an die siebziger Jahre denkt, als es fast unmöglich war, bequem bei sich zuhause zu arbeiten, und die Abhängigkeit von einer Institution die Regel war, so brachte diese Entwicklung natürlich eine Form der Teilzeitarbeit in den Studios mit sich, weil die Ressourcen auch zeitlich mit mehreren anderen Nutzern geteilt werden mußten. Da dies in den Institutionen auch heute noch der Fall ist, würde es erklären, warum die dort entstandenen Werke teilweise recht enttäuschend sind. Es ist nicht so sehr eine Geldfrage – was kostet die eine Maschine im Vergleich zu der anderen? –, sondern viel eher eine Zeitfrage: Wieviel Zeit kann ein Komponist auf die Beherrschung seiner Instrumente verwenden? Wieviel Zeit widmet er der Erarbeitung seines Werks? Es ist klar, daß unter diesem Gesichtspunkt die Werke aus privaten Studios, die mit viel bescheideneren Mitteln auszukommen haben – was nebstbei auch zu stärkerer Imagination zwingt –, eine gute Chance haben, interessanter zu sein. Historisch betrachtet, ist sicherlich der PC die wichtigste Entwicklung seit dem Aufkommen des Tonbandes, aber nicht so sehr, weil er ein Computer ist, sondern weil er dem Komponisten erlaubt, sein Instrumentarium von Grund auf zu beherrschen, unabhängig vom Diktat einer Institution.

Ich weiß nicht, wie meine Kollegen von der Computeranimation auf diese Feststellung reagieren, aber es ist offensichtlich, daß die Situation des unabhängigen Computer-Cineasten viel weniger beneidenswert ist als jene des unabhängigen Komponisten.

Und die Zukunft?

Daß die Stellung des Computers als Werkzeug zur Verbesserung der Klangmanipulation und -Bearbeitung unangreifbar ist, versteht sich von selbst, aber welche Entwicklungen

AND THE FUTURE?

There is no doubt that the current position of the computer as a tool for improving the manipulation and processing of sound is irrefutable, but what developments will the computer effect in the near future? Which profound changes can and will it trigger in the way musical creativity functions? Which perspectives will it open up in musical praxis that have not already been announced a hundred times in the last forty years? Here and now, we can already see the beginnings of real-time treatment, not only in the area of sound synthesis and interaction, but also in the highly sophisticated methods of treatment, such as are now being utilized after a time lapse. We can imagine interactive systems – more powerful and dependable than anything that exists today. The makers of the CAVE at the Ars Electronica Festival 96 have already given us a glimpse into this future.

Basically, the tasks of the future have already presented themselves to us, at least as far as the most important concepts deriving from musique concrète and from electroacoustic music thereafter have not yet all been electronically formulated. In addition, the majority of the composers in this field are still using working techniques that seem to have been directly inherited from instrumental music, as Bob Ostertag also quite rightly remarks in his article quoted above: "This is even more apparent when one considers that, formally speaking, the large majority of pieces involving computer response to live instrumentalists are simply variations in algorithmic composition." I believe it is possible to maintain with some certainty that even today, sound recording – which ought to be the foundation for a fundamental questioning of musical practice – still has only a very marginal effect in raising consciousness, even and especially within the institution of electroacoustic music. How else could we explain the fact that electroacoustic is still no more than an

wird der Computer in den kommenen Jahren bewirken? Welche tiefgreifenden Veränderungen kann und wird er in der Funktionsweise der musikalischen Schöpfung auslösen? Welche Perspektiven wird er der musikalischen Praxis eröffnen, die nicht schon hunderte Male in den letzten vierzig Jahren angekündigt worden sind? Hier und jetzt kann man schon die Echtzeitverarbeitung absehen, nicht nur im Bereich der Klangsynthese und der Interaktion, sondern auch bei äußerst ausgefeilten Bearbeitungsmethoden, wie man sie zeitverzögert schon heute einsetzt. Im Bereich der prozeßbasierten Musik kann man sich interaktive Systeme vorstellen – leistungsfähiger und zuverlässiger als alles, was heute existiert. Einen kleinen Blick in diese Zukunft haben uns die Macher des CAVE beim Ars Electronica Festival 96 ja schon gewährt.

Im Grunde liegen die zukünftigen Aufgaben schon vor uns, zumindest insofern, als die wichtigsten Konzepte, die sich aus der Musique Conctrète und in weiterer Folge der elektroakustischen Musik ableiten lassen, noch nicht alle elektronisch ausformuliert wurden und der Großteil der in diesem Bereich tätigen Komponisten sich noch heute Arbeitstechniken bedienen, die unmittelbar aus der Instrumentalmusik ererbt scheinen, wie das übrigens Bob Ostertag zu Recht in seinem oben erwähnten Artikel anmerkt: „Dies wird umso deutlicher, wenn man bedenkt, daß, formal gesprochen, der Großteil jener Werke, bei denen der Computer auf Live-Instrumentalmusiker reagiert, lediglich eine algorithmische Komposition von Variationen darstellt." Ich glaube, man kann mit einiger Sicherheit behaupten, daß selbst heute noch die Tonaufzeichnung – die doch Grundlage einer ganz fundamentalen Infragestellung der musikalischen Praxis sein müßte – nur extrem marginal bewußtseinbildend wirkt, auch und gerade innerhalb der Institution Elektroakustik. Wie anders wäre zu erklären, daß die letztere noch

appendage to the traditional music structures – the universities, conservatories, concert halls – which are all part of the heritage from the classical music of Europe. And that is what needs to be fundamentally changed. The computer will follow later.

heute nur ein Anhängsel der traditionellen musikalischen Strukturen ist – der Universitäten, Konservatorien, Konzertsäle -, die alle aus der musikalischen Klassik Europas übernommen wurden. Und das gilt es, von Grund auf zu verändern. Der Computer folgt dann später.

ROBERT NORMANDEAU

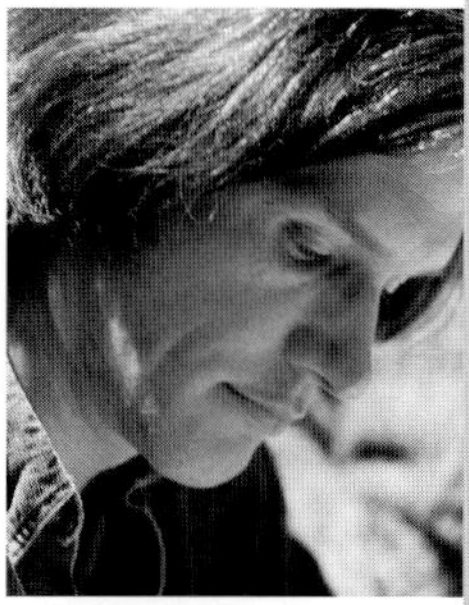

Robert Normandeau (CDN) is currently involved in acousmatic music. He was a member of the computer music jury.

Robert Normandeau (CDN) arbeitet zur Zeit im Bereich der akusmatischen Musik. Er war Mitglied der Jury für Computermusik.

[1] "When I first used the term "musique concrète" in 1948, I wanted to emphasize a reversal of musical work through the use of this adjective. Rather than coding musical ideas using the symbols of the notation system and leaving the actual implementation of these ideas to familiar instruments, in this case it was a question of gathering what was sonically concrete, regardless of its origin, abstracting from it the musical value inherent to it. This expectant attitude justifies the use of the term and marks the beginning of an entirely new direction in thinking and acting." Pierre Schaeffer, "Traité des objets musicaux", Seuil, Paris 1966, p. 23.

[2] The Acousmatics were pupils of Pythagoras, who received his teachings as he held his lectures from behind a curtain, so that they would not be distracted by his physical presence.

[3] As Michel Chion aptly defined it in his book "L'art des sons fixès", Èd. Métamkine, Fontaine, 1991.

[4] In this context, Schaeffer's astonishing book "À la recherche d'une musique concrète", published 1952 in Seuil, is to be highly recommended; in particular, it contains the "Premier Journal de la musique concrète 1948-49" and is full of the fundamental concepts, which were to be developed in the following fifty years.

[5] Bob Ostertag, p. 60

[1] „Als ich 1948 den Terminus 'Musique concrète' erstmals verwendete, wollte ich durch dieses Adjektiv eine Umkehrung der musikalischen Arbeit verdeutlichen. Anstatt die musikalischen Ideen durch die Symbole des Notensystems zu kodifizieren und ihre konkrete Ausführung bekannten Instrumenten anzuvertrauen, handelte es sich hier darum, das klanglich Konkrete zu sammeln, wo immer es auch herkommen mochte, und daraus jene musikalischen Werte zu abstrahieren, die ihm innewohnen. Diese Erwartungshaltung berechtigte zur Verwendung des Begriffs und markierte den Zugang zu völlig neuen Denk- und Handlungsrichtungen." Pierre Schaeffer, „Traité des objets musicaux", Seuil, Paris 1966, S. 23.

[2] Die Akusmatiker waren Schüler des Pythagoras, die seine Lehren erhielten, indem er seine Vorlesungen hinter einem Vorhang verborgen hielt, auf daß sie durch seine physische Anwesenheit nicht abgelenkt würden.

[3] So hat Michel Chion sie treffend in seinem Buch „L'art des sons fixés", Éd. Métamkine, Fontaine, 1991, definiert.

[4] In diesem Zusammenhang empfiehlt sich die Lektrüe von Schaeffers erstaunlichem Buch „À la recherche d'une musique concrète", erschienen 1952 in Seuil, das besonders das „Premier Journal de la musique concrète 1948–49" enthält und reich ist an jenen grundlegenden Konzepten, die im darauffolgenden halben Jahrhundert ausgearbeitet werden sollten.

[5] Bob Ostertag, S. 60

Phoenix Twilight at Turntable Dusk

Auferstehung aus der Asche des Vinyl

Naut Humon

Das Buch zum Prix Ars Electronica 96 enthielt ein spekulatives Szenario von Bob Ostertag.

The 1996 Prix Ars Electronica publication contained a speculative scenario by Bob Ostertag. He postulated that the Computer Music classification is headed for a dead end due to an inordinate preoccupation with "higher learning" and the tendency of the participants to be very insular, not attempting to include many of new, hybrid types of music. Because the Computer Music category is attracting only a marginal public audience outside its own creative cocoon, this purist concrete tradition is in need of additional sonic surgery. Having now been a member of the jury for this category, I can offer similar sentiments to Ostertag's, but with a quite different slant. He stated, "it appears that the more technology is thrown at the problem, the more boring the results." This does not have to be the case. The more advanced software & equipment in the right hands, can paint sonic wonders which illustrate timbral design that has significantly matured from the possibilities of past techniques. Do these improvements manifest a parallel jump in raw musical impact? Yes, sometimes, especially when

Darin postuliert er, daß die Klassifikation „Computermusik" in eine Sackgasse führen wird, und zwar wegen ihrer übertriebenen Ausrichtung auf „höhere Sphären" und wegen der Tendenz der Teilnehmer, ein Dasein im Elfenbeinturm zu führen und keinen Versuch zu unternehmen, neue, hybride Formen der Musik einzuschließen. Da die Computermusik auch nu einen marginalen Teil des Publikums außerhalb des eigenen kreativen Kokons anzieht, bedarf ihre purististisch-konkrete Tradition dringend der zusätzlichen akustischen Chirurgie von außerhalb. Als Mitglied der Musikjury kann ich Ostertags An sichten nachempfinden, wenn auch mit einem etwas anderen Ansatz. Ostertag sagt: „Je mehr Technologie, desto langwe liger das Ergebnis", aber das muß nicht unbedingt so sein. Die hochentwickelte Software und Ausrüstung kann – in den richtigen Händen – klangliche Wunder he vorbringen, die ein timbrisches Design bie ten, das sich weit über die Möglichkeiten der früheren Techniken hinausentwickelt hat. Aber manifestiert sich diese Weiterentwicklung auch beim schieren musikalischen Eindruck in einem ähnlichen großer Sprung? Ja, manchmal, besonders wenn ein Komponist, der sich des akusmatische Erbes bewußt ist, die Angelegenheit in di Hand nimmt und seine persönliche Vision bewußt und geschickt über dieses Erbe hinausführt.

Die elitäre elektroakustische Gemeinc ist relativ klein, und die ihr zugrunde lie-

the composer aware of the acousmatic heritage takes matters into their own hands and skillfully extends a personalized vision.

The elite electroacoustic community is relatively tiny and its underlying uniformity is a quality echoed in all other broader musical genres as well. But the fact that formal Computer Music has been underground for over fifty years is a testament to a specific style that won't be compromised even when its aesthetic development appears at a virtual deadlock. Stuck in a hall of mirrors, Computer Music looks at itself – and if it doesn't ever look for an exit, it may remain conveniently isolated. Reclusiveness of this nature, however, will continue to produce the occasional masterworks by artists whose audio novellas are unencumbered by the common clichés or equipment characteristics of their particular era and rise to a level of distinction. Transcending the stricter acousmatic computer predicament will mean more interaction with other sounds, methods and possibly even mechanisms outside the formal construct of the genre, possibly to even draw from fringe experimentalists and the edges of beatpop culture. The competition has arrived at the point where certain artists may not have to rely on just computerized interfaces for sonic performances at all – and yet the musical result would be closely analogous to any wild digital manifestations. To evaluate these fresh structures, the criteria for selective listening must be amended to include musical forums where computer involvement is ancillary. A major example of this is the transformation and development within the last decade or so of international DJ culture as it migrates from broadcast music selectors and dancefloor rhythm providers to the art of Turntablism and Digital Jockeys.

These vinyl and CD operators manipulate their media via a manual,

gende Uniformität hallt auch in anderen, breiteren musikalischen Genres nach. Die Tatsache, daß die formale Computermusik nun seit über fünfzig Jahren im Untergrund dahindämmert, ist Beweis für einen spezifischen Stil, der auch dann nicht aufs Spiel gesetzt wird, wenn seine ästhetische Entwicklung an einem toten Punkt angekommen zu sein scheint. Eingesperrt in ein Spiegelkabinett, betrachtet sich die Computermusik selbst – und wenn sie nicht bald nach einem Ausgang sucht, wird sie wohl ganz bequem isoliert bleiben. Eine (Ab-)Geschlossenheit dieser Art aber wird weiterhin eben nur dann und wann ein Meisterwerk hervorbringen, wenn ein Künstler – unbelastet von den allgemeinen Klischees und von der für eine spezielle Epoche charakteristischen Ausrüstung – mit einer Audio-Novelle über sich hinaus und auf eine vom Bisherigen unterscheidbare Ebene wächst. Über die strikten Vorgaben der Akusmatik hinauszugehen, bedeutet mehr Interaktion mit anderen Klängen, Methoden und möglicherweise sogar Mechanismen außerhalb des formalen Konstrukts des Genres, bedeutet vielleicht sogar, von randständigen Experimentalisten oder von den Grenzgängern der Beatpop-Kultur zu profitieren.

Der Wettbewerb ist an einem Punkt angelangt, an dem bestimmte Künstler nicht einmal mehr computergestützte Interfaces für ihre Aufführungen bräuchten – die musikalischen Resultate könnten auch analog erzeugt werden und dennoch kaum von den wildesten digitalen Manifestationen zu unterscheiden sein. Um diese frischen Strukturen zu evaluieren, müssen die Kriterien unseres selektiven Hörens so angepaßt werden, daß sie auch musikalische Formen umfassen, in denen der Computer nur ein Hilfsmittel ist. Eines der besten Beispiele ist die Veränderung und Entwicklung des letzten Jahrzehnts im Bereich der internationalen Disk-Jockey-Kultur, die sich von einfachen Tanzboden- und Hintergrundmusiklieferanten hin zu einer „Plattenteller-Kunst" und zum digitalen Jockeytum verschoben hat.

Diese Vinyl- und CD-Operatoren manipulieren ihr Medium mit Hilfe eines manuellen, analogen Mix-Stils, der erst kürzlich begonnen hat, die Möglichkeiten zu erahnen, die im Informationszugang über Modems und in computergestützen

analog mixing style that only recently has begun to recognize the possibilities inherent in information access modems and CPU workstation desks. Real-time software that stretches perception, position, and space will work to alter the DJ's role from disk playback mixer to sound diffusion projectionist, thereby blurring the worlds between storage/retrieval performers and their electroacoustic counterparts.

So why the DJ? As Paul D. Miller/a.k.a. DJ Spooky observes: "In the electronic milieu that we all move in today, the DJ is a custodian of aural history who acts as the cybernetic inheritor of the improvisational tradition of jazz, where various motifs would be used and recycled by the various musicians of the genre. The DJ combines the musical expression of other musicians with their own and in the process creates a seamless flow of music. In the mix, creator and re-mixer are woven together in the syncretic space of the text of samples and other sonic material to create a seamless fabric of sound that in a strange way mirrors the modern macrocosm of cyberspace where different voices and visions constantly collide and cross-fertilize one another. The linkages between memory, time, and place, are all externalized and made accessible to the listener from the viewpoint of the DJ who makes the mix. Thus, the mix acts as a continuously moving still frame, a camera lucida capturing moment-events. The mix, in this picture, allows the invocation of different languages, texts, and sound to converge, meld, and create a new medium that transcends its original components. The sum created from this audio-collage leaves its original elements far behind. Call the mixes and songs generated by the assembly process of DJing and sequencing etc. the social construction of memory. This shows the inexplicable mutability of sound as different people share the memories brought about by the same songs. It

Arbeitsplätzen steckt. Eine Real-Time-Software, die Perzeption, Position und Raum zu strecken vermag, wird dazu beitragen, die Rolle des DJ vom einfachen Wiedergabekünstler hin zu einem Projekteur der Klangverteilung zu verändern und dabei die Grenzen zwischen dem bloßen Sucher und Mixer gespeicherter Dinge und seinen elektroakustischen Gegenpart verwischen.

Warum also der DJ? Paul D. Miller alias DJ Spooky bemerkt: „Im elektronischen Milieu der Gegenwart ist der DJ ein Kustos der auditiven Geschichte, der als kybernetischer Erbe der Improvisationstradition des Jazz gelten kann, wo zahlreiche Motive von den unterschiedlichen Musikern des Genres immer wieder benützt und recycelt wurden. Der DJ kombiniert die musikalischen Ausdrucksformen anderer Musiker mit seiner eigenen und erzeugt in diesem Prozeß einen nahtlosen Fluß der Musik. Im Mix werden Schöpfer und Mixer in den synkretischen Textraum von Samples und anderem Klangmaterial verwoben, um ein nahtloses Klanggewebe zu erzeugen, das auf eine seltsame Weise den modernen Makrokosmos des Cyberspace widerspiegelt, in dem verschiedene Stimmen und Visionen ständig kollidieren und sich gegenseitig befruchten. Die Verknüpfungen zwischen Gedächtnis, Zeit und Raum werden externalisiert und dem Hörer aus dem Blickwinkel des DJs zugänglich gemacht, der den Mix erstellt. So gesehen, dient der Mix als ein sich ständig bewegender Bildkader, als eine camera lucida, die Momentereignisse einfängt. In diesem Bild erlaubt der Mix die Heraufbeschwörung unterschiedlicher Sprachen, Texte und Klänge, die sich zusammenfinden, verschmelzen und ein neues Medium bilden, das weit über seine Komponenten hinausgeht. Das Ganze ist auch in dieser Audio-Collage mehr als die Summe der Teile. Man könnte die Mixes und Songs, die durch den Assemblierungsprozeß des DJ mit Sequencern usw. entstehen, die soziale Konstruktion der Erinnerung nen-

demonstrates the uncanny power to metamorphosize, through audio alchemy, the passage of sound into a kind of unspoken story, that like its predecessor, the oral tradition, can pass on „tales" of songs."

Interpreting these "tales" by taking brief snatches of time and reconfiguring these as rhythms, soundscapes, and solos is the province of Turntablism. This is where DJ's manipulate vinyl platters with their hands, backtracking the LP record surface along with moving the crossfader mixer with speed and precision. When it hits a groove, a Turntable Jazz show offers a dazzling display of inspired animation and virtuosity. Synchronized jams fracture into bouts of individual turntable juggling. Beats are broken down and rebuilt; tempos shift. Vocal tracks are mutated into stuttering mantras delivered in a panoply of scratch styles (there are more than 300) – forwards, scribbles, backspins, flares, crabs, cuts, drags, chops, tears, chirps, and transforms. Any scratch DJ can interpret any instrument – be it drums, guitars, basses or sound effects – and transform these sonic foundations into a sophisticated and lyrical modern musical language. "We're really not just the normal DJ's anymore," Q-Bert, a world champion DJ extraordinaire, declares: "We're turntablists who play the turntable like a real musical instrument. We don't only just perform other people's music – we make our own". John Carluccio, director of the Battle Sounds documentary adds – "The Hip Hop DJ of the late 70's heard a break, one small portion of a musical composition, this break was repeated over and over by playing two of the same record (play one and cue two, then play two and cue one etc.) The mistake of the scratch, a sharp pull of a record, occurred, creating incredible noise, and when the DJ mastered this irregularity, it became a sound element of composition". Rob Swift of Brooklyn's notorious X-Men continues:

nen. Dies zeigt die unerklärliche Veränderbarkeit des Klanges, wenn verschiedene Menschen die Erinnerung teilen, die von den gleichen Songs hervorgerufen werden. Es demonstriert die geheimnisvolle Kraft der Metamorphose durch eine Audio-Alchimie, den Übergang des Klangs in eine Art unausgesprochene Geschichte, die wie ihr Vorläufer, die Erzähltradition, ganze 'Erzählungen' von Songs weitertradieren kann."

Die Interpretation dieser „Erzählungen" – indem kurze Zeitabschnitte genommen und als Rhythmen, als Klanglandschaften, als Soli neu konfiguriert werden – ist die Domäne des „Turntablism". Hier manipulieren DJs Vinylscheiben mit der Hand, verfolgen die Plattenspuren auf der LP und regeln gleichzeitig schnell und präzis das Mischpult. Wenn sie einschlägt, dann bietet eine Turntable Jazz Show eine faszinierende Darstellung inspirierter Animation und Virtuosität. Synchronisierte Jam-Sessions zerflattern in einen Wettstreit individuellen Tellerdrehens. Beats werden auseinandergenommen und neu aufgebaut, die Tempi werden verschoben. Vokalstücke mutieren zu stotternden Mantras, dargeboten in einer Fülle von Scratch-Stilen (es gibt über 300 davon) – Forwards, Scribbles, Backspins, Flares, Crabs, Cuts, Drags, Chops, Tears, Chirps und Transforms.

Jeder Scratch-DJ kann jedes beliebige Instrument interpretieren – Schlagzeug, Gitarre, Bass oder Klangeffekte – und diese klanglichen Grundlagen in eine ausgefeilte und lyrische moderne Musiksprache verwandeln. „Wir sind nicht mehr einfach nur die Plattenaufleger", erklärt Q-Bert, ein weltmeisterlicher DJ der Extraklasse: „Wir sind Plattentellerkünstler, die den Plattenteller spielen wie andere ein Instrument. Wir spielen auch nicht einfach die Musik anderer Leute ab – wir machen unsere eigene!" John Carluccio, Regisseur des Dokumentarfilms „Battle Sounds", fügt hinzu: „Der Hip-Hop-DJ der späten Siebziger hörte einen Break, eine kleine Portion einer musikalischen Komposition, und dieser Break wurde wieder und wieder abgespielt, indem zwei gleiche Platten liefen (eins spielen, zwei abhören, dann zwei spielen und eins abhören usw.). Der Fehler des 'Kratzens' – 'Scratch' – passierte, ein scharfer Ruck an der Platte, und ein unglaubliches Geräusch entstand, und wenn der DJ diese Unregelmäßigkeit

"You can incorporate the fader of a mixer and cut the sound off and on, but you can move it back and forth in rhythm, on beat to a another song that's playing. That's what the scratching motion consists of". Grandmaster Theodore (USA) is credited with first stumbling across this innovation in his home bedroom one day in the late seventies by accidentally rubbing the record while backspinning it to hold the beat he was on as his mother was complaining about the excessive loudness in the house. A short time later he and associate Grandmaster Flash were showcasing this crazy action at their Bronx music parties where they would go head to head setting the stage for a new cult of technique – the DJ Battle. This friendly olympic style competition features condensed routines that transfigure a pair of Technic SL -1200 decks under lightning fast stylus gouging to an expressionistic plateau similar in character to many historical tape splice or computer-generated pieces. Pushing the realm of speed, dexterity and re-arranging to a fine art, DJ's DXT, Cash Money, Jazzy Jeff, Roc Raida, Cutmaster Swift, Noise, Mixmaster Mike and the phenomenal Q-Bert etc. all share credit for furthering the founding Battle legacy.

Actual DJ teams have emerged where ensembles of ten or more separate turntables can all be heard at once. Crews such as the Invisible Skratch Piklz, Beat Junkies, X-Men, Fifth Platoon, Enforcers and others unveil group dynamics and interplay unheard of in Hip Hop.

From the more abstract and experimental zone, Christian Marcley, Otomo Yoshihide, DJ Spooky, Olive, and Project DARK, etc., deal in the domains of mood sculpture or layered sources which recombine as walls of impressionistic soundfields. By deconstructing, fragmenting and reinventing their aural pasts, these oblique architects act as analog filters who on occasion completely re-

zu beherrschen gelernt hatte, wurde sie z einem Klangelement der Komposition." Rob Swift von den bekannten X-Men aus Brooklyn ergänzt: „Man kann den Fader eines Mischpults einbauen und den Ton ein- und ausblenden, aber man kann die Platte auch im Rhythmus eines anderen gerade laufenden Stücks vor- und zurück bewegen – und das ist die Scratch-Bewegung."

Großmeister Theodore (USA) wird nachgesagt, er sei als erster auf diese Innovation gekommen, als er eines Tages in den späten Siebzigern in seinem Schla zimmer eher zufällig über eine Platte rub belte, als er sie zurückdrehte, um ihren Beat zu halten, und sich seine Mutter üb den höllischen Lärm im Haus beschwerte Kurze Zeit später zeigten er und Co-Groß meister Flash diese verrückte Aktion in ih ren Music Parties in der Bronx, wobei sie Kopf an Kopf arbeiteten – der DJ Battle war geboren. Diese friedliche olympische Disziplin umfaßt einige kondensierte Rou tinen, die zwei Technics SL-1200 Decks unter blitzschneller Bearbeitung von Ton-arm und Tellern in eine expressionistisch anmutende Plattform verwandeln, mit Klängen, die jenen früherer Tonband- od Computerarbeiten um nichts nachstehen Indem sie die Geschwindigkeit, Geschick lichkeit und die Neuarrangements auf ei geradezu künstlerisches Niveau hoben, h ben sich DJs wie DXT, Cash Money, Jazzy Jeff, Roc Raida, Cutmaster Swift, Noise, Mixmaster Mike und der phänomenale C Bert nebst weiteren in die Ehrenliste der „DJ-Battle"-Gründer und -Entwickler ein-getragen.

Ganze DJ-Teams sind entstanden, be denen Musik von zehn und mehr Platten tellern gleichzeitig gehört werden kann. Crews wie die Invisible Skratch Piklz, Bea Junkies, X-Men, Fifth Platoon, Enforcers und andere enthüllen eine gruppendyna-mische Arbeit und ein Zusammenspiel, d im Hip Hop bisher unbekannt war. Im noch abstrakteren und experimentelleren

customize the actual physical mediums of playback equipment and vinyl. Barry Schwartz invented a giant turntable apparatus with a metal disc surface that's six feet in diameter! An extended "tone arm" that drops a stylus of dry ice onto the huge plate emits an unearthly howl of thermal reactive tones accompanied by an upside-down phonograph player that is revolving underneath.

As the roles of the turntablist and the electroacoustician collide, the possibilities for real frontiers in sound and construction are up for grabs.

Wheels of steel & silicon will merge or splinter and remain the phoenix at the door of vinyl turntable obsolescence.

Bereich arbeiten Christian Marcley, Otomo Yoshida, DJ Spooky, Olive, Project DARK und andere an Stimmungsskulpturen, mit überlagerten Tonquellen, die sich zu Wänden impressionistischer Klangfelder verdichten. Durch die Dekonstruktion, Fragmentation und Neuerfindung ihrer auralen Vergangenheit fungieren diese schrägen Architekten als analoge Filter, die bei Bedarf auch das aktuelle physische Medium des Playback-Equipments samt Vinylscheibe neu konfigurieren und umbauen. Barry Schwartz erfand einen riesigen Plattenteller mit einer metallenen Plattenfläche von einem Meter achtzig Durchmesser! Ein verlängerter Tonarm mit einer „Nadel" aus Trockeneis, der diese enorme Platte abtastet, gibt ein außerirdisches Geheul thermischer Reaktionsklänge von sich, begleitet von einem darunter angebrachten, kopfüber montierten Phonographen ...

In dem Maße, in dem die Rolle des Plattendrehers und des Elektroakustikers kollidieren, öffnen sich neue Horizonte für Klang und Konstruktion – zum Greifen nah.

Und Scheiben aus Stahl und aus Silikon werden verschmelzen oder sich teilen – wie ein Phönix steigt der Plattenteller aus der Asche der scheinbaren Obsoletheit der Vinyl-LP.

NAUT HUMON

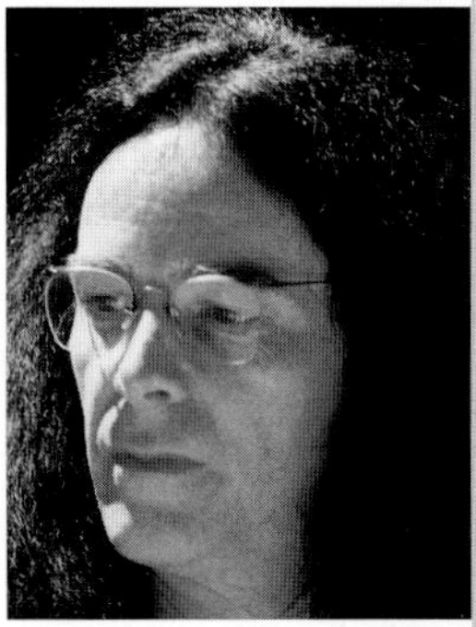

Naut Humon (USA), curator of "Sound Traffic Control" and director of operations for "Recombinant", was a member of the Computer Music Jury.

Naut Humon (USA), Kurator von „Sound Traffic Control" und operativer Leiter von „Recombinant", war Mitglied der Jury für Computermusik.

Computer graphics – where are we now ...

Computergraphik – eine Standortbestimmung

Mitch Mitchell

Die Computergraphik ist von der Welt der Film- und Unterhaltungsbranche gekidnappt worden. Was einst als hochintellektuelles Unterfangen in Wissenschaft und Kunst begonnen hat, ist zur hauseigenen Spielwiese einer Karnevals-Bühnenshow geworden – wenn auch zur teuersten weit und breit ...

Computer graphics have been hi-jacked by the worlds of film and entertainment. What started out as a highly intellectual endeavour in both science and art has been reduced to playing front of house at a carnival side-show – be it the most expensive one in town.

Computer graphics were originally developed to permit easier understanding of what was going on inside the machine. They were initially a direct view into the core of the computer. At first static blinking lights, the visual interface soon became moving dots on a cathode ray tube. These dots accelerated with time and eventually blurred into lines. The lines soon joined to become shapes, then the shapes became so-called wire frames of three-dimensional bodies and eventually these took on the appearance of real world objects. By hidden line removal they gave the impression of solid objects with solid surfaces.

Early on, computers were very expensive and only resided in major organisational locations such as research establishments and academic institutions. They were slow and processing took a long time. Specific research projects with carefully projected aims, and the operators' own experiments, in "down time", were the only activities usually possible. Thus most endeavours were largely of an intellectual nature

Ursprünglich wurde die Computergraphik entwickelt, um das, was im Inneren der Maschine vorgeht, leichter verständlich zu machen – ein direkter Blick in Herz des Computers. Waren es anfänglich noch statische, blinkende Lichter, so wurden daraus bald bewegte Punkte auf der Kathodenstrahlröhre, dem neuen Interfac und diese Punkte bewegten sich rasch fo zogen eine Spur nach sich und wurden zu Linien. Die Linien vereinigten sich zu Formen, dann wurden die Formen zu den so genannten Drahtgitter-Modellen dreidimensionaler Körper, und irgendwann nah men diese das äußere Erscheinungsbild von Körpern aus der realen Welt an. Durc Entfernung der dem Betrachter verborgenen Linien erweckten sie den Eindruck fester Objekte mit einer festen Oberfläche

In der Frühzeit waren Computer sehr teuer und fanden sich nur bei größeren Organisationen wie Forschungseinrichtungen und akademischen Instituten. Sie waren langsam, und die Verarbeitung der Daten dauerte und dauerte. Spezifische Forschungsprojekte mit sorgfältig geplanten Zielen – und außerhalb der Betriebszeiten noch die Experimente der Mitarbeiter – waren normalerweise die einzigen möglichen Aktivitäten. Und so blieben auch all Ansätze von weitgehend intellektueller N tur – zumeist naturwissenschaftlich -, unc wenn schon „künstlerisch", dann aber gleichfalls auf intellektueller Ebene. Dam hätten sich die Dinge in die verschiedensten Richtungen entwickeln können, und

either scientific in objective, or "artistic" with a capital "A". At this stage things could have gone in a number of directions – some large organisations established an artist in residence with the idea that these mammoth machines could become the tools of great art. Abstract and strange works were often created in these circumstances as the medium was explored in a cerebral sort of way. Access was allowed only to a very small number of acolytes – it was not available to the community in general.

The main technological characteristic of computer development has been the speed with which capacity has increased, processing speed has increased and price has fallen. This has resulted in an ever-increasing democratisation of computer power. Additionally, in the early days, massive energy was put into computer research and development work by military and space administrations throughout the world. Much of this effort was directed at the visualisation process for simulation and control purposes. The end of the cold war, and pretty well at the same time, the cost cutting in the space program, has meant that organisations engaged in computer research for these industries have had to seek out alternative markets. Thus graphics engines and software have become readily available for ever lower prices.

The use of computers in imaging has followed roughly two parallel paths. These are 3D and animation on the one hand, and image modification and manipulation on the other. The latter has provided the main push and the most popular applications. In the wider world of print and photography, the development of Photoshop and programs like it, beside computers capable of running it fast enough, has opened up a huge world of image "sweetening". Because visual data takes up so much space, the moving image has been more of a problem. It is easy to manipulate single images with a photo

einige große Organisationen leisteten sich einen „Artist in Residence" mit dem Hintergedanken, daß diese Mammutmaschinen vielleicht doch auch Werkzeuge zur Schaffung von großer Kunst sein könnten. Unter diesen Bedingungen entstanden oft abstrakte und seltsame Werke – das Medium wurde auf eine nicht minder kopfbetonte Weise erforscht. Der Zugang war nur wenigen Auserwählten möglicht, gewöhnlich Sterbliche blieben ausgeschlossen.

Die technologische Hauptcharakteristik der Computerentwicklung war die Geschwindigkeit, mit der sich die Kapazität vergrößerte, die Rechengeschwindigkeit steigerte und der Preis fiel. Dies hat zu einer immer stärkeren Demokratisierung der Rechnerleistung geführt. Und darüber hinaus wurde in jenen frühen Tagen viel Energie in Erforschung und Entwicklung von Computergraphik gesteckt, besonders seitens des Militärs und der Raumfahrtorganisationen der ganzen Welt. Viel davon floß in die Entwicklung von computergraphischen Prozessen zur Simulation und Steuerung. Das Ende des Kalten Krieges und die fast gleichzeitigen Budgetkürzungen im Raumfahrtbereich brachten es mit sich, daß immer mehr Firmen, die ursprünglich für diese Branchen im Bereich der Forschung und Entwicklung tätig gewesen waren, sich um neue Märkte umsehen mußten. Und daher wurden Graphikmaschinen und die zugehörige Software zu immer niedrigeren Preisen erhältlich.

Die Verwendung des Computers in der Bilddarstellung findet im wesentlichen auf zwei parallelen Ebenen statt, zum einen auf der 3D- und Animationsebene, zum anderen im Bereich der Bildbearbeitung und -manipulation. Gerade letztere stellt eine der Haupttriebfedern dar und bietet die bekanntesten Anwendungen. In der Welt des Drucks und der Photographie hat die Entwicklung von Programmen wie PhotoShop – samt einer Hardware, auf der sie schnell genug laufen – eine riesige Welt der „Bildverschönerung" eröffnet. Da aber Bilddaten viel Platz benötigen, blieben die bewegten Bilder immer etwas problematischer. Es ist einfach, ein einzelnes Bild mit einem Retuschierprogramm zu bearbeiten, aber bei rund einem Megabyte pro Kader stellen selbst gewöhliche Fernsehbilder schon große Ansprüche an die

retouch package, but at approximately one megabyte per frame, even ordinary television makes pretty severe hardware demands. Thus digital effects for television use were originally implemented through special dedicated hardware which was tremendously expensive. As the years went by, the capabilities of this equipment became more and more sophisticated whilst the prices ever spiralled downwards. At the time of writing, special software, running on so-called open-platform off-the-shelf computers, is just becoming the norm. In the area of 3D computer-generated imaging, titles for TV were initially about all that could be dealt with – flying logos as they were laughingly dubbed. CGI in television advertising and title design has been the subject of a series of visual fashions. First there were the green grids and wire frames, then the flying chrome logos, then morphing, and more recently, an entire style based on the ability to blur and soften whole areas of the screen, whilst keeping selected planes in focus. In the world of the television commercial CGI has become an essential tool used to some degree in almost every job.

But in the world of theatrical motion pictures, the quantity of data required for each frame has been a major obstacle. However, with personnel migrating between commercials and feature production, a requirement for many of the computer imaging techniques used in television has developed. From time to time there have been efforts to use computer graphics in film, but most of the early attempts were doomed, either by their expense, inflexibility, or the commercial failure of the movies themselves. Thus films like "Westworld", "Tron" and "The Last Starfighter" were unsuccessful early adopters which used massively oversized mainframes to do what could now be attempted on most home PC's. In the end it was "compositing" of

Hardware. Deswegen wurden digitale Effekte für das Fernsehen auch ursprünglich nur mit horrend teurer, spezieller Hardware eingesetzt. Im Lauf der Jahre wurden die Fähigkeiten dieser Ausrüstung immer besser, während gleichzeitig der Preis immer weiter sank. Gegenwärtig wird gerade eine Kombination aus spezieller Software und ganz „normalen" handelsüblichen Geräten – offenen Plattformen – der Regelfall. Im Bereich der 3D-computergenerierten Bilddarstellung war man früher gerade imstande, Fernsehtitel zu erzeugen – „flying logos" wurden sie scherzhaft genannt, und die Computergraphik im Fernsehen folgte einem strengen Modezyklus: Zuerst kamen die grünen Gittermodelle, dann die fliegenden Chrom-Logos, danach das Morphing und jetzt eben eine Mode, die auf unserer Fähigkeit basiert, alle Teile des Bildschirms weich und verschwommen darzustellen, bis auf eben die hervorzuhebenden Objektebenen. In der Fernsehwerbung wiederum ist die Computergraphik zu einem unerläßlichen Werkzeug geworden, das bis zu einem gewissen Grad in jedem Spot Anwendung findet.

Aber auf dem Gebiet des Kinos blieb die erforderliche Datenmenge pro Kader noch lange Zeit das Haupthindernis. Erst als das Personal zwischen den Werbe- und den Kinofilmen hin- und herzuwandern begann, entwickelte sich auch im Kinofilm ein wachsender Bedarf nach computergesteuerten Bildbearbeitungstechniken. Von Zeit zu Zeit gab es ja Versuche, Computeranimationen im Film einzusetzen, aber die meisten der frühen Ansätze waren zum Scheitern verurteilt, entweder aus Preisgründen oder vom Film als solchen her. So blieben Filme wie „Westworld", „Tron" und „The Last Starfighter" mutige, aber erfolglose Einzelgänger, die riesige Großrechner einsetzten, um das zu erzielen, was heute auf jedem besseren PC machb wäre. Letztlich war es die „Montage" der Special Effects, die als Ersatz für die altmo dischen optischen Methoden diente und

special effects as a replacement for the older "optical" methods that provided a means for the computer to slip into the film makers' toolbox. It was really as an extension to this that the 3D software made its reappearance. With the ever decreasing costs and increasing power of these systems CGI is now taking a major roll in the film industry.

Huge multi-million dollar productions based around very obvious computer generated images are now being announced on an almost weekly basis. The computer graphics are the stars of these films whilst the actors and even scripts appear to take a subservient role. Thus we have films based around the ability to show audiences things they have not been able to see realistically presented in the past. These projects are nothing more than side shows at carnivals – freak shows. "See the bearded woman, the Siamese twins, or nowadays, roll up, roll up, see the twister, see the dinosaurs, see the volcano erupt, see the dragon fly, the robot transform or just see an entire city wiped out by aliens". These films are no more than the early Lumière films. They show us objects or events with little attempt at story telling: workers arriving at the factory, a train leaving the station, or now, aliens invading earth, a dinosaur stomping by. In big budget films, computer imaging is being used in a very crude way; conceptually speaking, it really is back to the early days of filmmaking. Computer graphics are being used in the main to create stand-alone spectacles – a very crude dramatic model. This is at once the point when the audience in general becomes aware of computer imaging and at the same time, it is tragically being strangled as a medium.

Outside of all this stands "Toy Story" an entire computer generated piece, which is superb in all departments -- not least the script. "Toy Story" stands tall and sadly alone. But having said this, it is effectively a Disney animated film in their best

es dem Computer ermöglichte, auch im Werkzeugkasten der Filmemacher eine Heimat zu finden. Und nur als Erweiterung dieses Werkzeugs tauchte dann auch die 3D-Software wieder auf. Bei ständig sinkenden Kosten und gleichzeitg kräftig anwachsender Systemleistung spielt die Computergraphik in der Filmindustrie mittlerweile eine wichtige Rolle.

Riesige Produktionen mit Millionenbudgets, die alle auf sehr offensichtlich computergenerierten Bildern aufbauen, werden mittlerweile fast wöchentlich angekündigt. Die Computeranimationen sind die Stars dieser Filme, während die Schauspieler und häufig auch das Drehbuch sich mit einer Nebenrolle zufriedengeben müssen. So haben wir jetzt eben Filme, die in erster Linie darauf basieren, dem Publikum Dinge zu zeigen , die es in der Vergangenheit nicht realistisch dargestellt sehen konnte. Aber diese Projekte sind letztlich nichts anderes als Jahrmarktsshows – ein Kuriositätenkabinett, wenn man so will. „Kommen Sie, sehen Sie die Dame ohne Unterleib, die siamesischen Zwillinge ...", und heutzutage eben „Fahrt herzu, rollt herbei, seht die Windhose, seht die Saurier, seht die Vulkaneruption, seht den fliegenden Drachen, seht den Roboter sich verwandeln, seht, wie die Außerirdischen eine ganze Stadt auslöschen!".

All diese Filme sind nichts wesentlich anderes als die frühen Lumière-Filme. Sie zeigen uns Objekte oder Ereignisse, ohne überhaupt zu versuchen, uns dabei eine Geschichte zu erzählen. Was damals die Arbeiter waren, die die Fabrik verließen, ein Zug, der den Bahnhof verläßt, das sind heute die Außerirdischen, die auf der Erde einfallen, die vorbeistampfenden Dinosaurier. In den Filmen mit großem Budget wird die Computerbildwelt – konzeptuell gesehen – auf sehr krude Weise eingesetzt, ganz im Stile jener Jahre, als die Bilder laufen lernten. Computeranimationen werden einfach zur Erzielung des spektakulären Effekts verwendet, was nicht gerade für die Finesse des dramatischen Modells spricht. Und genau an diesem Punkt wird der Einsatz des Computers auch für das Publikum unmittelbar sichtbar, gleichzeitig aber wird er – und das ist das Tragische – als Medium stranguliert.

Jenseits all dessen steht „Toy Story", ein voll-

tradition. In terms of style, narrative, characterisation and dramatic progression it is no different to other Disney cartoon films. "Toy Story" is certainly a milestone and a marvellous entertainment, but is it a masterpiece?

After 100 years, film has produced many masterpieces, as have all the other creative arts. Even after its first half-century, film had already produced a number of generally acknowledged masterpieces such as "Citizen Kane" made well over 50 years ago. But how about computer graphics? Many milestones have been passed, but in its first 50 years computer graphics has produced no real masterpieces. This is not a bad thing in itself, but it does demonstrate the immaturity of the medium. CGI has a long way to go – it is in its earliest infancy and what we have seen is merely the smallest tip of the iceberg.

But CGI does have a major problem. Having been adopted by the motion picture industry as a new "eighth wonder of the world", huge amounts of money have been thrown at it. Thus work in this arena has outpaced most other areas of CG imaging and at seminars, competitions and festivals throughout the world, movie effects have come to dominate the proceedings and our view of what CGI might be. The problem is that the movies' use of computer imaging is a very limited subset of what is possible and the styles already available. Movies, in the main, demand photorealistic rendering and combination with original photography. They set out to mimic the real world even if that might be an imagined future or romanticised past.

Up until recently the actual equipment to realistically tackle computer animation was terribly expensive and often tricky to operate. Things are changing now and much cheaper workstations – really souped-up PC's – are becoming available as are really intuitive user-friendly modelling and animation programmes. These factors

ständig computergeneriertes Stück, das in jeder Hinsicht hervorragend ist, nicht zuletzt, was das Drehbuch betrifft. „Toy Story" steht großartig, aber allein da. Andererseits ist er „nur" ein Disney-Trickfilm in der besten Tradition des Hauses – hinsichtlich seines Stils, der Erzählweise, der Charakterdarstellung, der Dramaturgie unterscheidet er sich nicht von den anderen Disney-Zeichentrickfilmen. Sicherlich ist „Toy Story" ein Meilenstein und bietet großartige Unterhaltung, aber ist er ein Meisterwerk?

In den hundert Jahren seiner Existenz hat der Film zahlreiche hervorragende Arbeiten hervorgebracht, und schon nach fünfzig Jahren seiner Geschichte waren einige allgemein anerkannte Meisterwerke entstanden, etwa „Citizen Kane", der vor gut über fünfzig Jahren gedreht wurde. Und die Computergraphik? Manch ein Meilenstein wurde überwunden, aber in ihren ersten fünfzig Jahren hat die Computerbildwelt nichts hervorgebracht, was als ein die Zeiten überdauerndes Meisterwerk anzusehen wäre. Das ist per se genommen auch nichts Schlechtes, aber es zeigt, daß das Medium noch nicht wirklich ausgereift ist. Die Computerbildwelt hat noch einen weiten Weg vor sich, und was wir bisher gesehen haben, ist nicht mehr als die kleine Spitze eines Eisbergs.

Dennoch, ein Problem existiert: Nachdem die Kinoindustrie die Computerbilder als „Achtes Weltwunder" adoptiert hat, wurden enorme Summen hineingesteckt. So hat denn auch die Arbeit in diesem Bereich alle anderen überholt, und bei Seminaren, Wettbewerben und Festivals auf der ganzen Welt haben die Special Effects der Filmbranche die Veranstaltungen zu dominieren begonnen. Sie schreiben uns gleichsam vor, was und wie computergenerierte Bilder zu sein haben. Problematisch dabei ist, daß die filmische Verwendung der Computergraphik ja nur einen kleinen Teil dessen ausmacht, was möglich ist und was stilistisch vorhanden wäre.

will conspire to make the technology available to a whole new generation of computer artists, with much less financial resources behind them, and therefore, much less financial pressure demanding that they should conform. This hopefully will be the salvation of the medium and shall also open the doors to new operations working much more economically and therefore independently of the large concerns which have up until now ruled the roost. Smaller, leaner companies will eventually lead the way and displace the dinosaurs operating a business model based on enormously expensive specialist workstations manned by fat cats and charged out according to epic ratecards.

Today the aesthetic is in competition with the medium – the physical tools – and is in danger of loosing. The future could be defined as the point when the aesthetic wins and "ideas" take control.

Filme verlangen im wesentlichen nach photorealistischem Rendering und einer Einbindung der Computerbilder in die Filmaufnahmen. Sie versuchen, die reale Welt zu imitieren, und sei es in einer erträumten Zukunft oder einer romantisierenden Vergangenheit.

Bis vor kurzem war ein modernes Equipment, mit dem man ernsthafte Computeranimationen machen konnte, schrecklich teuer und häufig auch mühsam zu bedienen. Das ändert sich, und immer billigere Workstations (oder sagen wir, aufgemotzte PCs) werden ebenso verfügbar wie benutzerfreundliche Modellier- und Animationsprogramme. Diese Faktoren werden zusammenwirken, um die Technologie einer völlig neuen Generation von Computerkünstlern zugänglich zu machen, mit wesentlich weniger finanziellen Ressourcen und deswegen auch viel weniger finanziellem Druck, der nach Anpassung schreit. Dies wird hoffentlich die Rettung des Mediums sein und sollte auch neuen Unternehmungen die Türen öffnen, die viel ökonomischer und deswegen auch unabhängig von den großen Konzernen operieren können, die bisher Herr im Hause waren. Kleinere, schlankere Firmen werden irgendwann die Führung übernehmen und die Dinosaurier verdrängen, die ein Unternehmenskonzept durchziehen, das auf riesigen, enorm teuren Workstations basiert, in dem der Big Boss das Sagen hat und nach höchst üppigen Preislisten abrechnet.

Heute steht die Ästhetik im Wettstreit mit dem Medium – dem physischen Werkzeug – und ist in Gefahr, den Kampf zu verlieren. Die Zukunft könnte als jener Punkt definiert werden, an dem die Ästhetik siegt und die Ideen die Herrschaft übernehmen.

MITCH MITCHELL

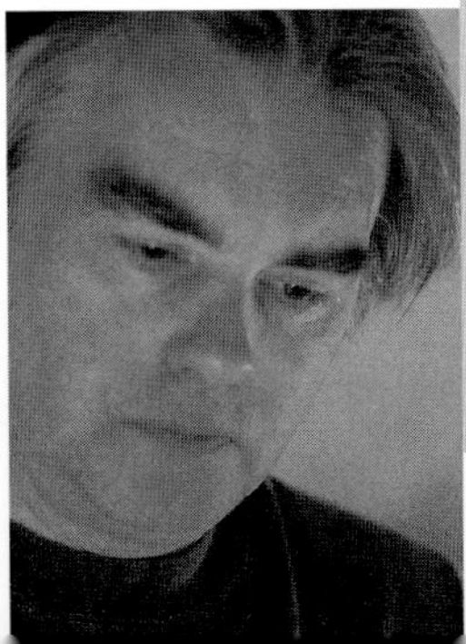

Mitch Mitchell (USA) is director of Printed Picture Company. He was a member of the computer animation jury.

Mitch Mitchell (USA) ist Direktor der Printed Picture Company. Er war Mitglied der Jury für Computeranimation.

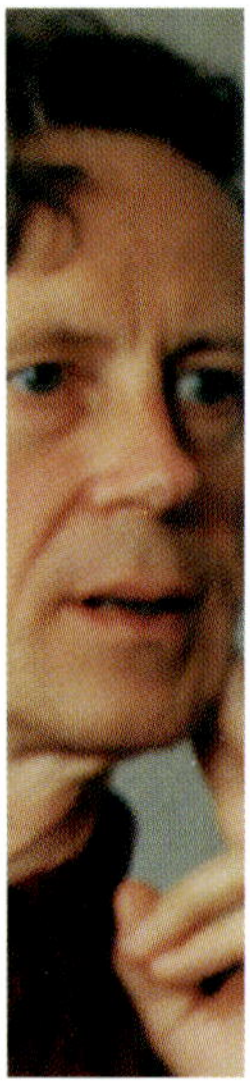

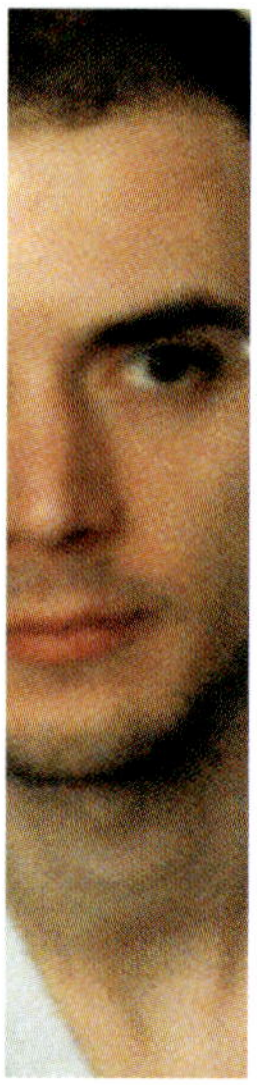

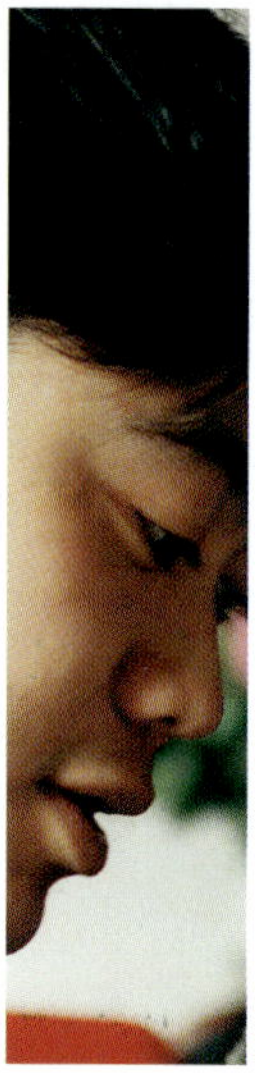

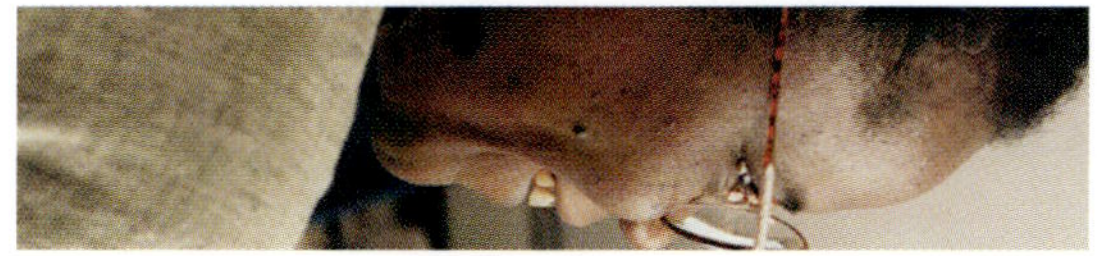

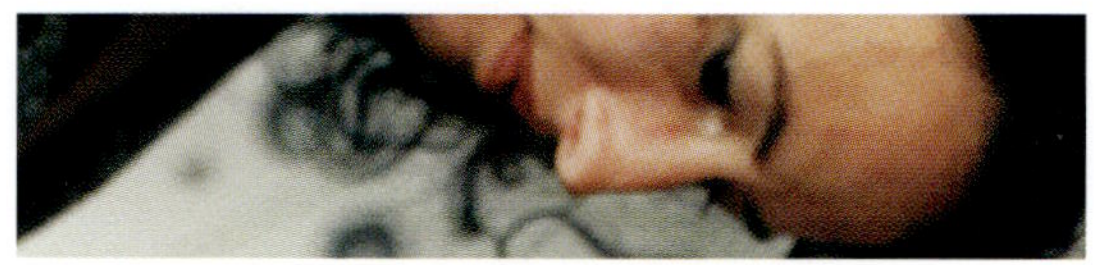

.NET JURY

CONNECTIVITY!
VER-BINDUNGEN SCHAFFEN!

There is a perpetual debate about art: about its definitions, boundaries, function, aims and means; about accessibility, politics, status; about high culture versus popular culture; about inclusion versus exclusion; about funding and money; about art and anti-art.

There's an abundance of camps in these discussions, and they tend to fight fiercely. But such discussions are not the fodder of the net jury: we are, as of yet, far too much engaged in attempting to define what net art might amount to.

Net art does not equal taking whatever is on your gallery's walls, converting it into something the computer can digest and making that accessible via the net. Net art deals with the consequences of what it means to be on the net and the implications of the choice of making that particular technology one's medium. For instance, film as a medium has its own rules and restrictions: it works within certain technical possibilities (montage, editing, camera points) and can't deal with others (you can't play back, the public can't interfere with the story-line nor select or influence the shots), it responds to specific aesthetic conventions, it is part of a history of visual display (focusing means singling out a character; black & white footage connotes "historical material") and narrative recep-

Es gibt eine immerwährende Debatte um die Kunst – um ihre Definition, Grenzen, Funktion, Ziele und Mittel, um Zugänglichkeit, Politik und Status, um Hochkultur versus „Populär-Kultur", um Einbeziehung und Ausgrenzung, um Finanzierung und Unterstützung, um Kunst und Anti-Kunst.

Es gibt mehr als genug unterschiedliche Lager in diesen Diskussionen, und sie neigen dazu, einander heftig zu bekämpfen. Aber solche Auseinandersetzungen sind keineswegs Sache der Netz-Jury: Wir sind bisher viel zu sehr damit beschäftigt, eine Definition für das zu finden, was Net Kunst sein könnte.

Netz-Kunst bedeutet nicht, einfach alles, was an Kunst in der eigenen Galerie herumhängt, in irgendetwas zu konvertieren, das der Computer verdauen kann, ur das Ergebnis über das Netz zugänglich zu machen. Netz-Kunst beschäftigt sich mit den Konsequenzen dessen, was es bedeu tet, im Netz zu sein und welche Implikatic nen es hat, sich ausgerechnet dieser Tech nologie als Medium zu bedienen. Der Film beispielsweise hat seine eigenen Regeln und Grenzen: Er arbeitet innerhalb bestimr ter technischer Möglichkeiten (Montage, Schnitt, Kameraposition) und kann sich nicht mit anderen auseinandersetzen (ma kann ihn nicht zurückspielen, das Publiku kann weder die Story noch die Einstellungen aussuchen), er entspricht bestimmter ästhetischen Konventionen, er ist Teil eine Geschichte der visuellen Darstellung (eine Großaufnahme dient immer dazu, einen Charakter hervorzuheben, schwarz-weiß-Material signalisiert „historische Aufnahme") und einer narrativen Rezeption. Kur um: Das Medium bestimmt die Kunst, in ihrer Form wie in ihrem Inhalt. Das gleich gilt für andere Medien – Romane, Gemäl de, Photographien usw. haben alle ihre

tion. In short, the medium shapes the art, both in form and content. The same goes for other media: novels, paintings, photographs etc. have their own rules and (dis)advantages. Having made an artwork, one may very well convert it to a digital format and put it on the net. But that transformation – and we grant that it may sometimes be a very complicated one – does not of its own accord constitute "net art".

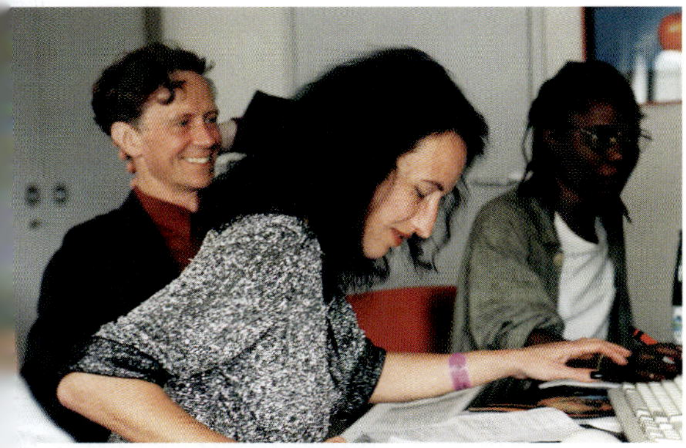

Museums or galleries where art is on show are clearly not themselves art – unless the building has a very special architecture, but then it would be the building that would earn this definition. And the documentation of a project is hardly ever a piece of art in itself. Nevertheless, the misunderstanding that reproducing or exhibiting analog art on the net will magically produce net art is a very common one, and far too many contestants insist on submitting homepages that merely "contain" art. After browsing them briefly, we invariably dumped them. (In fact, at some point we became rather vicious about it: "Oh, another gallery," we'd say, at that point using the term "gallery" as a generic term for homepages that were basically exhibitions of art.)

Since every medium has its own rules, we will attempt to describe the aesthetics and structure that the net engenders, in order to make artists more aware of them. And just as important, we aim to identify

eigenen Regeln, ihre Vor- und Nachteile. Wenn man ein Kunstwerk gemacht hat, dann kann man es natürlich in ein digitales Format konvertieren und ins Netz stellen. Aber diese Transformation – und wir geben zu, daß sie bisweilen recht kompliziert werden kann – macht daraus nicht a priori Netz-Kunst.

Museen oder Galerien, in denen Kunst gezeigt wird, sind auch nicht per se Kunst, sofern nicht das Gebäude eine ganz spezielle Architektur hat, aber dann wäre es eben auch das Gebäude, dem diese Definition zukäme. Und die Dokumentation eines Projektes ist nur in den seltensten Fällen selbst ein Kunstwerk. Dennoch, das Mißverständnis, daß die Reproduktion oder Ausstellung analoger Kunst im Netz wie von Zauberhand Netzwerk-Kunst erzeugt, ist weit verbreitet, und viele Einreicher bestehen darauf, Homepages einzusenden, die nur Kunst „enthalten". Nach kurzer Durchsicht haben wir sie ohne Ausnahme ausgeschieden. (Ab einem gewissen Zeitpunkt wurden wir da schon richtig bösartig: „Ach, schon wieder eine Galerie!", wobei der Begriff „Galerie" eben als Sammelbegriff für alle Homepages stand, die im wesentlichen Kunstausstellungen waren.)

Nachdem jedes Medium seine eigenen Regeln hat, werden wir versuchen, die Ästhetik und Strukturen zu beschreiben, die das Netz mit sich bringt, um Künstler auf seine Besonderheiten aufmerksam zu machen. Und – genauso wichtig – wir zielen darauf ab, neue Entwicklungsgebiete zu identifizieren und etwas Aufmerksamkeit darauf zu lenken. Wir wollen Künstler auf diese neuen, oft noch jungfräulichen Regionen der Kreativität hinweisen und sie auffordern, diese doch zu erforschen.

Das Neue am Netz – und seine wichtigste Eigenschaft – ist, daß es verbindet. Konnektivität macht seine Struktur und seine Mittel aus. Das Netz „verbindet Computer, Leute, Sensoren, Vehikel, Telephone und so ziemlich alles andere in einem globalen Netzwerk, das schnell und billig ist", wie Joichi Ito das ausge-

new areas of development on the net and to generate some attention towards these areas. We wish to point out these new, often virginal regions of creativity to artists and to thereby encourage artists to explore them.

The novelty of the net and its prime importance is that it links. Connectivity is its structure and its means. The net "connects computers, people, sensors, vehicles, telephones, and just about anything together in a global network which is fast and cheap," as Joichi Ito put it; this interconnectedness is the context. "The fun and substance of the [net] is that it is meant to connect living minds at work in all manners of complex and purposeful configurations," Derrick de Kerkhove wrote in the 1995 "Prix Ars Electronica" catalogue. Using this connectivity, showing an awareness of this connectivity and making sense of all the archived and realtime information that abounds in this context is what net art should be dealing with. Anything stable – "stable" as in: done, finished, not growing, unchanging – is not net art. What we are looking for is places on the net that reflect, stimulate, enhance this connectivity, that draw from it and thrive on it – that need the net in order to be able to exist.

We came up with a short list of criteria that we think are indispensable to define this elusive yet all-pervasive "connectivity":

Grammar

Just as a novelist uses narrative devices, a sculptor uses spatial arrangements and a director draws from cinematographic conventions, people on the net are working with and continue to develop a grammar. Basically, a grammar defines the lay-out of a homepage, it's accessibility and its embeddedness. This grammar can be parsed into how one uses links, the transparency of the rules of navigation, the use of frames and backgrounds, et cetera.

drückt hat, und diese Verbindung – Vernetztsein – ist der „Kontext". „Das Unterhaltsame und das Wesentliche des [Netzes liegt darin, wache Geister bei der Arbeit in den verschiedensten komplexen und sinnstiftenden Konfigurationen miteinander in Verbindung zu bringen", schrieb Derrick de Kerkhove im Buch zum Prix Ars Electronica 95. Und diese Konnektivität auszunützen, ein Bewußtsein für sie zu zeigen und all den in diesem Zusammenhang so zahlreichen archivierten und in Echtzeit verfügbaren Informationen einen Sinn abzugewinnen – damit sollte sich die Netz-Kunst beschäftigen. Alles Stabile – „stabil" im Sinne von fertig, beendet, nicht weiter wachsend –, ist keine Netzkunst. Wir suchen Orte im Netz, die diese Konnektivität reflektieren, stimulieren, unterstreichen, die davon leben, sie hegen und pflegen. Die das Netz brauchen, um existieren zu können. Wir haben eine kurze Liste von Kriterien zusammengestellt, die wir für unabdingbar halten, um diese kaum greifbare und doch alles durchdringende „Konnektivität" zu definieren:

Grammatik

Wie ein Romancier narrative Hilfsmittel benutzt, ein Bildhauer räumliche Arrangements erstellt und ein Regisseur aus den kinematographischen Konventionen schöpft, so arbeiten Leute im Netz daran, eine eigene Grammatik zu entwickeln. Solch eine Grammatik definiert das Grundlayout einer Homepage, ihre Zugänglichkeit und ihre Einbettung. Diese Grammatik kann daraufhin untersucht werden, wie Links eingesetzt werden, wie transparent die Navigationsregeln sind, welche Rahmen und Hintergründe verwendet werden und so fort.

Struktur

Alle Medien, die wir vor dem Netz gekannt haben, waren linear, basierten auf der Zeit. Es gibt eine sequentielle Ausrichtung, eine schrittweise Chronologie, bei

Structure

All media we knew before the net are linear, time-based. There is a sequential orientation: a step-by-step chronology where one step is the (indispensable) building block for the next, and where the author of art is by necessity, as Canadian filmmaker David Cronenberg put it, imposing "a benevolent dictatorship". On the net, linearity may be abandoned. You can keep previous versions of your work online, you can offer people a range of routes to go through your story, allow archives to accumulate and mix them with your current work. Users should always be allowed to retrace their steps, to go through homepages in a variety of ways, none of them "better" or more meaningful than the other.

Public service and net-awareness

Does a page, program or interface provide the net with a public service, with some kind of community information? Does a homepage or interface enhance our knowledge (or sensation) that we are part of a giant, global network? Is it able to acknowledge what else is happening on the net and use that as a treasure, as a joke, a metaphor, as something ironical, as something to dwell upon or fantasize about?

Co-operation

Does a project reflect that it is embedded in the richness and vastness of the net, or is it insulated? Is it something various people work on or with? Are there any outside links? Is material from other places on the net being incorporated, commented upon, drawn from? Does it stimulate cooperation and contributions? Can one interact, or is one supposed to just click a few buttons and consume?

Community and identity

The net allows people to meet, mix and mingle in new ways: people from various backgrounds who live in widely varying

der jeder Schritt die (unabdingbare) Voraussetzung für den nächsten darstellt und wo der Autor der Kunst aus Gründen der Notwendigkeit eine – wie der kanadische Filmemacher David Cronenberg es ausdrückt – „gutwillige Diktatur" errichtet. Im Netz ist Linearität kein Thema: Man kann frühere Versionen der Arbeit online halten, man kann den Benutzern eine Vielzahl von Routen anbieten, auf denen sie durch die Story reisen können, man kann Archive einrichten und sie mit der gegenwärtigen Arbeit vermischen. Die Benutzer sollten immer in die Lage versetzt werden, ihre Spuren zurückverfolgen und auf unterschiedliche Weisen durch die Homepages wandern zu können, und keine davon ist „besser" oder bedeutsamer als eine andere.

Öffentliche Dienste und Netz-Bewusstsein

Bietet eine Seite, ein Programm oder ein Interface dem Netz einen öffentlichen Dienst an, liefern sie irgendeine Art von für die Gemeinschaft relevanter Information? Erhöht eine Seite oder ein Interface unser Bewußtsein (oder Gefühl), daß wir Teil eines gigantischen globalen Netzwerks sind? Erkennt sie an, was sonst noch im Netz vor sich geht, und benützt sie das als Schatz, als Scherz, als Metapher, als etwas Ironisches, als etwas, worüber man nachgrübeln oder phantasieren kann?

Kooperation

Läßt ein Projekt erkennen, daß es in den Reichtum und die Weite des Netzes eingebunden ist, oder ist es isoliert? Können verschiedene Leute daran oder damit arbeiten? Gibt es Verknüpfungen nach außen? Wird Material von anderen Orten im Netz eingebunden, kommentiert, mitbenützt? Stimuliert es die Mitarbeit, regt es zu eigenen Beiträgen an? Kann man mit ihm interagieren, oder wird von einem erwartet, daß man bloß ein paar Knöpfe drückt und konsumiert?

Gemeinschaft und Identität

Das Netz erlaubt den Menschen, sich auf neuartige Weise zu treffen, über Grenzen jeder Art hinweg: Leute mit dem unterschiedlichsten

places can meet and swap information, create communities, build worlds and develop a sense of belonging. Does a program, interface or homepage promote this sense of community? Does it allow you to exchange information? Has this community any influence on the analogue life of its participants? Does it allow people to take on various (virtual) identities? Does it allow people to overcome some of the disadvantages they may encounter in analogue life, such as sex, age, colour, disability, lack of social status?

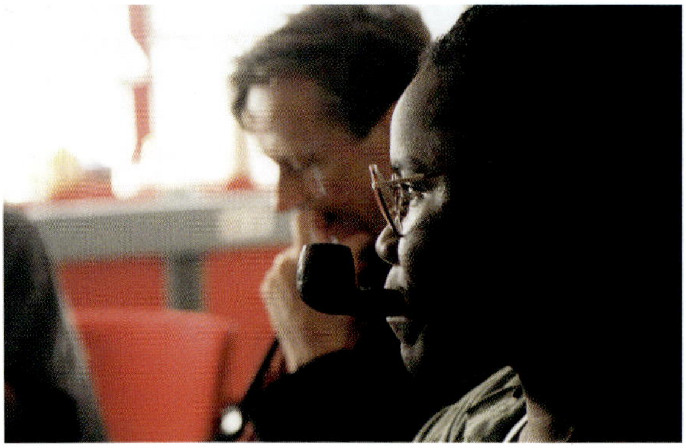

Openness

Is the protocol or programming language that is being used, open and transparent for others, or is it proprietary software? Can participants or users share it or modify it? We are looking for people who come up with memes: ideas that are structured to survive. "Survival" on the net is often about being copied: loading a copy of a webpage in your browser's cache, giving out freeware or shareware (that is, copies of one's program), offering people the possibility to copy, save, cut, paste and use information, pictures, programs, or other chunks of distilled ideas and creativity. Being copied on the net is indeed a measure of success: the more one's page is linked to, the more one's program or interface is used by others, the more people subscribe to one's memes by replicating it, and the more viable and vivid the meme becomes, the better adapted it is. Does the

Hintergrund, die an weit voneinander entfernten Orten leben, können einander begegnen, Gemeinschaften bilden, Welten erschaffen und ein Zugehörigkeitsgefühl entwickeln. Unterstützt ein Programm, ein Interface oder eine Homepage dieses Gemeinschaftsgefühl? Erlaubt es Informationsaustausch? Hat diese Gemeinschaft irgendeinen Einfluß auf das (analoge) Leben ihrer Mitwirkenden? Erlaubt sie den Menschen, mehrere (virtuelle) Identitäten anzunehmen? Erlaubt sie den Menschen, zumindest einige der Nachteile zu überwinden, denen sie im analogen Leben ausgesetzt sein könnten – Geschlecht, Alter, Hautfarbe, Behinderung, fehlender soziale Status?

Offenheit

Ist das eingesetzte Protokoll oder die verwendete Programmiersprache für andere transparent und offen oder handelt es sich um eine geschützte Software? Können die Mitwirkenden oder Benutzer daran teilhaben oder sie modifizieren? Wir suchen nach Leuten, die mit Memen arbeiten – mit Ideen, deren Struktur auf Überleben eingerichtet ist. „Überleben" im Netz bedeutet häufig, kopiert zu werden: Eine Kopie einer Web-Page in den Zwischenspeicher des Browsers laden, Freeware oder Shareware anbieten (d. h. Kopien de eigenen Programms), anderen die Möglichkeit zum Kopieren, Speichern, Ausschneiden und Weiterverwerten von Informationen, Bildern, Programmen oder in anderer Form Anteil an den eigenen destillierten Ideen, an der eigenen Kreativität einräumen. Kopiert zu werden ist im Netz tatsächlich ein Gradmesser für den Erfolg. Je öfter auf die eigene Seite zugegriffen wird, je öfter ein Programm oder Interface von anderen benutzt wird, umso mehr Leute unterstützen die eigenen Meme durch Replikation und umso eingängiger und verbreiteter wird das Mem, umso besser angepaßt ist es an das Medium. Klammert sich der Erzeuger an die Originalfor

maker hold on to its original form and guard this memetic baby, or does he or she allow for mutation and variation to occur? Are they willing to let go and see how this particular idea evolves? (Interestingly, in last year's forum about the net prize, all winners expressed what we labelled a "willingness to be copied".)

While considering each submitted homepage and interface in terms of the criteria listed above and discussing our own affiliations on the net, we couldn't help but notice that Virtual Homesteads are becoming more prominent. Much to our surprise, hardly any homesteads had been submitted – presumably, we conjectured, because their makers and inhabitants mistakenly do not consider these homesteads to be net art.

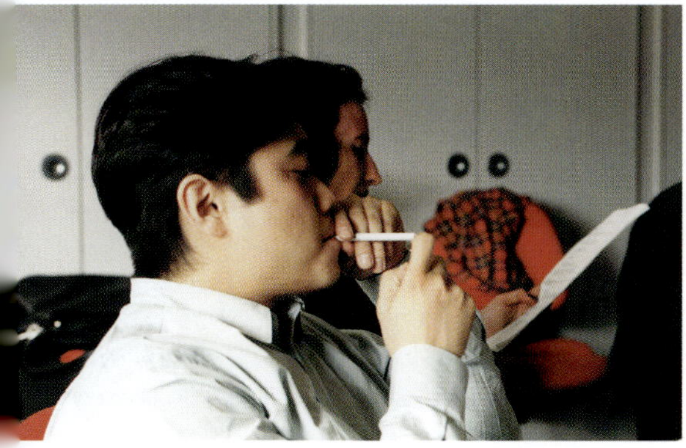

Virtual Homesteads are communal "worlds". One can walk through them and encounter (see, read about) various people and objects there. There are things to do, things to see or read, adventures to be enjoyed. People who visit these worlds often become regulars, and help to build and expand their world.

These worlds and homesteads take on a life of their own: unpredictable things happen that startle even their makers. They grow a history, stories circulate, urban legends abound, and the concerted effort of their inhabitants even reaches into the future. (Why else would one build, if not for the future?) And apart from this binding sense of community that is being pro-

und bewacht er sein memetisches Baby eifersüchtig, oder erlaubt er, daß Mutationen und Variationen stattfinden? Ist er bereit loszulassen und zu beobachten, wohin sich diese spezielle Idee entwickelt? (Interessanterweise haben beim Künstlerforum des vergangenen Jahres alle Preisträger das ausgedrückt, was wir als „Wunsch und Wille, kopiert zu werden" bezeichnet haben.)

Während wir jede eingereichte Homepage und jedes Interface mit den oben angeführten Kriterien verglichen und unsere eigenen Beziehungen zum Netz diskutierten, ist uns aufgefallen, daß Virtuelle Siedlungen (Virtual Homesteads) offenbar immer prominenter werden. Zu unserer Überraschung sind aber kaum solche Siedlungen eingereicht worden – wahrscheinlich, so haben wir gefolgert, weil die meisten ihrer Erzeuger und Bewohner diese Siedlungen fälschlicherweise nicht als Netz-Kunst betrachten.

Virtuelle Siedlungen sind kommunale „Welten". Man kann sie durchwandern und die unterschiedlichsten Personen und Objekte treffen (oder sehen bzw. von ihnen lesen). Es gibt Dinge zu tun, zu sehen oder zu lesen, Abenteuer zu genießen. Wer eine solche Welt besucht, wird häufig zum Stammgast und hilft selbst mit, die Welt auf- und auszubauen.

Diese Welten und Siedlungen entwickeln ein Eigenleben: Unvorhersehbare Dinge passieren, die selbst ihre Schöpfer erstaunen, sie entwicklen eine Geschichte, „Geschichten" werden über sie erzählt, „urban legends" tauchen zuhauf auf, und die konzertierte Anstrengung ihrer Bewohner reicht bis in die Zukunft (wofür sonst sollte man bauen, wenn nicht für die Zukunft?). Diese virtuellen Gemeinschaften lassen aber nicht nur diesen bindungsstiftenden Gemeinsinn entstehen, sondern lassen den Menschen auch Raum zum Atmen: Man kann mit Charaktereigenschaften herumspielen, die normalerweise als invariabel und Teil der Kernpersönlichkeit angesehen werden: Geschlecht, Hautfarbe, Alter, soziale Schicht.

Bis vor kurzem waren die Unterschiede zwischen diesen verschiedenen Typen von Welten noch groß, aber das ändert sich gerade. Ebene um Ebene wird hinzugefügt, neue Protokolle und Programme werden entwickelt. IRC (Internet Relay Chat) ist ein textgestütztes Medium, seit der Erfin-

duced, these virtual communities tend to give people some air to breath. One can play with characteristics that are usually perceived to be stable and part of one's "core personality": sex, colour, age, class.

The differences between these types of worlds used to be vast. But that is changing. Layers upon layers are being added and new protocols and programs developed. IRC (Internet Relay Chat) is a text-based medium, but once "Palace" was invented, we had a graphic IRC of sorts as well. MUDs and MOOs are experimenting with graphic appendages. "Active Worlds" allow for the same building capabilities as MUDs and MOOs, but are based on graphics and use text only for conversations. There are now some VRML-worlds that accommodate inhabitants and to which one can contribute.

Each of these worlds has its own particularities and advantages. MUDs and MOOs don't need fat computers. Since they are text-based, they can be run on almost any kind of computer. And the programs are free. Some argue that these text-based worlds, like novels, leave more room for the imagination and incite you to fantasize, to make your own visual representation of your co-inhabitants and your environment. "Palace" and "Active Worlds" are proprietary software and need license keys, devoted servers and lots of computer memory. On the up side, these worlds provide their inhabitants with shared pictures: you can build whatever is in your mind and present that to others, and let them use it. Also, some people and some cultures are better versed in pictures than in text.

How should these various interfaces and programs that all create worlds be judged? Are text-based worlds and graphic-based worlds two versions of the same phenomenon, or should we consider them to be different "media", just like books and television are different media?

What is interesting is that these worlds are solid: they persevere. The inhabitants

dung von „Palace" verfügen wir aber auch über eine Art graphisches IRC. MUDs und MOOs experimentieren mit graphischen Anhängseln, Active Worlds erlauben dieselben Bau-Eigenschaften wie MUDs und MOOs, aber basieren auf graphischen Grundstrukturen und verwenden den Text nur für die Konversation. Und es gibt mittlerweile einige VRML-Welten, die Bewohner aufnehmen und zu denen man aktiv beitragen kann.

Jede dieser Welten hat ihre Besonderheiten und Vorteile. MUDs und MOOs brauchen keine fetten Computer, weil sie text-orientiert sind: Sie laufen auf so ziemlich jeder Art von Rechner, und die Programme gibt's umsonst. Einige behaupten, daß die Text-Welten – ähnlich wie ein Roman – der Phantasie mehr Raum lassen und einem erlauben, sich seine Vorstellung von den Mitbewohnern und seiner Umwelt zu machen. „Palace" und „Active Worlds" wiederum sind geschützte Software und benötigen einen Lizenz-Schlüssel, eigene Server und riesige Speicher. Auf der Haben-Seite liefern diese Welten ihren Bewohnern gemeinsame Bilder – man kann bauen, was man gerade im Kopf hat, kann es anderen präsentieren und es ihnen zur Verfügung stellen. Außerdem können einige Menschen und Kulturen mit Bildern besser umgehen als mit Texten. Wie also sollte man solch unterschiedliche Interfaces und Programme bewerten, die alle Welten schaffen? Sind Text-Welten und Graphik-Welten zwei verschiedene Ausprägungen desselben Phänomens, oder sollten wir sie eher als zwei verschiedene „Medien" betrachten, wie ja auch Bücher und Fernsehen zwei verschiedene Medien sind?

Das Interessante an diesen Welten ist ihre Dauerhaftigkeit. Die Einwohner tendieren stark dazu, Permanenz zu konstruieren: Sie listen ihre Vergangenheit auf, archivieren sie, sie bauen und erweitern (und was in diesen Welten gebaut wird, kann nicht rückgängig gemacht werden),

are bent on creating permanence: they list and archive their history, they build and add (and whatever is built in these worlds cannot be undone), they invest time and energy, they gather skills, they build levels, they develop friendships and affinities. In short, inhabitants put their stakes in these worlds.

We are looking forward to more open protocols and platforms to create worlds like these. As of now, MUDs/MOOs, IRC, WebChats and VRML and a few games (Doom, Quake, Diablo) are the only open platforms. We look forward to seeing more of those and we predict that in the near future, Virtual Homesteads will combine text, VRML, avatars, games, sound and graphic into multi-media surroundings. It would seem that the media or technologies

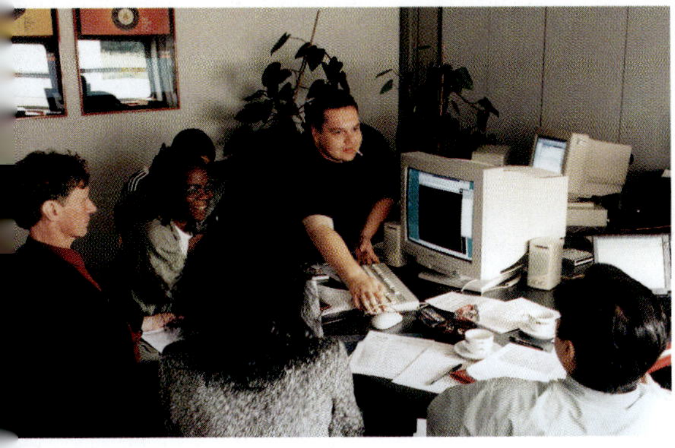

that these Virtual Homesteads make use of, are converging, while their originals are still available too. When these different modes of presentation do converge, the impact of Virtual Homesteads will be huge. After all, the World Wide Web only came into existence when text and pictures were combined via an open protocol – and look at the impact that created.

THE PRIZES

The problematics of sensory life is squarely the object of the prizewinner, "sensorium" <www.sensorium.org>. While it does perhaps suffer from a rather austere if not completely disincarnate style, it attempts to give us the sense of connecting

sie investieren Zeit und Energie, erwerben Fähigkeiten, erstellen neue Ebenen, entwickeln Freundschaften und Zuneigung – kurzum, die Bewohner setzen auf ihre Welten.

Wir freuen uns auf mehr offene Protokolle und Plattformen, um Welten wie diese zu erschaffen. Bisher sind MUDs/MOOs, IRC; Web-Chats, VRML und einige wenige Spiele (Doom, Quake, Diablo) die einzigen offenen Plattformen. Wir hoffen, in Zukunft mehr davon zu sehen, und wir prophezeihen, daß zukünftige Virtuelle Siedlungen Text, VRML, Avatars, Spiele, Klang und Graphik in multimediale Umgebungen einbauen werden. Er scheint, daß die Medien und Technologien, auf denen die Virtuellen Siedlungen aufbauen, sich aufeinander zu entwickeln, während ihre Originale weiter bestehen bleiben. Wenn diese unterschiedlichen Präsentationsmodi verschmelzen, werden die Virtuellen Siedlungen einen tiefgehenden Eindruck hinterlassen. Denn schließlich kam ja auch das World Wide Web erst zustande, als Text und Bilder durch ein offenes Protokoll kombiniert wurden – und man sehe sich nur an, was das für Folgen hatte!

DIE PREISTRÄGER

Die Problematik des Sinnenlebens ist – kurz ausgedrückt – das Anliegen des Gewinners der Goldenen Nica, „sensorium" <www.sensorium.org>. Wenn es auch ein wenig unter einem ziemlich nüchternen, ja, geradezu „fleischlosen" Stil leidet, so versucht es doch, uns das Gefühl zu geben, in Echtzeit mit dem Leben im Netz verknüpft zu sein, gerade wenn es den Fluß der ausgetauschten Packets in Umweltklänge übersetzt. Entscheidend an dem Werk ist, daß es ein wenig von jenem reflektiert, was William Butler Yeats „die Emotion der Vielfalt" genannt hat, ein Gefühl des Erstaunens und Respekts im Moment der Erkenntnis, daß es so viele gleichzeitige Präsenzen im Netz gibt. Mit seinem zwar spröden, aber dennoch effizienten Design legt „sensorium" Zeugnis ab vom vernetzten Leben auf dieser Welt und ist so ein Sinnesorgan dieser Welt selbst. Ein Teil dieser sich entwickelnden Kunst (und Psychologie) wird vom Satelliten inspiriert. Selbst der

in realtime with the life of the net, even as it translates the flow of packet switching into ambiance-like musical sounds. What is important about the piece is that it reflects something of what William Butler Yeats called "the emotion of multitude", a sense of awe at the moment of recognition of so many all-at-once simultaneous presences. "sensorium", in rather dry but nevertheless effective design, also gives us evidence of connective life on earth, and hence a sense of the earth itself. It is part of this developing art (and psychology) which is inspired by the satellite. Even the name "sensorium" deserves a mention. What it offers is quite literally a "common sense", hence it is not an absolutely necessary for it to be tied to each one of us personally, although that could be nice too.

We continued to keep "public mindedness" or "public service" in mind. The net is a social service. It carries social obligations and it can stimulate many creative services that benefit everybody. "Make Money Fast" <www.clark.net/pub/rolf/mmf/> fitted this criterium perfectly. That was the factor that consciously or otherwise brought the jury to agree to award a prize to a site that, at first glance, doesn't "look like art" at all. It also seems too "specialized", focusing, as it does, on a rather exotic variety of Web ills. That is part of it. But "Make Money Fast" is unique to the Net. It is a publicly accessible platform which can help you trace the origin of the annoying chain-mail that sometime clutters your e-mail. The chain-mail is a disease of the Web, an automated malfunction of human intelligence once it is accelerated by networks. Finding antidotes that are neither brutal nor legalistic is a public service. The human dimension of "Make Money Fast" is also strong, warm and funny.

Humour was also present to our minds when looking into "TechnoSphere" <www.technosphere.org.uk/>, the other ex-aequo second prize. "TechnoSphere" allows one to build a creature out of a va-

Name „sensorium" ist als solcher erwähnenswert – er trägt die Konnotation „gemeinsamer Sinn" in sich und muß auch deswegen nicht persönlich an einen von uns gebunden sein, obwohl auch das nett sein könnte.

Wir haben auch weiterhin nach Arbeiten mit „Gemeinsinn" oder „Allgemeinnutzen" Ausschau gehalten. Das Netz ist ein sozialer Dienst, es trägt soziale Verantwortung und kann viele kreative Dienstleistungen stimulieren, von denen jedermann profitieren kann. „Make Money Fast" <www.clark.net/pub/rolf/mmf/> entsprach diesem Kriterium perfekt, und dies war auch der Grund, warum die Jury einer Web-Site eine Auszeichnung zuerkannt hat, die auf den ersten Blick überhaupt nicht nach „Kunst" aussieht. Sie scheint auch zu spezialisiert zu sein, denn sie konzentriert sich auf eine eher exotische Variante der Netzkrankheiten. Das ist die eine Seite. Aber „Make Money Fast" ist einzigartig im Netz. Es handelt sich dabei um eine öffentlich zugängliche Plattform, die dem Benutzer erlaubt, die lästigen Kettenbriefe, die manchmal die eigene Mailbox verstopfen, bis zum Ursprung zurückzuverfolgen. Diese Kettenbriefe sind eine Seuche des Web, eine automatisierte, durch das Netzwerk beschleunigte Fehlfunktion der menschlichen Intelligenz. Ein Gegenmittel zu erfinden, das weder brutal ist noch auf der Rechtsstreit-Ebene funktioniert, ist ein Dienst an der Allgemeinheit. Außerdem besitzt „Make Money Fast" eine starke menschliche Dimension, es ist warm und fröhlich.

Der Humor kam auch nicht zu kurz, als wir „TechnoSphere" <www.technosphere.org.uk/> betrachteten, den Gewinner der zweiten Auszeichnung. „TechnoSphere" erlaubt einem, aus einer Vielzahl möglicher Köpfe, Körper und Glieder eine Kreatur zusammenzubasteln, ihr einen Namen zu geben und sie in den digitalen Raum loszulassen. Das Kleintier sieht anfänglich recht hübsch aus, aber dann sieh

riety of choices for heads, bodies and legs, name it and let it loose in digital space. The critter looks rather cute at first, but then you never see it again. However, you can receive a regular report on its condition by calling up its ID: a page will show you its fitness, age and activity. One can trace its location and family tree. More serious events – such as death, attacks and pregnacies – are announced via e-mail. This semi-random but endlessly recurring activity allows us to populate the Net with new varieties of e-pets or "net pets" if you prefer. It is an interesting concept, if still a bit disincarnate for the voracious sensory craving of many Webfreaks.

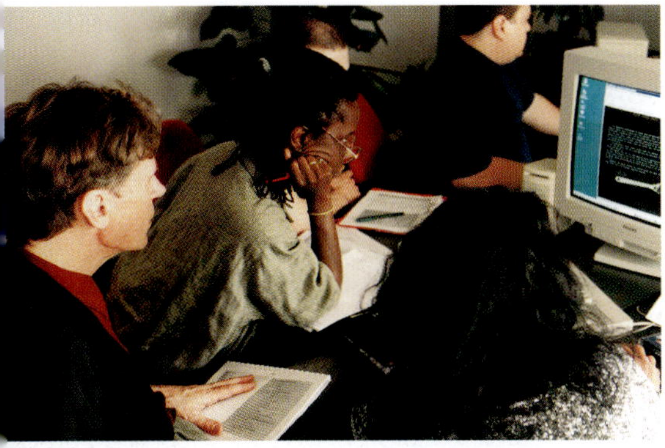

The honorary mentions

Hypernarratives: the Web is full of interesting stories told with adroit use of the technology. However, many are linear and hence fail to make use of one of the net's greatest resources, the instant accessibility to anything in any order. Others don't work very well: either they are too top-heavy and take too much time to download, hence interrupting the flow of the narrative structure, or they are too disjointed to allow a user to make sense of them. "Cyberpoetry" <student.uq.edu.au/~s27 1502> avoided such pitfalls and tells its many stories with exciting voice and choice alternatives.

VRML: A special category had to be created for this rising technology of the Web. We were surely all tempted at one

man es nie wieder. Allerdings kann man regelmäßig einen Bericht über sein Befinden erhalten, wenn man seine ID-Nummer aufruft: Eine Seite zeigt dann seinen Gesundheitszustand an, sein Alter und was es gerade tut. Man kann seinen derzeitigen Aufenthalt herausfinden und seinen Familienstammbaum. Ernsthaftere Ereignisse – wie Herzinfarkt, Tod oder auch Schwangerschaft – werden über E-Mail angekündigt. Diese halb-zufällige, aber endlos rekursive Aktivität ermöglicht es, das Netz mit zahlreichen neuen Varianten von elektronischen Haus- oder besser gesagt „Netz-Tieren" zu bevölkern. Es ist ein interessantes Konzept, wenn auch vielleicht noch nicht genug aufgemotzt für den alles verschlingenden sensorischen Hunger mancher Web-Freaks.

Die Anerkennungen

Hypererzählungen: Das Web ist voll von interessanten Geschichten, die unter geschicktem Einsatz der Technologie erzählt werden. Allerdings sind viele davon linear aufgebaut und lassen den Einsatz der tiefgehenden Ressourcen des Netzes vermissen – die sofortige Zugänglichkeit aller Teile in jeder beliebigen Reihenfolge. Andere wiederum sind zu kopflastig – es dauert zu lange, die einzelnen Teile zu laden, was den Fluß der narrativen Struktur unterbricht – oder aber zu disjunkt, als daß ein Benutzer hinter den Sinn kommen könnte. „Cyberpoetry" <student.uq.edu.au/~s271502> hat all diese Fallgruben vermieden und erzählt seine vielen Geschichten mit aufregender Stimme und erstklassigen Wahlmöglichkeiten.

VRML: Eine eigene Kategorie mußte für diese im Web nun zunehmende Technologie geschaffen werden. Früher oder später war wohl jeder von uns in Versuchung, „Ocean Walk" <www.atom.co.jp/VRML2/ocean/index .html> – dem schönsten und besten VRML-Stück, das die meisten von uns im Web gesehen haben – einen Preis zuzuerkennen. Abgehalten hat uns eine von uns aufgestellte Faustregel: Wenn eine Web-Site – egal wie schön sie ist – ebenso gut und deswegen wesentlich schneller auf einer CD-ROM funktionieren würde (wie in diesem Fall), dann verdient sie nicht

point or another to hand a prize to "Ocean Walk", <www.atom.co.jp/VRML2/ocean/ index.html> the most beautiful and effective VRML many of us have seen on the Web. What stopped us was the rule of thumb that, however beautiful or effective the site, if what it shows would be just as much fun as a CD-ROM and consequently quite a bit faster, it would not qualify for more than a mention. This VRML piece only worked well for us on a Silicon Graphics high-speed processor. "Container-City" <www.mesch.ch/container-city> was selected for its imaginative use of grids and blocks with great artistry. Both pieces, of course, are gratifyingly interactive.

The rule of the CD-ROM also applies to documentary sites such as "Harappa" <www.harappa.com>. What struck us with that one is its sheer beauty and ease of execution, glorious colours and design, fascinating and complex information made simpler than browsing in a specialised library.

Virtual Homestead: worlds are communities on line which today number above 400,000 registrants (called "immigrants" in the right circles), complete with different varieties of avatars capable of different varieties of functions from live and spatially referenced speech to limb and facial expression movements. Reaching a new warp speed in connectivity, worlds such as "Sherwood Forest" <www.ccon.org/events/sherwood.html> and "Virtual Roof" <www. hack-roof.de/> are beyond the Web altogether, residing on a server and providing the kind of permanence and stability normally associated with the real. Indeed you can build and unbuild anything you want, but you cannot unbuild someone else's property, nor can anyone else unbuild yours.

Code as Medium: The idea of this category might not have come to us, were it not for two delightfull sites which bring relevance to "art for art's sake" on the Web. "Form" <www.c3.hu/hyper3/form/> is design that has humour as its only content;

mehr als eine Anerkennung. Dieses VRML-Werk lief bei uns nur auf einer schnellen Silicon-Graphics-Maschine wirklich zufriedenstellend. „Container-City" <www.mesch.ch/container-city> wurde wegen seiner imaginativen und geradezu meisterlichen Verwendung von Gittern und Blöken gewählt. Beide Werke sind selbstverständlich erfreulich interaktiv.

Die CD-ROM-Faustregel gilt auch für dokumentarische Sites wie „Harappa" <www.harappa.com>. In diesem Fall aber haben uns die reine Schönheit und Einfachheit der Ausführung, die wunderbaren Farben und das Design überzeugt, die faszinierende und komplexe Informationen viel leichter zugänglich machen, als es eine spezialisierte Bibliothek könnte.

„Virtual Homesteads": Welten sind Online-Gemeinschaften, denen heute über 400.000 Mitglieder angehören (in den richtigen Kreisen auch „Immigranten" genannt), bevölkert von verschiedenen Avatars mit den unterschiedlichsten Fähigkeiten, die von der Live-Sprache und räumlichen Bezügen bis hin zu Bewegungen von Gliedern und Gesicht reichen. Welten wie „Sherwood Forest" <www.ccon.org/events/sherwood.html> und „Virtual Roof" <www.hack-roof.de/> liegen in ihrer Geschwindigkeit und Struktur schon fast jenseits des Web, sie residieren auf eigenen Servern und bieten jene Permanenz und Dauer an, die man normalerweise mit der Wirklichkeit assoziiert. Und tatsächlich kann man darin alles auf- und abbauen, was man will, aber das Eigentum eines anderen bleibt vor der eigenen Bauwut geschützt, wie auch die eigenen Schöpfungen vor der Demontage durch andere sicher sind.

Code als Medium: Eine derartige Kategorie wäre uns gar nicht in den Sinn gekommen, gäbe es nicht zwei ausgezeichnete Sites, die das Element des „L'art pour l'art" ins Web einbringen. „Form" <www.c3.hu/hyper3/form/> ist ein Design, dessen einziger Inhalt Humor ist, „Jodi" <www.

"Jodi" <www.jodi.org> is most elegantly coded. (Try to find the rocket inside.)

Webcam: Ever since we dipped into "Mike's FishTank" and threw snowballs with "The Snowball Throwing Machine," there have been enough experiments with Webcams to warrant a specific category. "ShootingBack" <n1nlf-1.media.mit.edu/shootingback.html> pokes technofun at the user by showing constantly changing conditions and angles chosen by the artist.

Collage/collider: "The Multi-Cultural Recycler" <shoko.calarts.edu/~alex/recycler.html>, manages to let cultures clash on the net, recycles art and builds something new from this collision.

Public Service: "Energized Gaming Culture" <www.farm.de/x/> received a mention for that group of sites, because it is a beautifully crafted, well researched and passably interactive site. It serves as a record of development of digital games with a special emphasis on their appearance online, thus acting as a living archive. Archiving the history of developments of specific genres on-line is a valuable public service.

At the end of what was unquestionably the longest and perhaps also the most arduous jurying process, we found that we had only eleven mentions instead of the usual twelve. We decided to leave it that way, thereby marking the fact that this year's selection may not have given us all the choices we were expecting. We would definitely like to see less technology and less transported art, and more net art.

jodi. org> hingegen ist äußerst elegant codiert (Versuchen Sie einmal, die Rakete zu finden!).

Webcam: Seit wir das erste Mal in Mike's FishTank eingetaucht sind und mit der „Snowball Throwing Machine" Schneeballschlachten veranstaltet haben, hat es genug Experimente mit Webcams gegeben, um eine eigene Kategorie zu rechtfertigen. „Shooting Back" <n1nlf-1.media.mit.edu/shootingback.html> schürt die Lust an der Technologie beim Anwender, indem es immer wieder je nach Lust des Künstlers geänderte Blickwinkel und Bedingungen bietet.

Collage/collider: „The Multi-Cultural Recycler" <shoko.calarts.edu/~alex/recycler. html> läßt Kulturen im Netz aufeinanderprallen, betreibt Kunst-Recycling und baut aus diesen Kollisionen etwas Neues auf.

Öffentliche Dienstleistung: „Energized Gaming Culture" <www.farm.de/x/> erhielt eine Anerkennung für diese Gruppe von Sites, weil es eine wunderbar konzipierte, gut recherchierte und passabel interaktive Web-Site ist. Es zeichnet die Entwicklung von digitalen Spielen auf, wobei es deren Auftreten im Netz besonders berücksichtigt, es ist also ein lebendes Archiv. Und die Archivierung der Geschichte von spezifischen Online-Genres ist ohne Zweifel ein wertvoller Dienst an der Netzöffentlichkeit.

Am Ende der zweifellos längsten und möglicherweise schwierigsten Entscheidungsfindung einer Jury im diesjährigen Prix mußten wir feststellen, daß wir nur elf Anerkennungen statt der eigentlich vorgesehenen zwölf hatten. Wir beschlossen, es dabei zu belassen, auch um zu dokumentieren, daß die diesjährige Auswahl uns vielleicht nicht in allem das geboten hat, was wir erwartet hatten. Wir würden uns für die Zukunft wünschen, eindeutig weniger Technologie und „transportierte" Kunst zu sehen und dafür mehr Netz-Kunst.

"sensorium's" key thematic concept is "sense." "sensorium" seeks to create a Web site that allows one to sense the world in a variety of ways.

As a tool via the net, "sensorium" is designed to expand the meaning of the term "sense" from various viewpoints. The true capabilities of the net do not just lie in mundane scenarios, such as virtual shopping at home, transmitting information or as a medium for expression, but rather in helping us sense such aspects of human activity as how and what we eat, the air we breathe, what are the relationships that we share with our family, other people and creatures. We hope that the net allows the audience to see the more important things in our lives.

BREATHING EARTH

We tend to believe the ground beneath us is stationary, yet it moves all the time. Indeed, it seems to be alive. The "Breathing Earth" offers a visualization of compressed earthquake data. Data is included from earthquakes that have occurred worldwide in the past 14 days. The contents take the form of a WWW site filmed with time-lapsed photography. Currently, on-line data is used from the International Database Center (IDC).

WEB HOPPER

"Web Hopper" is an interWeb site that enables you to visualize your Web hopping and that of other Net users. The interface shows your trail crossing the world as a red line and those of other users as blue lines.

The "Web Hopper" translates http packet information about the IP address into latitude and longitude. When the http packets pass through the IWE Network Operation Center (NOC), the location information is visualized via a Java applet real-time base. To translate from the host name to the location information, a

Thematischer Schwerpunkt im Konzept von „sensorium" ist der Begriff „Sinn". „sensorium" versucht, eine Web-Site zu schaffen, die es erlaubt, die Welt auf verschiedenste Art „sinnlich" zu erfassen.

Als Werkzeug im Netz ist „sensorium" darauf ausgerichtet, die Bedeutung des Begriffs „Sinn" unter unterschiedlichen Aspekten zu erweitern. Die wahren Möglichkeiten des Netzes erschöpfen sich nicht bloß in mondänen Szenarien wie virtuellem Einkaufen von zuhause aus, in einfacher Informationsübermittlung, es ist auch nicht nur Ausdrucksmittel, sondern bietet vielmehr die Möglichkeit, uns beim sinnlichen Erfassen diverser Aspekte des menschlichen Verhaltens zu helfen. Wie und was essen wir, welche Luft atmen wir, welche Beziehungen haben wir zu unserer Familie, zu anderen Menschen und Lebenwesen? Wir hoffen, daß das Netz dem Publikum ermöglicht, die wichtigen Dinge im Leben zu erkennen.

BREATHING EARTH (ATMENDE ERDE)

Wir neigen dazu, die Welt unter unseren Füßen als stationär zu empfinden, und doch bewegt sie sich die ganze Zeit, ja, sie scheint beinah lebendig zu sein. „Breathing Earth" bietet eine Visualisierung komprimierter Daten von Erdbeben. Es werden Daten aus allen jenen Erdbeben einbezogen, die weltweit innerhalb der jeweils letzten 14 Tage registriert wurden. Der Inhalt wird in Form einer WWW-Site dargestellt, in zeitversetzter Photographie gefilmt, wobei Online-Daten aus dem International Database Center (IDC) verwendet werden

WEB HOPPER

„Web Hopper" ist eine Inter-Web-Site, die es erlaubt, Ihre Web-Besuche und -Reisen ebenso zu visualisieren wie jene anderer Netzwerkbenutzer. Das

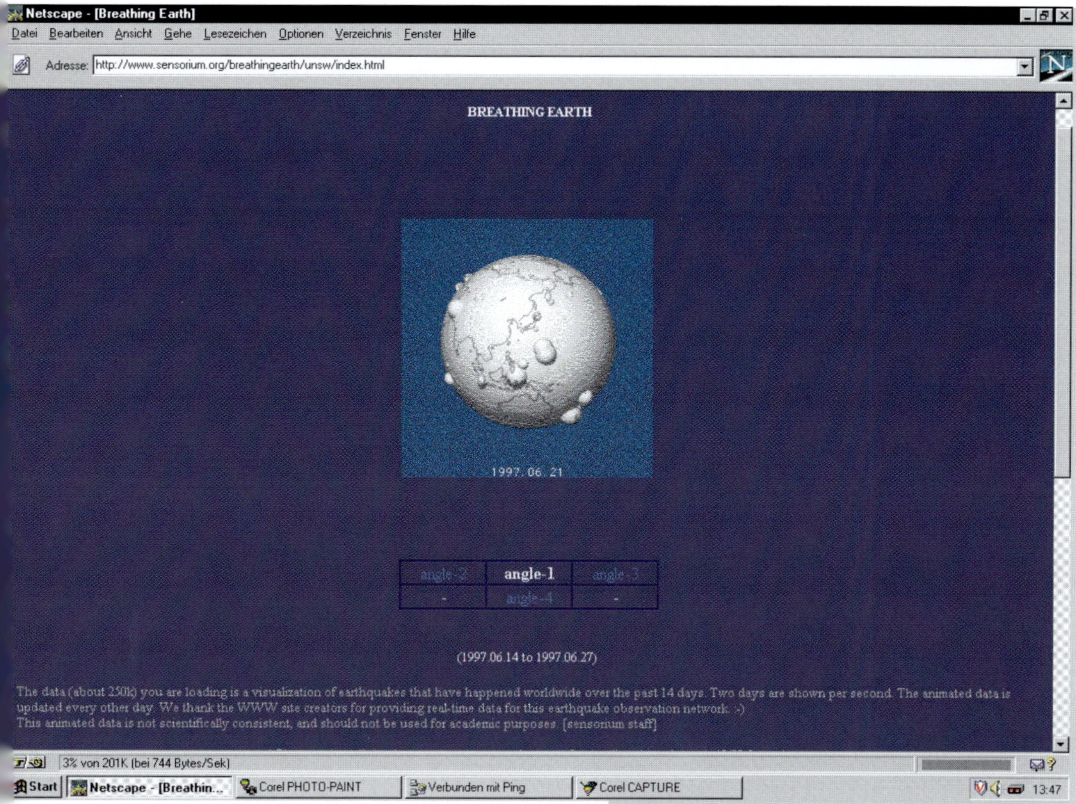

database on the host name to Latitude/Longitude page at Illinois University (Matt Gardner at the Pablo VR Group) was used.

This interface shows countless visualizations of "just connected moments." You might be able to grasp the Internet status quo.

Net Sound

"Net Sound" translates nine different communications protocols into sound data based on network traffic data at the lab. (As ICMP request, RIP, NTP, FTP, HTTP, NNTP, NFS, X-window.)

"Net Sound" is an outlet for the stetho system. This real-time system translates different communications protocols into MIDI sounds based on network traffic data

Interface zeigt Ihre Spur durch die virtuelle Welt als rote Linie, jene von anderen Usern als blaue Linien. Der „Web Hopper" übersetzt die http-Packet-Information über IP-Adressen in Längen- und Breitengradinformationen. Wenn die http-Packets durch das IWE Network Operation Center (NOC) geschleust werden, wird die Ortsinformation über eine kleine Java-Applikation in Echtzeitdaten umgewandelt. Für die Übersetzung des Host-Namens in eine geographische Information wird eine (von Matt Gardner von der Pablo VR Group) an der Illinois University installierte Datenbank mit Längen- und Breitengradinformation verwendet.

Dieses Interface zeigt unzählige Visualisierungen von „Verbunden mit"-Momenten – vielleicht eine Möglichkeit, den Status quo des Internet festzuhalten.

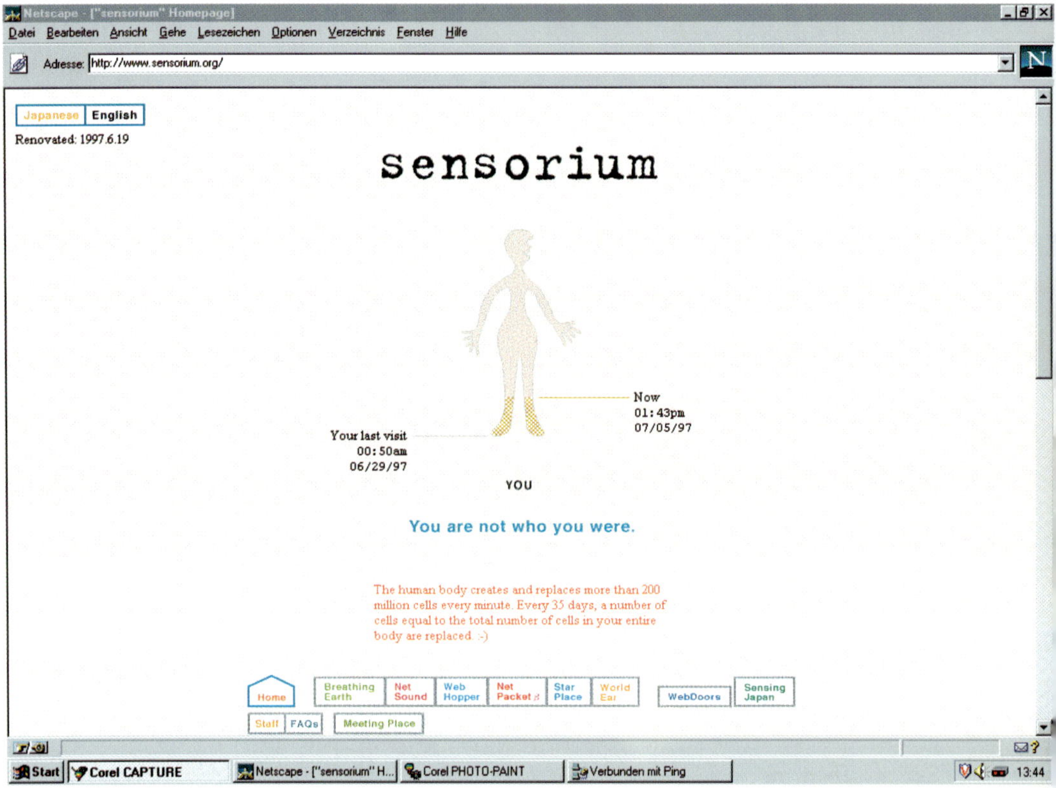

(the TCP dump command). The original purpose of the stetho system was to support the management of networks. The system translates network traffic so that that sounds can be heard via telephone even in a non-computer environment.

Net Packet

This companion to "Net Sound" uses visual data instead of sound. It is Java programmed for real-time visualization.

Star Place

"Star Place" was launched with the idea of creating a site where visitors could experience the universal sense of time and space in their lives as opposed to cyberspace.

As we began to develop the site, we became aware of the knowledge common to everyone on the planet: the earth is fly-

Net Sound

„Net Sound" übersetzt auf der Basis des Datendurchsatzes im Laboratorium neun verschiedene Kommunikationsprotokolle wie ICMP-Request, RIP, NTP, FTP, HTTP, NNTP, NFS und X-window in Klangdaten. „Net Sound" ist im wesentlichen ein Ausgang für das Real-Time-Stetho-System, das die Netzwerkdurchsatzdaten unterschiedlicher Protokolle (die TCP-Dump-Kommandos) in MIDI-Klänge umwandelt. Ursprünglich wurde Stetho zur Unterstützung von Netzwerkmanagement-Aufgaben entwickelt. Das System übersetzt den Verkehr im Netz so, daß Klänge telephonisch abrufbar und auch in einer computerlosen Umgebung nachvollziehbar sind.

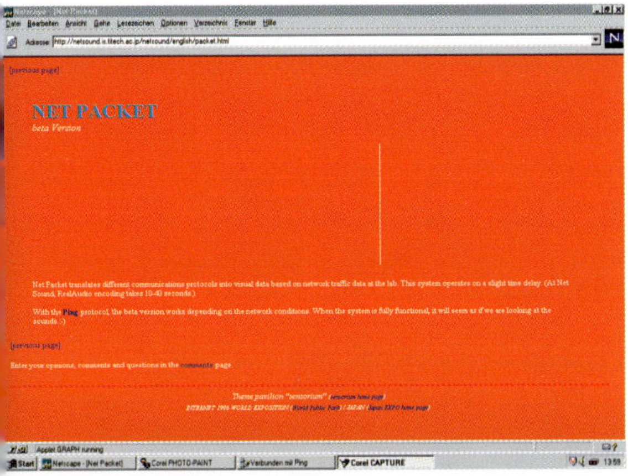

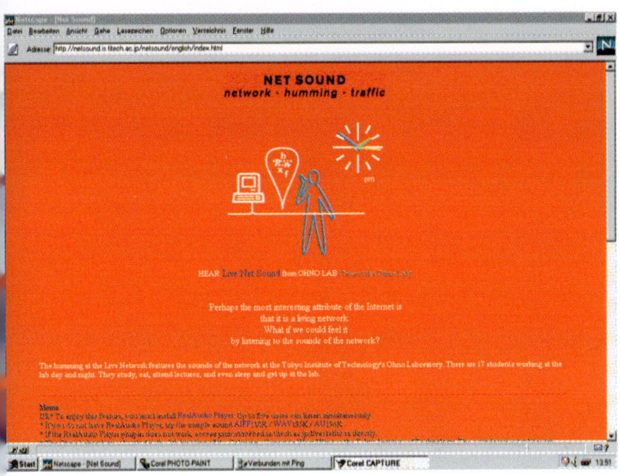

ing through the universe. We do not experience the sensation of flying, but we know the Earth spins around the sun because of night and day and the seasons. The Earth's speed — 30 kilometers per second — is much faster than anything we can feel.

"sensorium" is the theme pavilion of the Japan zone in the Internet 1996 World Exposition (IWE '96). Now it will be going into a new phase as a non-profit Web site maintained by the team members with help from Japan's WIDE project (widely Integrated Distributed Environment – the early and largest inter-academic network

Net Packet

Das Schwester-Package zu „Net Sound" verwendet visuelle Daten statt Klänge und wurde als Echtzeitanwendung in Java programmiert.

Star Place

„Star Place" entstand aus dem Gedanken, eine Web-Site zu schaffen, in dem die Besucher, im Gegensatz zum Cyberspace, ein universelles Gefühl von Raum und Zeit in ihrem Leben erfahren können. Als wir begannen, diese Site zu entwicklen, wurde uns bald bewußt, welches Wissen allen auf diesem Planeten gemeinsam ist: Die Erde fliegt durch das Universum. Wir fühlen diese Flugbewegung nicht, aber wir wissen aus dem Ablauf von Nacht und Tag, aus der Folge der Jahreszeiten, daß die Erde um die Sonne kreist. Die Geschwindigkeit der Erde – 30 Kilometer pro Sekunde – ist viel höher als alles, was wir mit den Sinnen erfassen können.

„sensorium" war der Themenpavillion Japans bei der Internet 1996 World Exposition (IWE '96). Jetzt wird es mit Unterstützung von Japans WIDE project (Widely Integrated Distributed Environment), dem ältesten und weitestverbreiteten interakademischen Netzwerk Japans unter der Leitung von Jun Murai, als Non-Profit-Web-Site von den Teammitgliedern neu eingerichtet.

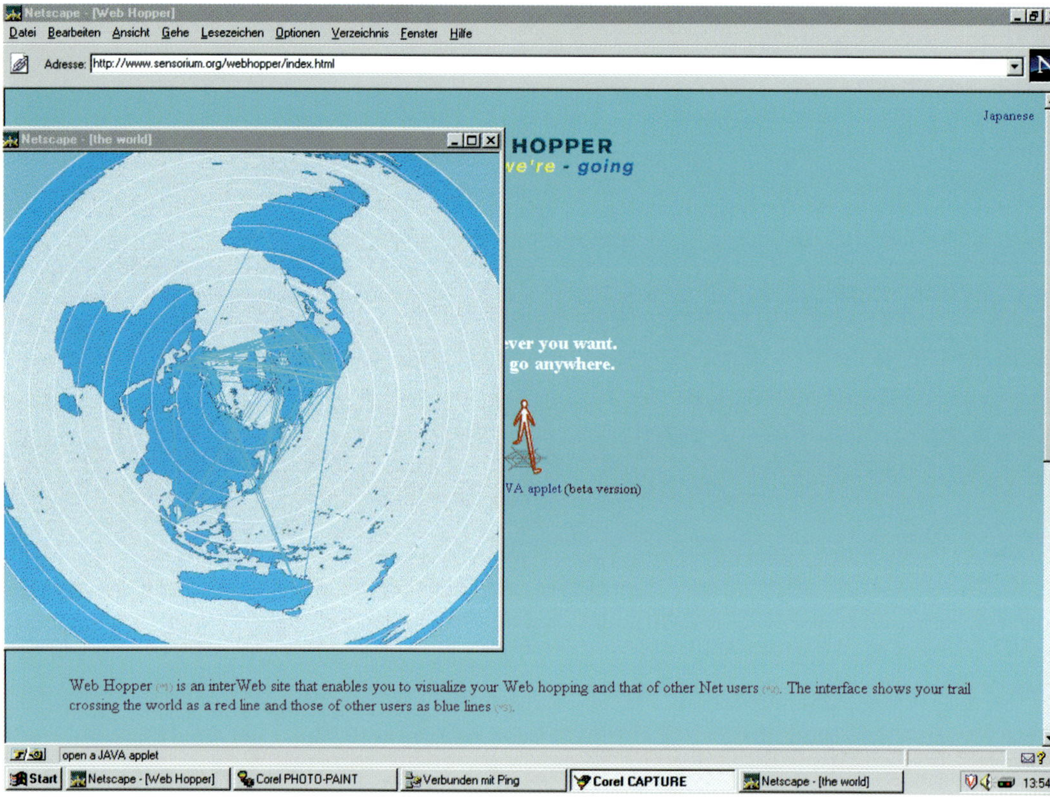
Netscape - [Web Hopper]
Datei Bearbeiten Ansicht Gehe Lesezeichen Optionen Verzeichnis Fenster Hilfe
Adresse: http://www.sensorium.org/webhopper/index.html
Japanese
Netscape - [the world]
HOPPER
ve're - going
ever you want.
go anywhere.
VA applet (beta version)
Web Hopper is an interWeb site that enables you to visualize your Web hopping and that of other Net users. The interface shows your trail crossing the world as a red line and those of other users as blue lines.
open a JAVA applet
Start
Netscape - [Web Hopper]
Corel PHOTO-PAINT
Verbunden mit Ping
Corel CAPTURE
Netscape - [the world]
13:54

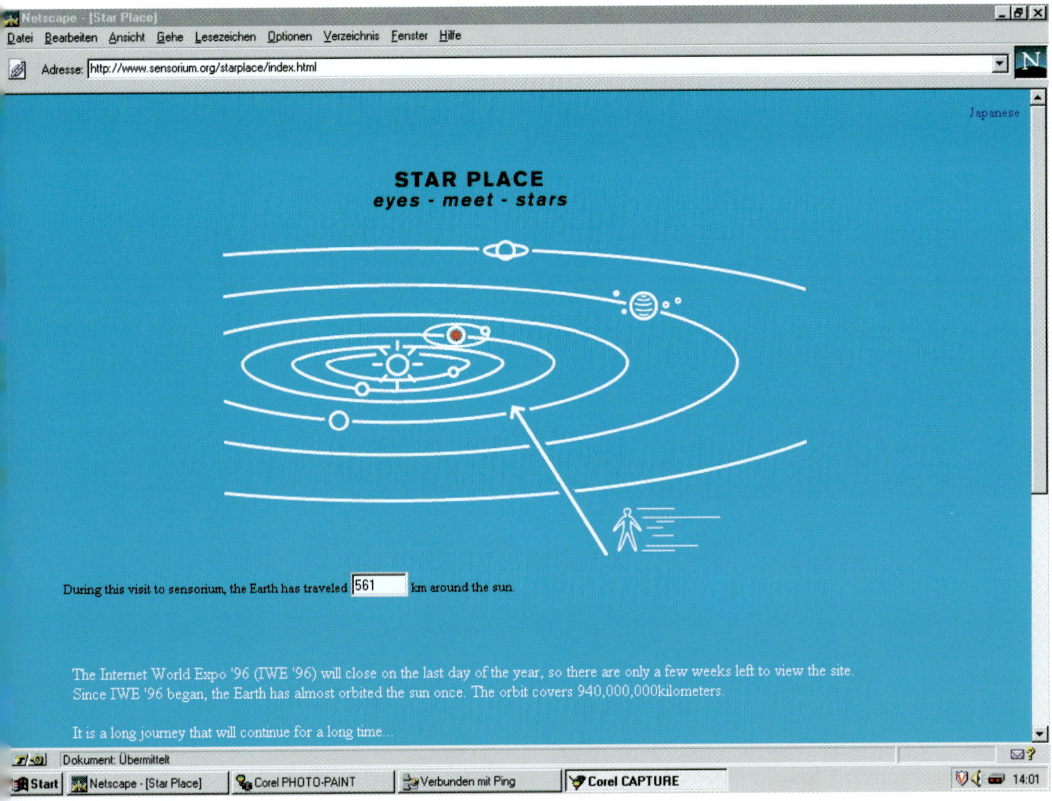

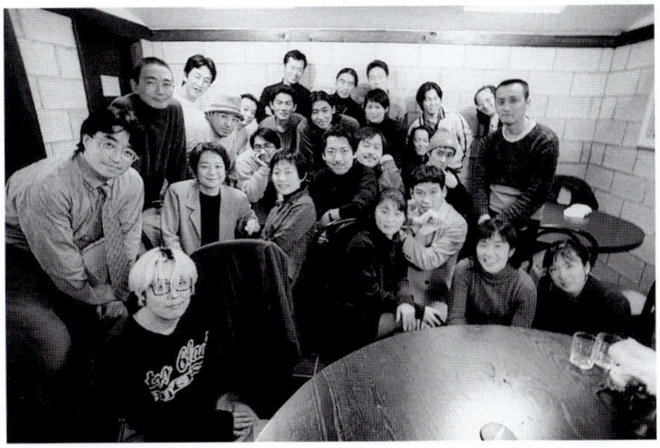

Project Taos is a group of various professionals led by Shin'ichi Takemura (http://www.mediamatic.nl/Doors/Doors2/Takemura/Takemura-Doors2-E.htm), an anthropologist and an associate professor of Tohoku University of Arts and Design. Project Taos is involved in activities ranging from urban design and product planning for large projects, design and production operations and corporate consulting. It is located in Minami Aoyama, Tokyo.

Project Taos ist eine Gruppe verschiedener Professionisten unter der Leitung von Shin'ichi Takemura, einem Anthropologen und Associate Professor an der Tohoku University of Arts and Design. Project Taos beschäftigt sich mit Aufgaben von Stadtplanung und Produktplanung für Großprojekte über Produktionsdesign und -überwachung bis hin zu Konsulententätigkeit für Unternehmen. Der Sitz der Gruppe ist in Minami Aoyama, Tokyo.

TechnoSphere

Gordon Selley / Rycharde Hawkes / Jane Prophet

"TechnoSphere" is an Internet-based artificial life simulator allowing people to create their own creatures and communicate with them as they grow, evolve, and die in a virtual 3D environment. The WWW interface allows users to make a new life form, register it with the system, and retrieve information about their creatures.

The Artificial Life Environment is a program that stores the artificial life forms and calculates their interactions with each other and the environment (modelled in 3D). It supports the reproduction and evolution of life forms. User requests made from the Web Server are processed here and the relevant information passed onto the rendering and email engines. The artificial life forms – which can be either herbivores or carnivores – grow, seek food and mates, chase and evade each other, and produce offspring that are variants on their parents.

The simulation includes the environment (a 16 km-square mountainous terrain with substantial plains), animals, and vegetation. It is possible to create many different types of vegetation within "TechnoSphere", but only one type has been provided thus far. The two basic classes of animals are herbivores and carnivores. There is scope for many specialisations of these two classes, but only two have been made available to the general public thus far: grazer and prowler. Grazers spend their lives wandering the plains, searching for food and companionship, whilst avoiding the insatiable appetites of prowlers. Companionship is sought in the form of other grazers who, for the most part, share a similar appearance. A prowler's existence is similar except that they prey on the grazers for food rather than plants.

Animals can be created by user, or by the simulation as offspring of creatures

„TechnoSphere" ist ein im Internet an gesiedelter 3D-Simulator künstlichen Lebens und erlaubt es Menschen, ihre eigenen Geschöpfe zu erschaffen, mit ihnen z kommunizieren, sie beim Heranwachsen, bei ihrer Entwicklung und schließlich bei ihrem Sterben zu beobachten. Über das WWW-Interface können neue Lebensformen geschaffen und im System registriert werden, außerdem können Daten über diese Lebewesen eingeholt werden.

Das Artificial-Life-Environment ist ein 3D-Programm, das die künstlichen Lebens formen speichert und ihre Interaktion mit anderen Lebewesen und der Umwelt berechnet (und in 3D modelliert). Es unterstützt die Reproduktion und Entwicklung von Lebensformen. Vom Web-Server kommende Benutzeranfragen werden hier bearbeitet, und die relevanten Daten werden an die Rendering-Maschine sowie den E-Mail-Server weitergeleitet. Die künstlichen Lebensformen – sie können Fleisch- oder Pflanzenfresser sein – wachsen, suchen nach Futter und Partnern, jagen einander und flüchten, sie bringen Nachkommen hervor, die Varianten ihrer Elterntiere sind.

Die Simulation umfaßt die Umwelt (e 16 km² großes gebirgiges Terrain mit größeren Ebenen), die Tiere und die Vegetation. Es ist möglich, innerhalb von „Techn Sphere" verschiedene Arten von Vegetatic zu schaffen, derzeit ist allerdings nur eine installiert. Die Tiere teilen sich in Fleisch- und Pflanzenfresser. Auch hier gäbe es di Möglichkeit, mehrere spezialisierte Untergruppen zu erstellen, derzeit sind nur zwe für das Publikum zugänglich. Die grasenden Tiere verbringen ihr Leben auf der Su che nach Nahrung und Partnern und flüch ten vor der unstillbaren Gier der Raubtiere Sie suchen sich andere grasende Tiere als Partner, die ja auch in den meisten Fällen eine gewisse äußerliche Ähnlichkeit aufweisen. Das Leben eines Raubtiers verläuf ähnlich, nur mit dem Unterschied, daß es

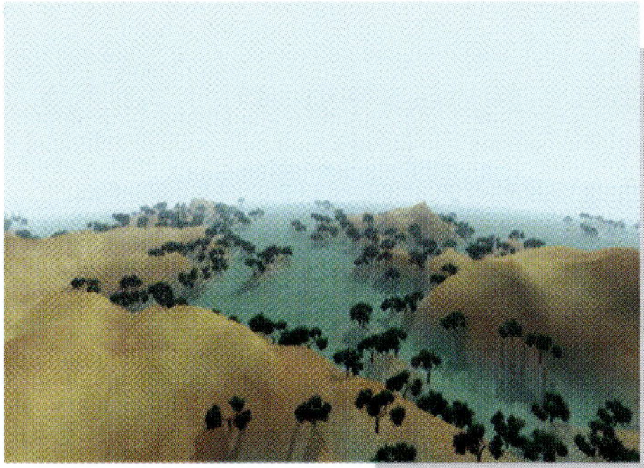

created by the users. Creatures start life as children, progress through adulthood, old age and eventually die, becoming another food supply for the carnivores. The artificial life environment allows creatures to mate, the offspring taking a mixture of physical and functional attributes from both of its parents. The animals that are the result of this mating start their existence in "TechnoSphere" as a foetus, which gestates over a number of weeks, until the animal carrying them gives birth. When an animals carcass has either been picked clean or has rotted away to nothing, the animals information is removed from the simulation state.

Each animal created by a user is given that user's email address. All children born of that animal will inherit that email address, as will their children, and their children's children, etc. This means that users are kept in contact with their immediate family without the need for further action.

Our aim has been to produce an online environment where users can experiment and engage with artificial life forms for fun or more serious purposes. "TechnoSphere" is popular with most Internet users ranging from children to scientists working in the area, its appeal is global and not limited to certain age groups.

One goal is to bring the user closer to the virtual world with the use of interface tools. Users receive email from their creatures at key moments in their lives which prompts the user to investigate the situation by using the online tools. Creatures may live up to three real world months emailing the person who created them roughly once a day. This long term, but at arms distance, relationship between

versucht, Grasfresser als Beute zu erlegen.
Die Tiere können entweder direkt von den Usern erstellt werden, oder sie entstehen als Nachwuchs von bereits existierenden Lebewesen. Sie beginnen ihr Leben als Jungtier, altern und sterben schließlich, wobei sie wiederum den Fleischfressern als Nahrung dienen. Das intelligente Environment ermöglicht auch die Fortpflanzung, wobei der Nachwuchs eine Mischung der Eigenschaften beider Elternteile annimmt. Die so entstehenden Tiere beginnen ihren Zyklus als Föten, werden nach einer bestimmten Tragezeit geworfen und entwickeln sich dann weiter. Wenn der Kadaver eines gestorbenen Tieres entweder aufgefressen oder verrottet ist, wird die Information über das Tier aus dem System gelöscht.

Jedes von einem User geschaffene Tier wird nach der E-Mail-Adresse des Schöpfers benannt. Diese vererbt sich auf die Jungen, deren Nachwuchs und so weiter. Auf diese Weise werden die User über ihre Lebewesen und deren Nachkommenschaft auf dem laufenden gehalten, ohne daß ein weiteres Eingreifen erforderlich wäre.

Unser Ziel war es, ein Environment zu Verfügung zu stellen, in dem die Benutzer zum Vergnügen oder zu ernsthafteren Zwekken mit künstlichen Lebensformen experimentieren können. „TechnoSphere" ist den meisten Internet-Usern vom Kindesalter bis zu Forschern ein Begriff, es ist weltweit attraktiv und auf keine Altersgruppe beschränkt.

Ein Anliegen ist es, durch die Verwendung der Interface-Werkzeuge den User näher an die virtuelle Welt zu bringen. Die Benutzer erhalten in den Schlüsselmomenten im Leben ihrer Kreaturen über E-Mail die Aufforderung, sich doch mittels der Online-Tools die Situation näher anzusehen. Die Kreaturen leben bis zu drei Monate lang, wobei sie durchschnittlich eine E-Mail am Tag an ihren Schöpfer senden. Diese langfristige und doch unmittelbare Beziehung

the user and their creations has proved very popular.

"TechnoSphere" has a large user base of 100,000 people from around the world During a typical 10 day period "TechnoSphere" has 750,000 WWW hits, and serves 2Gigabytes of information to 40,000 users.

zwischen Schöpfer und Geschöpf hat si[...] als sehr populär erwiesen.

„TechnoSphere" hat weltweit 100.0[...] User. In einer typischen 10-Tage-Periode wird im WWW auf „TechnoSphere" run[d] 750.000 Mal zugegriffen, es liefert rund zwei Gigabytes an Information an 40.00[0] Benutzer.

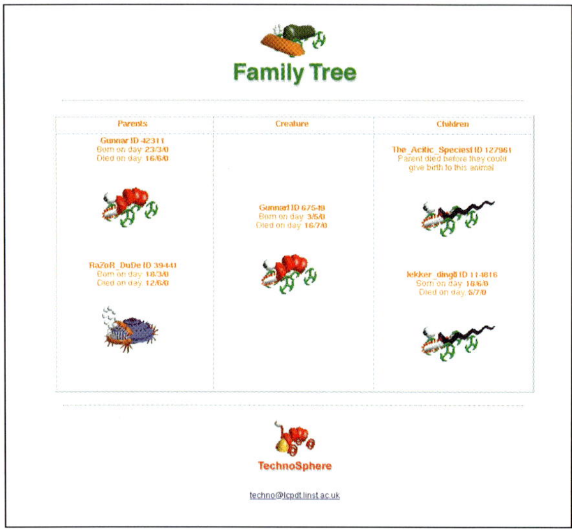

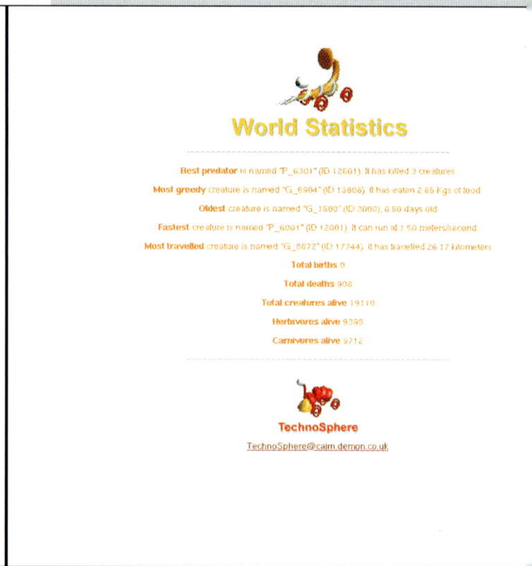

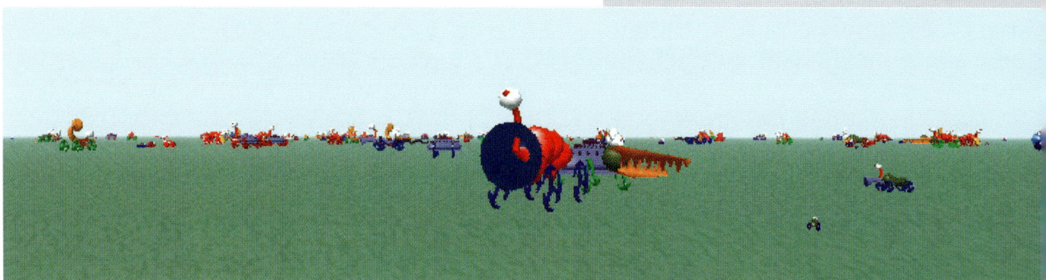

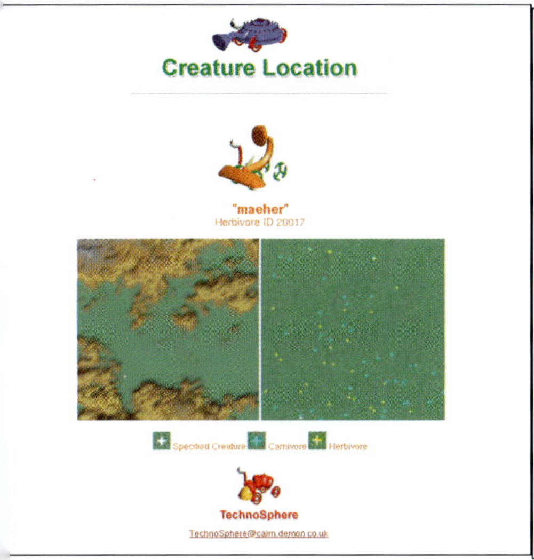

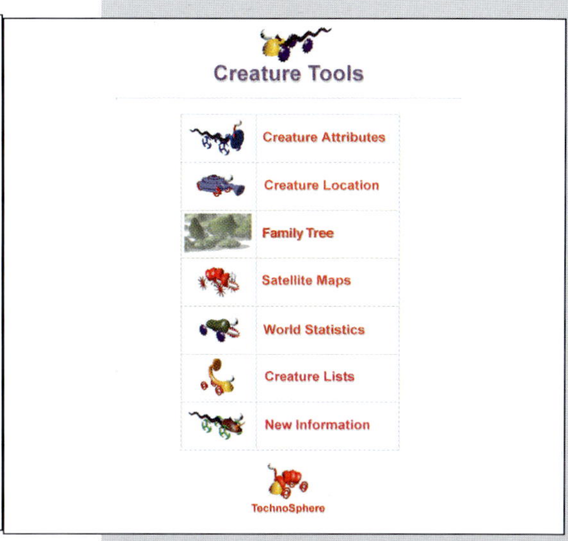

Gordon Selley (GB) is the technical manager for TechnoSphere. He is currently researching networked environments and information spaces as Senior Research Fellow at London College of Printing, School of Media.

Gordon Selley (GB) ist der technische Manager von TechnoSphere. Derzeit arbeitet er als Senior Research Fellow am London College of Printing School of Media und beschäftigt sich mit vernetzten Umgebungen und Informationsräumen.

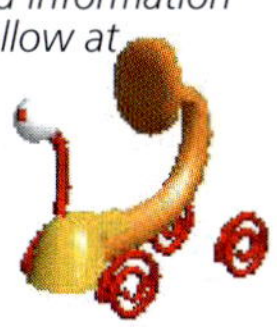

Rycharde Hawkes (GB) has developed the GetALife artificial life engine for TechnoSphere version II. He graduated from Coventry Polytechnic, UK in Computer Science. He is now employed as a Member of Technical Staff at Hewlett-Packard Laboratories, Bristol, UK.

Rycharde Hawkes (GB) hat die „GetALife" Artificial Life Engine für TechnoSphere Version II entwickelt. Er hat das Coventry Polytechnic im Fach Computerwissenschaften absolviert und arbeitet derzeit als Mitglied des technischen Stabes der Hewlett-Packard Laboratories, Bristol.

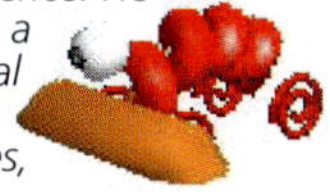

Jane Prophet (GB) is an artist working with video and digital media. She studied Fine Art to complete an MA in Electronic Graphics at Coventry University. Her interest in digital systems and artists' use of computer imaging became the focus of a Ph.D. at Warwick University.

Jane Prophet (GB) ist Künstlerin und arbeitet mit Video und digitalen Medien. Sie hat an der Coventry University zum MA aus Elektronischer Grahik graduiert. Ihr Interesse an digitalen Systemen und der Verwendung von Computerbildern in der Kunst wurde zum Schwerpunkt ihrer Dissertation an der Warwick University.

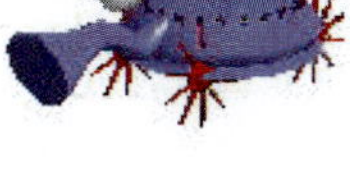

MMF / Make Money Fast

Rolf Schmidt

MMF stands for "Make Money Fast". If you've read Usenet or owned an e-mail account for any amount of time, you've run into these messages that pollute newsgroups and fill your mailbox. "Read this twice...then do the math!" they exhort. "I almost let this slip through my fingers!". And so on. The text tells you to send cash to the people on the list of names, add your name to the bottom, and spam it yourself via Usenet or e-mail. If you do this, the letter claims, you'll be rich beyond your wildest dreams.

Of course, they're nothing but old-fashioned chain letters. They're illegal virtually everywhere, they don't work, they're annoying, they waste network bandwidth, they insult your intelligence, and perhaps worst of all, they've upset the lives of their hapless victims who have found themselves mailbombed, kicked off of their ISPs, hassled by government authorities, and worse.

Experienced netters refer to this garbage as MMF because an early, popular variant suggested that its victims spam it with the title "Make Money Fast".

The MMF Hall of Humiliation contains plenty of pointers to other sites which combat MMF by carefully explaining why MMF doesn't work, and the penalties suffered by participants, in a calm, rational manner. But, despite the hundreds of resources available and the fact that chain letters are expressly prohibited by most ISPs, the problem is worse than ever. So, I've decided to take another approach: severely embarrassing the would-be scammers who waste my time and yours.

MMF steht für „Make Money Fast". Jedem Usenet-Leser oder E-mail-Besitzer sind schon jene Nachrichten begegnet, die Newsgroups verseuchen und die Mailbox anfüllen. „Lesen Sie das zweimal ... und dann rechnen Sie nach!" und so fort. Der Text fordert dazu auf, Personen auf der beigefügten Namensliste Geld zu senden, den eigenen Namen am unteren Ende der Liste dazuzuschreiben und die Nachricht selbst über Usenet oder E-Mail weiterzuverbreiten. Und wer dies tut, so behauptet der Brief, wird weit über seine wildesten Träume hinaus reich.

Natürlich ist das nichts anderes als die guten alten Kettenbriefe. Sie sind praktisch überall verboten, funktionieren nicht, nerven, verschwenden Netzwerkleistung, sie beleidigen unsere Intelligenz, und – was vielleicht am schlimmsten ist – sie haben das Leben der unglücklichen Opfer gestört, die sich selbst mit Mails bombardiert sehen, die von ihren Providern hinausgeworfen und von Behörden verfolgt werden oder noch Schlimmeres erleiden. Erfahrene Netz-Surfer nennen diesen Mist kurz MMF, weil eine frühe und recht populäre Variante ihre Opfer mit der Kurzformel „Make Money Fast" auf den Leim geführt hat. Die „MMF Hall of Humiliation" – eine „Un-Ruhmeshalle" sozusagen – enthält zahlreiche Verweise auf andere Sites, die MMF bekämpfen, indem sie genau erklären, warum MMF nicht funktioniert, welche Strafen auf die Mitwirkenden warten, und das auf ruhig und rationale Weise. Aber trotz hunderte verfügbarer Ressourcen und trotz der Tatsache, daß Kettenbriefe von den meisten Providern ausdrücklich verboten sind ist das Problem schlimmer denn je. Und deswegen habe ich beschlossen, einen anderen Ansatz zu wählen: Jene Möchtegern-Schwindler ernsthaft in Verlegenheit zu bringen, die meine und Ihre Zeit verschwenden.

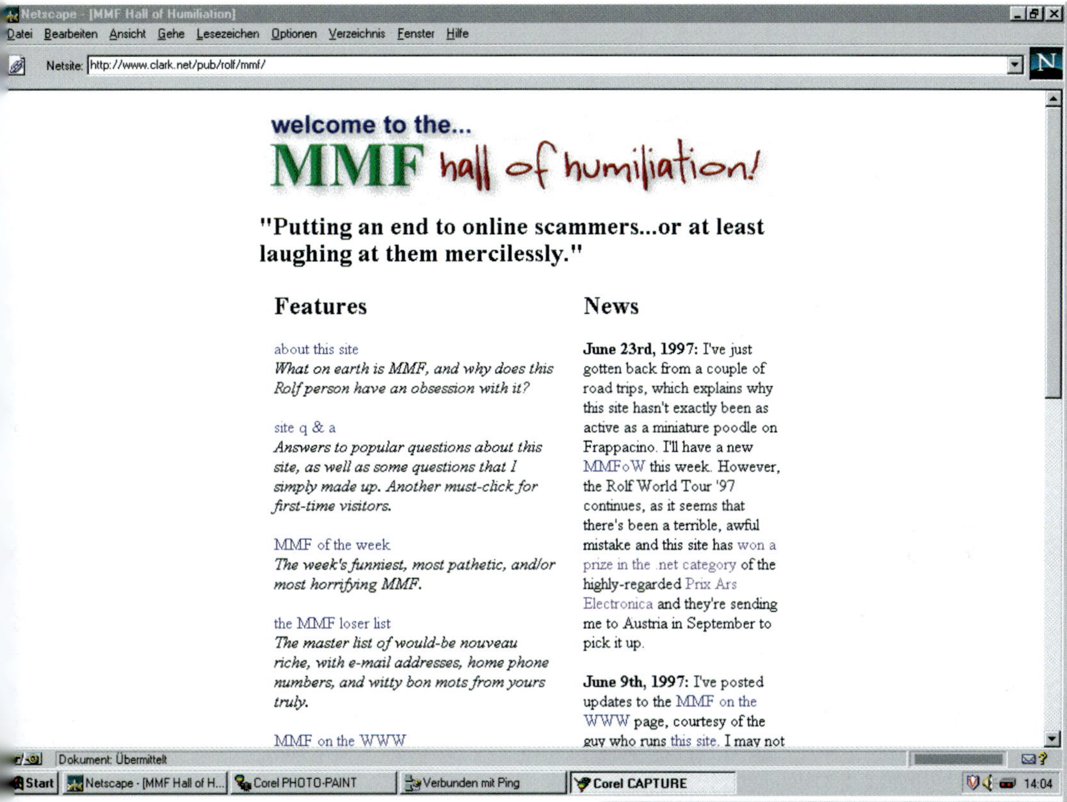

Rolf Schmidt (D) lives in the United States of America and is a marketing professional for a well-known computer peripheral company. The 30-year-old has been running The MMF Hall of Humiliation since 1996.

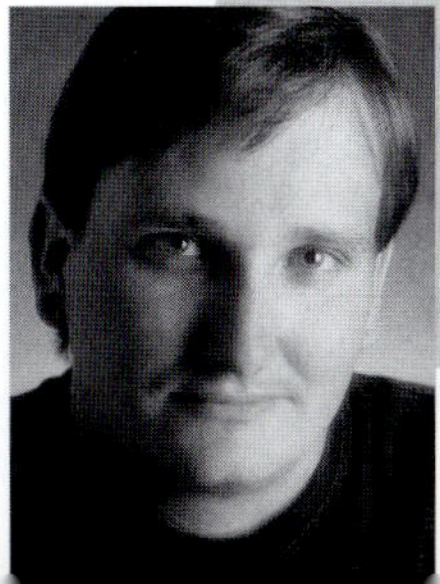

Rolf Schmidt (D), geb. 1967, lebt in den USA und ist Marketing-Manager bei einem führenden Hersteller von Computerperipherie. Er betreibt „The MMF Hall of Humiliation" seit 1996.

THE MULTI-CULTURAL COLLIDER

AMY ALEXANDER

There has been a recent proliferation of video cameras on the World Wide Web. These cameras provide documentation, surveillance, and a very specific representation of the camera owners and their surroundings. Ordinary people and places around the world are instantly subject to becoming part of the mass culture - and are potentially also subject to cultural recycling. "The Multi-Cultural Recycler", in addition to its tongue-in-cheek attempt at performing cultural recycling on ordinary situations, also examines the meetings and collisions of all of these disembodied representations out in "cyberspace."

When a visitor accesses "The Multi-Cultural Recycler", s/he is presented with two options for how to "start" the "Recycler":

Option 1: The default version. The "Recycler" selects two or three camera Web sites from anywhere in the world at random, and captures the live or latest image from their cameras. The Recycler then performs digital image processing on these images to "recycle" them into a new image - a new object of "Web art." Since the actual process used is also selected at random, each access to the Recycler site produces a unique image.

Option 2: "Make your own Cultural Compost!" If the visitor selects this option, s/he is presented with menus from which s/he can select the source cameras to be recycled. The visitor may also select, in place of a camera, the immediately previous "recycled" image, or one of the recycled images from the Gallery page. Since these previously recycled images were created by other visitors to the site, visitors have the opportunity to perform "cultural recycling" on each other's work.

Whichever camera selection option the visitor chooses, the live or latest images will be captured from their Web sites, and

In letzter Zeit ist eine wahre Flut von Videokameras über das World Wide Web hereingebrochen. Diese Kameras liefern Dokumentationen, Überwachungsbilder und eine sehr spezifische Darstellung ihrer Eigentümer und deren Umgebung. „Gewöhnliche Leute" und Orte auf der ganzen Welt werden so augenblicklich Teil einer Massenkultur – und sind auch potentiell einer kulturellen „Wiederverwertung" ausgesetzt. „The Multi-Cultural Recycler" untersucht – neben dem augen-zwinkernden Versuch eines kulturellen Recyclings aus alltäglichen Situationen – die Begegnungen und Kollisionen all dieser entkörperten Darstellungen draußen im Cyberspace.

Wenn ein Besucher den „Multi-Cultural Recycler" aufruft, werden ihm zwei Startoptionen angeboten.

Bei Option 1, der Standardeinstellung, wählt der „Recycler" nach dem Zufallsprinzip zwei oder drei Kamera-Web-Sites irgendwo auf der Welt aus und lädt deren letzte oder Live-Bilder. Dann bearbeitet er sie digital und „recycelt" sie zu einem neuen Bild – eben zu einem Objek der "Web-Kunst". Nachdem auch der Bearbeitungsprozeß von Zufallsparametern gesteuert wird, erzeugt das Programm bei jedem Aufruf ein einzigartiges Bild.

Option 2: "Mach deinen eigenen kulturellen Kompost!" Wenn der Besucher diese Option wählt, erscheinen mehrere Menüs, aus denen er die zu recyclenden Bildquellen wählt. Statt einer Videokamera kann auch ein soeben „wiederverwertetes" Bild aus einer Galerie-Seite gewählt werden, und nachdem diese Bilder ja von anderen Besuchern der Web-Site geschaffen wurden, können Besuch der Site ihre Werke gegenseitig „kulture recyclen".

Welche Kamera-Option auch immer gewählt wird, die letzten oder die aktuellen Live-Bilder werden auf ihren Web-

processed through one of roughly twenty image processes. All of the processes result in both juxtaposition and merging of the original images, creating a collage which is a document of their relationship as fragments of Web culture, and of their chance meeting in cyberspace.

Sites „eingefangen" und in rund zwanzig Arbeitsgängen bearbeitet. All diese Prozesse resultieren in einer Gegenüberstellung und gleichzeitigen Verschmelzung der Originalbilder und schaffen eine Collage, die ihre Beziehung als Fragmente der Netz-Kultur sowie ihre zufällige Begegnung im Cyberspace dokumentiert.

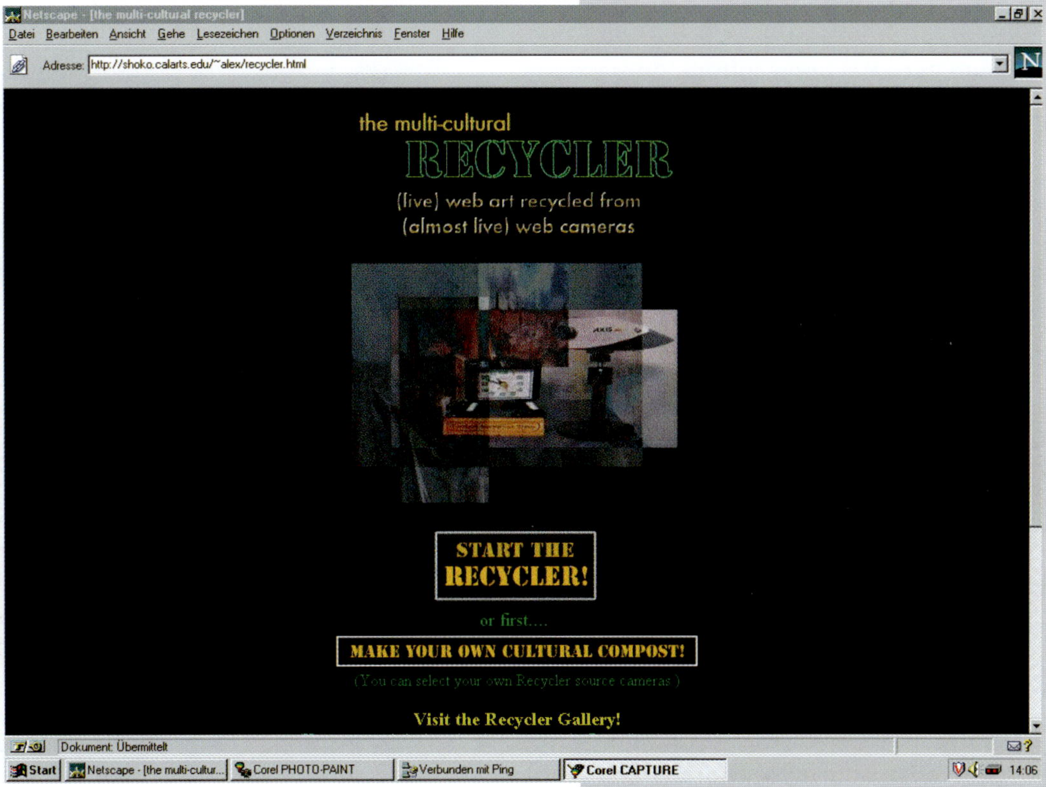

Amy Alexander (USA) has worked in film, video, computer animation and interactive media and has taught at California Institute of the Arts and the University of Southern California. She received a BA from Rowan College of New Jersey and an MFA from California Institute of the Arts. She is active in Internet art and is interested in continuing to explore the integration of interactivity with temporally-based visual art forms.

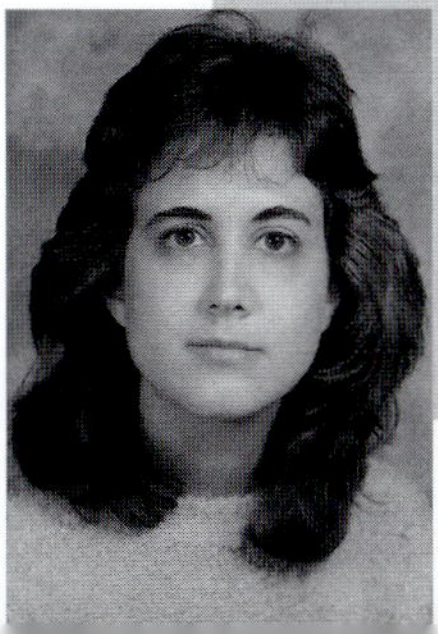

Amy Alexander (USA) arbeitete in den Bereichen Film, Video, Computeranimation und Interaktive Medien. Sie hat am California Institute of the Arts und an der University of Southern California gelehrt, graduierte am Rowan College of New Jersey zum BA und am California Institute of the Arts zum MFA. Derzeit arbeitet sie als Internet-Künstlerin und interessiert sich vor allem für die Erforschung der Einbindung von Interaktivität in die zeitlichen Formen visueller Kunst.

Sherwood Forest

Bruce Damer

"Sherwood Forest" Community was an experiment in virtual community building and culture in a constructivist cyberspace environment called AlphaWorld, one of the cityscapes in the Active Worlds environment. "Sherwood Forest" was built and populated by members of the Contact Consortium, an organization dedicated to studying, promoting and enriching the Internet-hosted virtual worlds as a new space for human contact and culture. Consortium members have years of experience in designing and running MUDs, MOOs and in computer graphics world building exercises and applied this to the "Sherwood Forest" Community Project.

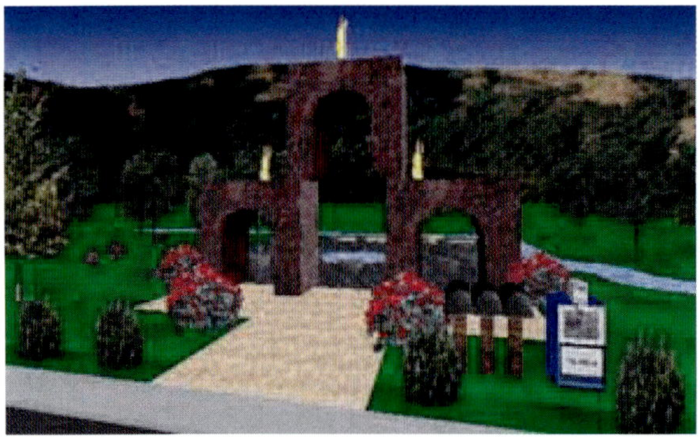

The purpose of "Sherwood Forest" was to design a very natural, attractive setting with woodlands, flowers and flowing water and then attract a community of users to build a village community in that space. A unique feature of Active Worlds is that it allows users all over the Internet with nothing more than a Windows PC and a modem connection to navigate and build in a large virtual space while interacting with others through "avatars" (virtual representations of other users). AlphaWorld represented a large scale extension of the early work of Farmer and Morningstar on Habitat (1986-92). Using this capability, "Sherwood" community planners recruited builders from some of the 100,000 registered citizens of AlphaWorld.

Die „Sherwood Forest"-Community war ein Experiment über den Bau virtueller Gemeinschaften und Kulturen innerhalb eines konstruktivistischen Cyberspace-Environments namens AlphaWorld, einer der Städte im Active-Worlds-Environment. „Sherwood Forest" wurde von Mitgliedern des Contact Consortium gebaut und bevölkert, einer Organisation, die sich dem Studium, der Verbreitung und Bereicherung der virtuellen Welten im Internet als neuem Raum für menschliche Kontakte und Gemeinschaften widmet. Zu den Mitgliedern des Konsortiums zählen sowohl Freiberufler als auch Spezialisten der Industrie, Forscher an Universitäten und Mitarbeiter von Firmen und öffentlichen Einrichtungen. Die meisten Mitglieder verfügen über jahrelange Erfahrung in Entwicklung und Betrieb von MUDs, MOOs und in Computergraphik-Projekten und haben diese in das „Sherwood Forest"-Community-Projekt eingebracht.

Ziel von „Sherwood Forest" war es, eine möglichst natürliche, attraktive Umgebung mit Wäldern, Blumen und Gewässern zu entwickeln und dann eine Gemeinschaft von Usern einzuladen, in diesem Raum ein Dorf zu bauen. Eine Besonderheit der Active Worlds ist, daß User aus dem gesamten Internet nichts als einen Windows-PC und ein Modem brauchen und damit in einem großen virtuellen Raum navigieren und Dinge bauen und gleichzeitig miteinander über Avatare (virtuelle Darsteller anderer User) kommunizieren können. AlphaWorld war eine groß dimensionierte Erweiterung der frühen Arbeit von Farmer und Morningstar in „Habitat" (1986-92). Unter Ausnützung dieser Möglichkeiten rekrutierte die Gemeinschaftsplaner von „Sherwood" potentielle „Bauherren" aus einigen der 100.000 registrierten Bewohner von AlphaWorld.

Why did we pick the theme of "Sherwood Forest"? Apart from the attractive fable of Robin Hood (which supplied some creative roles), it turns out that the Luddite movement against technology began in the "Sherwood Forest" region of Britain. We felt that if there was a rebellion against life in this new virtual worlds technology, it might as well happen inside a virtual "Sherwood Forest"!

Und warum „Sherwood Forest"? Abgesehen von der attraktiven Fabel von Robin Hood (die einige recht kreative Rollen inspiriert hat), hat auch die Bewegung der maschinenstürmerischen Ludditen ihren Anfang in der Gegend des Sherwood Forest in England genommen. Und wir dachten, wenn es denn eine Rebellion gegen das Leben in dieser neuen Technologie der virtuellen Welten geben soll, dann kann sie ihren Ausgang auch gut und gerne in einem virtuellen „Sherwood Forest" nehmen.

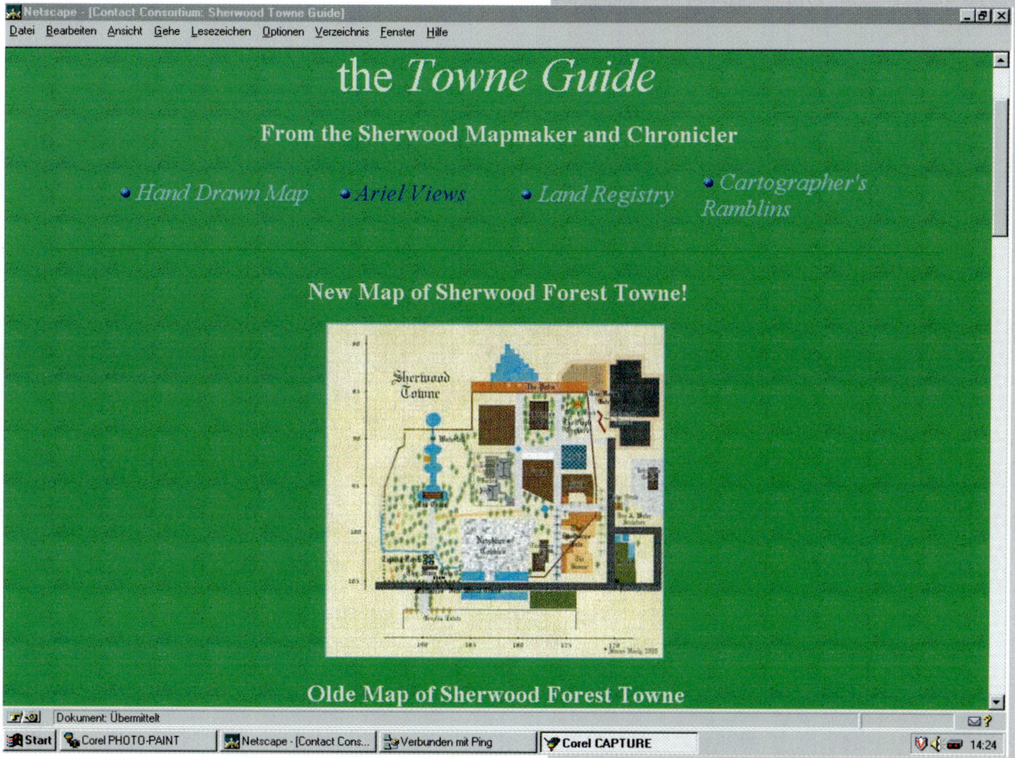

Bruce Damer (USA) holds a BSc in Computer Science from the University of Victoria, Canada and an MSEE from the University of Southern California. Currently he is a member of the faculty at San Francisco State University Multimedia Studies Program. He formed DigitalSpace Corporation to serve as a catalyst for applications of virtual worlds in business and education and he co-founded the Biota.org SIG to seed digital biota in virtual worlds.

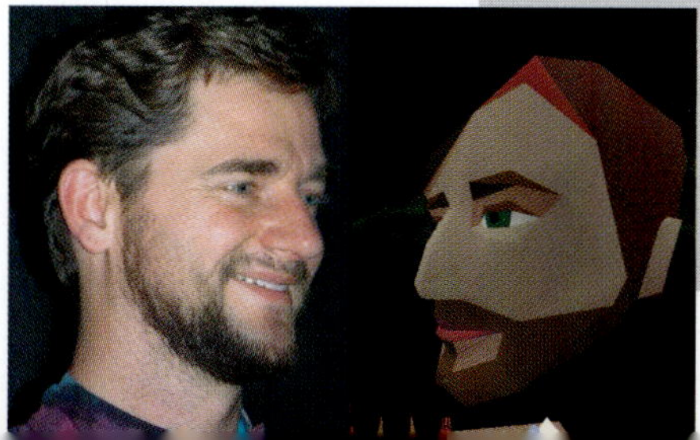

Bruce Damer (USA) graduierte an der University of Victoria, Kanada, zum BSc aus Computerwissenschaften und an der University of Southern California zum MSEE. Derzeit ist er Fakultätsmitglied am San Francisco State University Multimedia Studies Program. Er gestaltete die als Katalysator für Anwendungen virtueller Welten im Business- und Bildungsbereich fungierende DigitalSpace Corporation und war Mitbegründer der Biota.org SIG zur Verbreitung von digitalen Bioten (Pflanzen und Tieren) in virtuellen Welten.

OCEAN WALK

RYOICHIRO DEBUCHI

I created "Ocean Walk" as an art piece. In other words, it is a world that aims to excite and move people.

Dive into the seas of Yonaguni Islands of Okinawa to explore the ruins that have been there for an unknown number of centuries. Discover the enormous pyramids and the "Stone Circle", the playground of those who live there: dolphins, turtles, and various other creatures. Click the fish and animals and they'll let you ride them.

There's also a little "game" in this world. If played well, it'll lead you to the secret "Dragon Palace". The Dragon Palace is an underwater paradise described in the Japanese legend "Urashima Taro". This Dragon Palace has been recreated in Okinawa fashion. The buildings, dancers' movements, and their costumes have been designed to realisitically resemble Okinawa tradition.

Our policy when creating VRML data is to keep file sizes as small as possible while building high quality worlds. The environments of VRML viewers are not universal: direct connection, dial-up using modems, speedy workstations, slow personal computers.

To reach the largest common denominator, we need to set our target to the low end machines. Worlds with complex objects and textures can look pretty; but I purposely chose not to do so. I also aim to have people on different platforms (i.e., SGI, Windows) enjoy our worlds. For this reason, we chose not to use any functions that are only available on specific platforms.

Ich habe „Ocean Walk" als ein Kunstwerk geschaffen, mit anderen Worten als Welt, die Menschen anregen und bewegen soll. Tauchen Sie ein in das Meer bei den Yonaguni-Inseln von Okinawa und erforschen Sie die Ruinen, die dort seit unzähligen Jahrhunderten ruhen. Entdecken Sie die riesigen Pyramiden und den „Steinkreis", den Spielplatz derer, die dort leben: der Delphine, Schildkröten und zahlreicher anderer Geschöpfe. Klicken Sie auf Fische und andere Tiere, und sie laden Sie zu einem Ritt ein.

Diese Welt enthält auch ein kleines „Spiel", und wenn es gut gespielt wird, führt es den Besucher zum geheimen „Drachen-Palast". Dieser Palast ist eine Unterwasserwelt, die in der japanischen Legende von „Urashima Taro" beschrieben wird. Hier wird der Palast im Stil von Okinawa neu geschaffen, die Gebäude, die Bewegungen der Tänzer und ihre Kostüme wurden nach realen Vorbildern aus Okinawa gestaltet.

Wenn wir VRML-Daten konzipieren, wollen wir bei möglichst hoher Qualität der virtuellen Welten möglichst kleine Datensätze zu erzeugen. Die Computer-Umgebung der Benutzer ist ja nicht überall gleich, sie reicht von Standleitungen zu Einwahlverbindungen über Modem, von schnellen Workstations zu langsamen Personal Computern.

Um eine gemeinsame Basis zu finden, müssen wir unser Augenmerk auf die Low-End-Geräte der Benutzer richten. Welten mit komplexen Objekten und Strukturen mögen zwar hübsch aussehen, aber ich habe bewußt darauf verzichtet. Ich möchte, daß Leute mit verschiedensten Plattformen unsere Welten benutzen können – von SGI bis Windows –, deshalb habe ich auch auf plattformspezifische Effekte verzichtet.

HTTP://WWW.ATOM.CO.JP/VRML2/OCEAN/INDEX.HTML

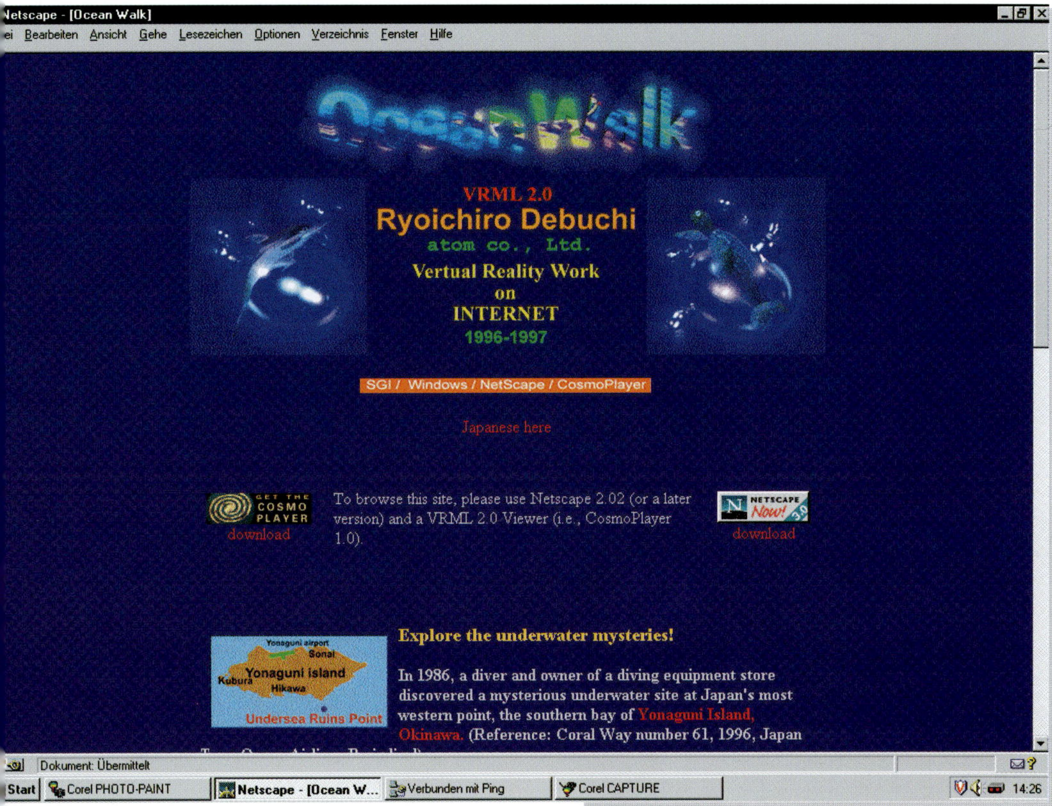

Ryoichiro Debuchi (J) studierte Visuelle Kommunikation und Design am Kyushyu Institute of Design. 1981 begann er seine Forschungs- und Produktionstätigkeit in der Computergraphik mit Programmentwicklung und hat seither ständig im Bereich der Computeranimation gearbeitet. Seine Arbeiten wurden beim SIGGRAPH Electronic Theater 1991 und 1994 prämiert und unter anderem in der SIGGRAPH Art Show gezeigt. Derzeit ist er in der VRML-Forschung bei atom co., Ltd. tätig.

Ryoichiro Debuchi (J) studied Visual Communication of Design at Kyushyu Institute of Design. Began research and production work on computer graphics in 1981 with programming development and has continuously produced animated computer graphics.
He has received an award at the SIGGRAPH Electronic Theater in 1991 and 1994 and has also entered the SIGGRAPH Art Show and many others. Now works in research on VRML contents at atom co., Ltd.

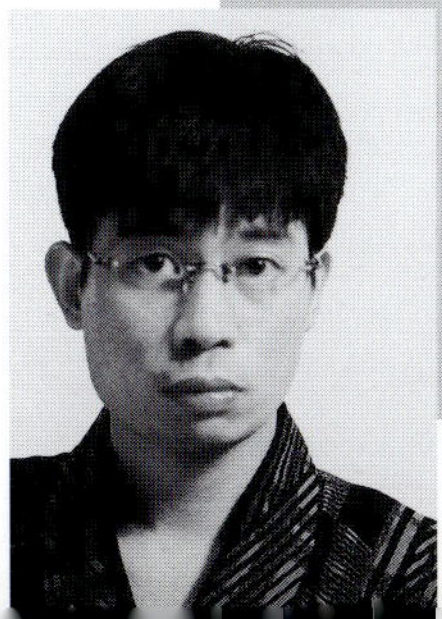

Virtual Roof

Hermann-Josef Hack

The basic idea of "Virtual Roof" is to provide multimedia for everyone – including those that no one else thinks of. Rather than being driven by the euphoria of technical developments and being degraded to a playground and experimental laboratory for the high-tech industry with its expensive show environments, the art project "Virtual Roof" intends to demonstrate new directions. If media competency represents a ticket into the information society, then those whose numbers are constantly growing may not be left out in the cold:

People without work, without money, without a home ...

Instead, people who are disadvantaged are put into a position where they can use new media to articulate and improve their own situations, to help others. The intention is to effectively help them to reclaim their dignity and self-respect. The project "Virtual Roof" provides a virtual home and thus at least an electronic address to those who no longer have a home. This is the starting point for actions and events in various cities to encourage even more initiatives for helping people to help themselves.

As a work-in-progress, "Virtual Roof" provides street newspapers ("asphalt magazin hannover") with a platform for presenting their work, for example, and encourages institutions like these to make themselves heard in cyberspace through virtual presence. In this way, informal contacts under the cover of anonymity lead to actual encounters with people who would have had nothing to do with one another in a normal city setting (as it actually happened in Hannover in March this year, when young students met with newspaper vendors).

Over a thousand visions have already come under the shelter of the virtual roof of the Ruhr region in Germany, the first

Multimedia für alle – auch für diejen[i]gen, an die sonst keiner denkt – möglich zu machen, ist die Grundidee des „Virtua[l] Roof". Anstatt sich rein von der Euphorie der technischen Entwicklung treiben zu lassen und zu Spielwiese und Versuchslabor für die High-Tech-Industrie mit teure[n] Show-Environments zu verkümmern, wil[l] das Kunstprojekt „Virtual Roof" neue We[]ge weisen. Wenn Medienkompetenz die Eintrittskarte in die Informationsgesellschaft darstellt, dann dürfen diejenigen nicht ausgeschlossen werden, deren Zah[l] ständig zunimmt: Menschen ohne Arbeit[,] ohne Geld, ohne Wohnung ...

Stattdessen werden Benachteiligte i[n] die Lage versetzt, durch Nutzung neuer Medien ihre Situation selbst zu artikulieren und zu verbessern, anderen zu helfen, ihnen gezielt zu helfen bzw. wieder Würde und Selbstwert zurückzugewinnen. Das Projekt „Virtual Roof" bietet denjenigen eine virtuelle Heimat und damit wenigstens eine elektronische Adresse, die nirgends mehr zu Hause sind. Vo[n] hier aus werden Aktionen und Events in verschiedenen Städten gestartet, um noc[h] mehr Initiativen zu fördern, mit denen sich Hilfe zur Selbsthilfe vermitteln läßt.

Als Work-in-Progress bietet „Virtual Roof" z. B. Straßenzeitungen (asphalt magazin hannover) eine Plattform zur Darstellung ihrer Arbeit und animiert diese Einrichtungen, sich durch die virtuelle Präsenz Gehör im Cyberspace zu verschaffen. Es entstehen (so geschehen in Hannover im März dieses Jahres, wo Schüler sich mit Zeitungsverkäufern ihrer Stadt trafen) aus unverbindlichen Kontakten im Schutz der Anonymität reale Begegnungen von Menschen, die sich im normalen Stadtbild nicht miteinander befaßt hätten.

Bereits über tausend Visionen sind schon zum virtuellen Dach Ruhrgebiet, der ersten Besiedelungsregion, eingegan-

homesteading region. These are passed on to the responsible politicians and evaluated by scientists and the communal association of the Ruhr region. As virtual empirical data, they form a basis for improving living conditions in this region.

gen. Diese werden den zuständigen Politikern überreicht und von Wissenschaftlern und dem Kommunalverband Ruhrgebiet ausgewertet. Als virtuelle Empirie bilden sie Grundlagen für eine Verbesserung der Lebensumstände dieser Region.

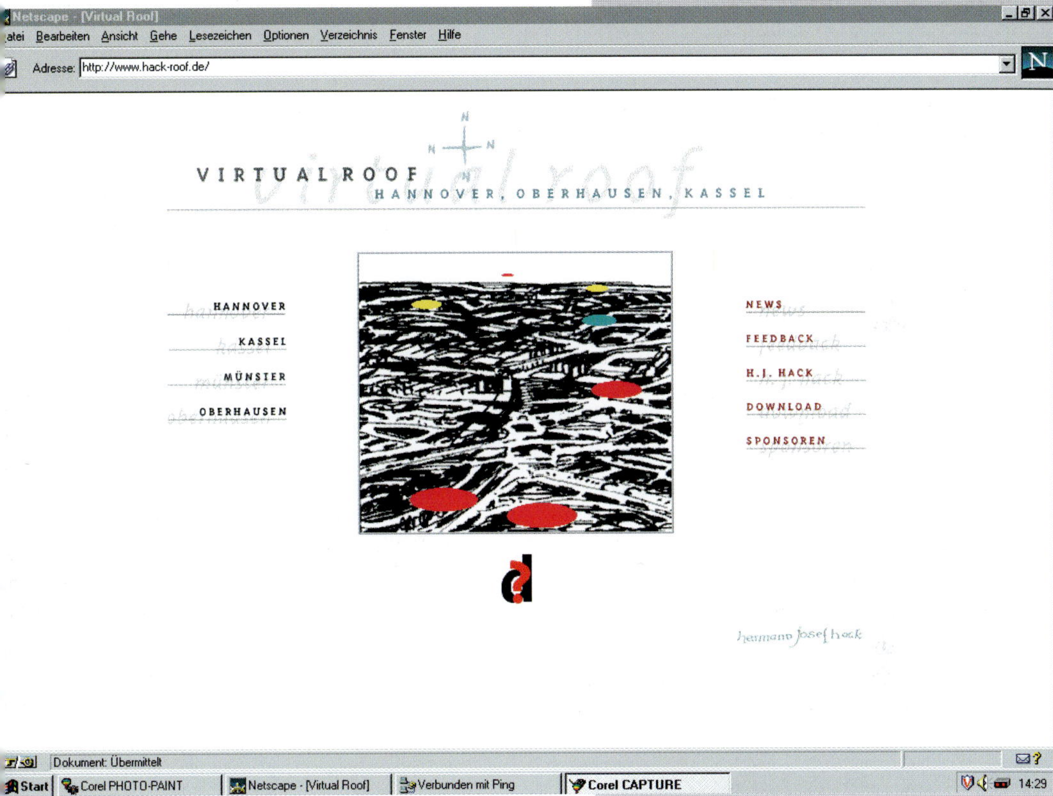

Hermann Josef Hack (D), studied free arts with Josef Beuys. Free-lance artist since 1990. Founder of the "Global Brainstorming Project", an initiative for improving the interdisciplinary communication of important global concerns. He has worked with Van Gogh TV – piazza virtuale – for documenta IX, and initiated projects such as "Global Switch". In 1996 Hack established the "Hackhypermedia virtual institute for the future", a future studies institute that works in the Internet and therefore without the infrastructural ballast of conventional institutes.

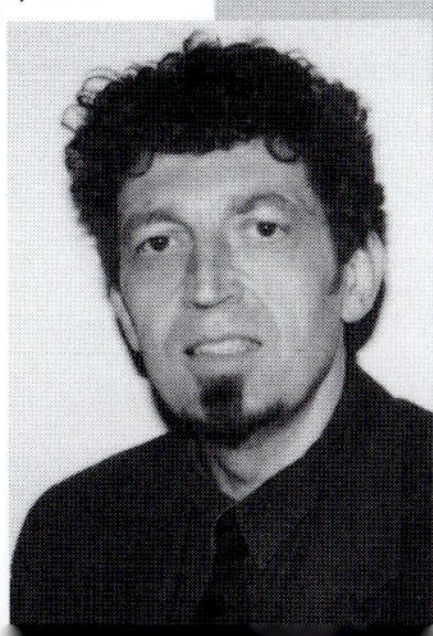

Hermann Josef Hack (D). Studium der freien Künste bei Josef Beuys. Ab 1990 freier Künstler. Gründer des „Global Brainstorming Project", einer Initiative zur besseren interdisziplinären Vermittlung wichtiger globaler Anliegen. Mitarbeiter bei Van Gogh TV – piazza virtuale – zur documenta IX sowie Initiator von Projekten wie „Global Switch". 1996 rief Hack das „Hackhypermedia virtual institute for the future" ins Leben, ein Zukunftsinstitut, das im Internet und damit ohne den infrastrukturellen Ballast herkömmlicher Institute arbeitet.

Harappa

Omar Khan

Harappa is a gateway to South Asia before 1947. The site has two main objectives. One is to start telling the story of the actual ancient city of Harappa and the civilizations of the Indus Valley, about which very little information is available outside scholarly circles. The second objective is to showcase pre-independence (1947) media from the Indian subcontinent, including film, sounds and a wide range of pictures.

An underlying theme is to explore the common history of South Asia (Bangladesh, Bhutan, Maldives, Nepal, India, Pakistan, and Sri Lanka), a history which goes far deeper than current political divisions may suggest.

The story of the actual, literal Harappa (now in Punjab, Pakistan) starts in the early 19th century, with the rediscovery of a city which flourished in 2,500 B.C. Its discovery was the first sign of the ancient Indus Valley civilizations. This story continues today with the first major excavations by a joint US-Pakistani team at Harappa since the 1940's.

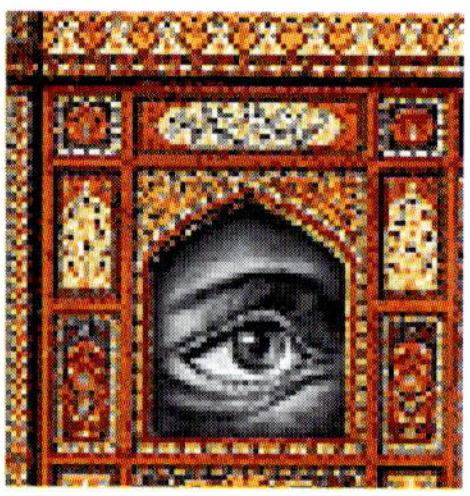

The most basic questions about Harappan civilization remain unanswered. The Indus Valley script has not been deciphered. All we know is that it was the largest ancient civilization, highly sophisticated in its management of trade and resources. Given the very latest research pushing the origin of Indus Valley cultures to 6,000 B.C., they may be the lost origin of many common elements of today's world, including

Harappa ist ein Tor zum Süd-Asien der Zeit vor 1947. Diese Site hat zwei Ziele: Einerseits will sie die Geschichte der tatsächlichen alten Stadt Harappa und der Zivilisation des Indus-Tals erzählen, über die außerhalb wissenschaflicher Kreise nur wenig Information erhältlich ist. Andererseits ist sie eine Darstellung der Medien auf dem Indischen Subkontinent vor der Unabhängigkeit 1947 unter Einbeziehung von Film, Klängen und einem breiten Spektrum von Bildern.

Zugrundeliegendes Thema ist die Erforschung der gemeinsamen Geschichte der Länder Südasiens (Bangladesch, Bhutan, Malediven, Nepal, Indien, Pakistan und Sri Lanka), einer Geschichte, die viel tiefer reicht, als die gegenwärtigen politischen Unterteilungen es vielleicht glauben machen.

Die Geschichte des tatsächlichen Harappa (heute im Pandschab, Pakistan) beginnt im frühen 19. Jahrhundert mit der Wiederentdeckung einer Stadt, die um 2500 v. Chr. ihre Blüte hatte. Damals wurden zum ersten Mal Spuren der antiken Indus-Zivilisation zutage gefördert. Heute führt ein pakistanisch-amerikanisches Team zum ersten Mal seit den 40er Jahren diese Ausgrabungen weiter.

Die grundlegenden Fragen zur Zivilisation von Harappa bleiben freilich noch unbeantwortet: Die Schrift der Indus-Kultur ist noch nicht entziffert, und wir wissen nur, daß sie die bedeutendste Zivilisation ihrer Zeit war und Handel und Gewerbe enorm entwickelt waren. Wenn diese Kultur tatsächlich um 6000 v. Chr. aufgekommen ist – wie jüngst veröffentlichte Untersuchungen annehmen lassen –, so könnte die Kultur des Indus-Tals der verlorene Ursprung vieler gemeinsamer Elemente der heutigen Welt sein: der Schrift, der Tierkreiszeichen-Symbolik, verschiedener Aspekte der Mathematik, der Textilien, Toiletten und vieler anderer.

writing, various signs of the Zodiac, mathematics, textiles, toilets and much more.

The Web site includes a 90 slide tour of the ancient Indus by Jonathan Mark Kenoyer, one of its foremost scholars. The site tries to weave together the relevant information in ways that broad audiences today can enjoy and find stimulating. The site shows actual objects and images the Harappans used, presents detailed analyses of objects like a unicorn seal, or offers a walk through the ancient Indus city of Mohenjodaro.

Die Web-Site führt uns anhand von 90 Bildern von Jonathan Mark Kenoyer, einem der bedeutendsten Forscher auf diesem Gebiet, auf eine Reise durch diese alte Indus-Kultur. Die Site versucht, die wichtigsten Informationen so zu verknüpfen, daß ein breites Publikum sie als anregend und unterhaltsam empfindet. Die Web-Site zeigt Objekte und Bilder, die von den Harappanern verwendet wurden, sie präsentiert eine detaillierte Analyse einzelner Gegenstände wie etwa eines Einhorn-Siegels und bietet einen Rundgang durch die alte Indus-Stadt Mohenjodaro.

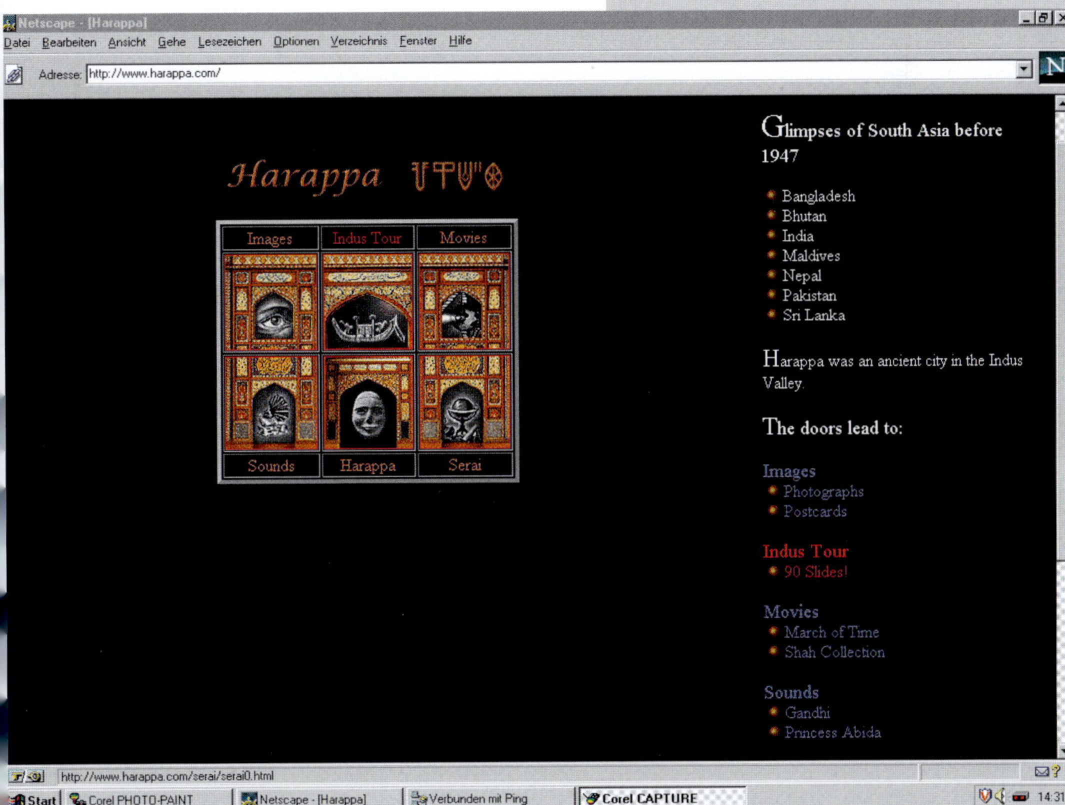

Omar Khan is Pakistani who now resides in San Francisco and holds an MA in the philosophy of art from Columbia University. He works as a multimedia producer and business consultant. Previous work includes directing a pioneering series of video documentaries on democracy for Pakistan's Parliament on behalf of The Asia Foundation in Islamabad, Pakistan.

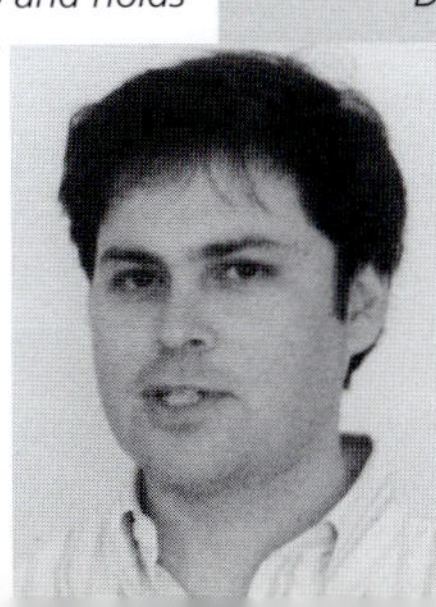

Der gebürtige Pakistani Omar Khan lebt in San Francisco und graduierte an der Columbia University zum MA aus Kunstphilosophie. Er arbeitet als Multimedia-Produzent und Konsulent. Zu seinen früheren Arbeiten zählt die Regie bei einer Serie von Videodokumentationen über Demokratie für das Pakistanische Parlament im Auftrag der Asia Foundation in Islamabad, Pakistan.

Cyberpoetry

Zervos Komninos-Kostant

"Cyberpoetry" is the name I gave to the poetry created on my Macintosh lc630avcd. My goal was to create a poetry that was computer generated, that was moving, that was two and three dimensional and that incorporated voice and sound. It was a poetry that could not be published as a traditional print publication. Whilst it resembled film-making, the text created the images and evoked moods and was always the principle object of every animation and movie.

CYBERPOETRY

poetry that can not be published in the traditional print media,
poetry that moves in time and space,
poetry that requires new ways of reading,
poetry that allows interaction,
poetry that is moving digital, aural, visual, and animated.

My cyberpoetry, indeed, is not the only "cyberpoetry" in existence. Many fine hypertext poems exist on the net, as well as text animations and combinations of sound, text and image. Randomly generated poems utilizing cgi scripting are adding an opportunity for interactivity with the reader. The Java language is providing new ways of reading poetry.

In 1996 I established a Web presence, developing cyberpoetry for the net and a gallery of cyberpoetry from around the globe. I want the site to be as active as possible, a celebration of words/text/ sounds, entertaining, informative and fun. Hope you enjoy it.

„Cyberpoetry" nenne ich die auf meinem Macinstosh LC360avcd entstandene Poesie, wobei es mein Ziel war, Dichtung zu schaffen, die computergeneriert ist, die sich in Bewegung befindet und zwei- und dreidimensional ist und Stimme und Klang miteinbezieht. Dies ist eine Art von Poesie, die durch traditionelle Druckmedien nicht wiedergegeben werden kann. Wenn das Werk auch der Filmgestaltung ähnelt, so hat doch der Text die Bilder geschaffen und Stimmungen hervorgerufen, d. h. der Text war immer das Hauptobjekt jeder Animation und jedes Filmabschnitts.

CYBERPOETRY

Dichtung, die in traditionellen Print-Medien nicht wiedergegeben werden kann,
Dichtung, die sich in Raum und Zeit bewegt,
Dichtung, die neue Lesarten erfordert,
Dichtung, die Interaktion erlaubt,
Dichtung, die digital, aural, visuell, animiert und bewegt ist.

Meine Form der Cyberdichtung ist natürlich nicht die einzige existierende Cyberpoesie – zahlreiche gute Hypertext-Gedichte ebenso wie Textanimationen und Kombinationen von Klang, Text und Bild – existieren im Netz. Zufallsgenerierte Gedichte auf der Basis von CGI-Skripts erweitern die Interaktionsmöglichkeiten beim Lesen, und Java eröffnete neue Wege beim Lesen von Poesie.

Seit 1996 bin ich im Web präsent, entwickle Cyberpoesie für das Netz und eine Galerie der Cyberdichtung aus aller Welt. Ich möchte meine Web-Site so aktiv wie möglich erleben, als ein Zelebrieren von Worten, Texten und Klängen, unterhaltsam, informativ und fröhlich. Viel Spaß!

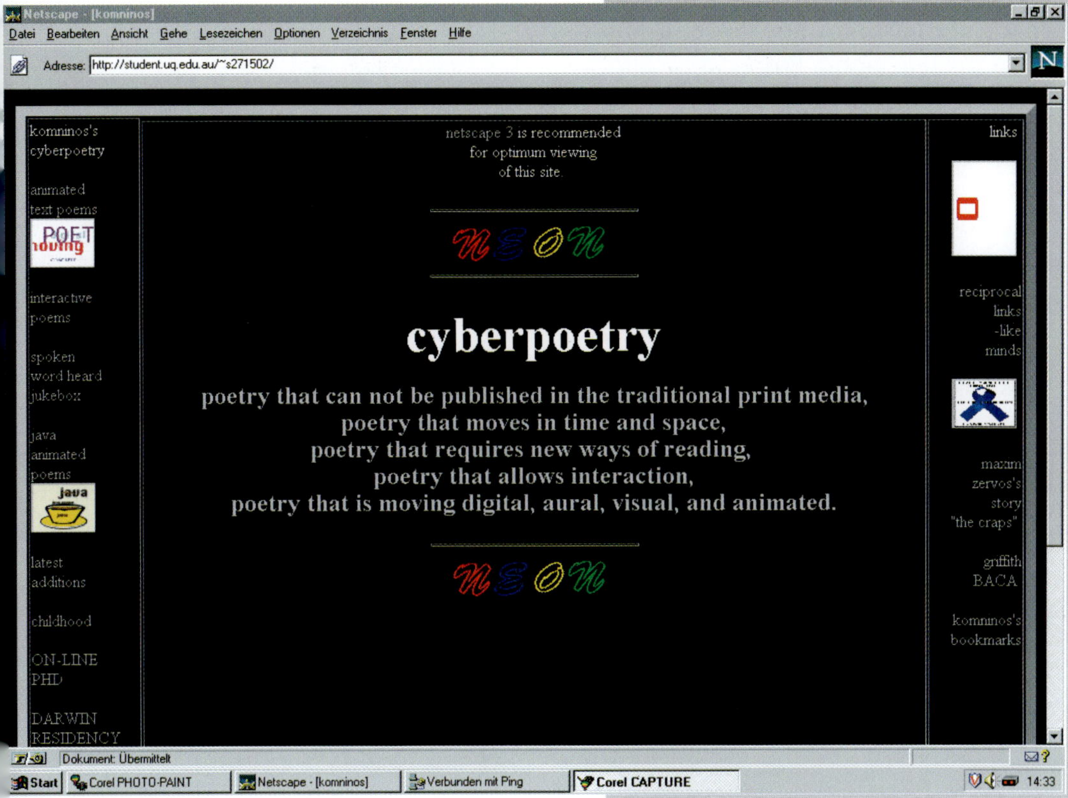

Zervos Komninos-Kostant (AUS) has been a performance poet since 1985 and holds a Master of Arts in creative writing from the University of Queensland. He authored a CD-ROM of cyberpoetry for his dissertation and now teaches effective writing to arts students at the Gold Coast campus of Griffith University. In 1992 he received the Australian human rights award for literature and in 1991 was awarded the Australia council's Ros Bower award for outstanding achievement in community arts. He has published collections of poetry for adults and for children. His works have been performed in Sydney and Brisbane.

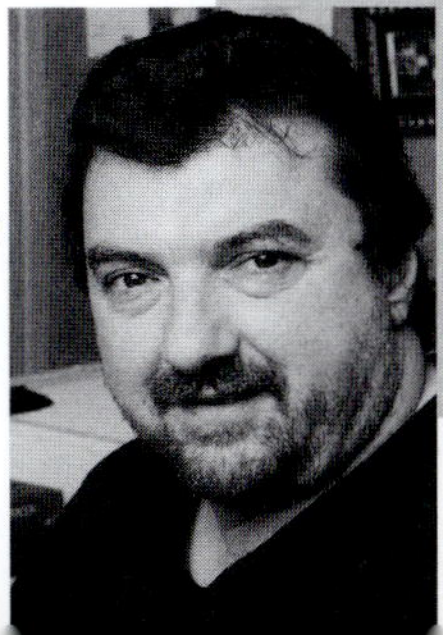

Zervos Komninos-Kostant (AUS) ist seit 1985 Performance-Poet und graduierte an der Universität Queensland zum MA aus Schriftstellerei. Als Dissertation hat er eine CD-ROM der Cyberpoesie herausgebracht und unterrichtet mittlerweile Kunststudenten am Gold Coast Campus der Griffith University in Effektivem Schreiben. Für seine herausragenden Verdienste um die Gemeinschaftskunst erhielt er 1992 den australischen Menschenrechtspreis für Literatur. Er hat Gedichtsammlungen für Erwachsene und Kinder veröffentlicht, seine Werke wurden in Sidney und Brisbane aufgeführt.

Shooting Back

Steve Mann

With Personal Imaging I have attempted to define a new form of interaction between humans and technology.

I have equipped a pair of ordinary sunglasses with miniature spatial light modulators, CCD sensor arrays, and appropriate optics, so that when I put them on, I see a computer screen. On my computer screen I see a video image of what is actually present in the real world. Rather than allow the light to simply pass through, as would be the case with ordinary sunglasses, the apparatus absorbs and quantifies incoming light, processes it computationally, and then sends the processed result on toward my eyes.

After several years of adaptation, the apparatus has begun to function as a visual prosthetic of sorts – a true extension of my mind and body that allows me to record exactly what I see. Clearly I don't need to carry a camera because I am a camera, but in "Shooting Back", I do anyway. Thus the act of merely making a documentary with an ordinary camcorder (while wearing my prosthetic camera) allows one camera to see through the other, and thus allows the audience the unique first-person perspective, as though being inside my eye, while I am shooting with a camcorder.

In "Shooting Back", I confront representatives of the "Surveillance Superhighway" (establishments such as department stores where video surveillance is used extensively, yet photography by customers is strictly prohibited). I begin with my camcorder held down at my side, pointing away from a representative of the SS. Then, I ask the representative "What are those mysterious ceiling domes – those dark hemispheres ..." or "Is that a video camera? Why are you taking pictures of me without my permission?". After the representative tells me that I am paranoid and that only criminals are concerned about cameras, I raise the camcorder up to my eye (the vantage point

Mit Personal Imaging habe ich versucht, eine neue Form der Interaktion zwischen Mensch und Technologie zu definieren.

Ich habe eine ganz gewöhnliche Sonnenbrille mit räumlichen Lichtmodulatoren, CCD-Sensorgruppen und passenden Optiken in Miniaturformat ausgestattet, so daß ich beim Tragen der Brille einen Computerbildschirm sehe. Auf diesem Bildschirm sehe ich ein Videobild dessen, was in der realen Welt um mich vorgeht. Anstatt einfach das Licht gefiltert durchzulassen, wie dies normalerweise bei Sonnenbrillen geschieht, absorbiert und quantifiziert dieser Apparat das einfallende Licht, bearbeitet die anfallenden Daten und sendet das Ergebnis der Datenbearbeitung an meine Augen.

Nach zahlreichen Jahren des Experimentierens ist das Gerät mittlerweile zu einer Art visueller Prothese geworden – einer echten Erweiterung meines Körpers und Geistes, die mir erlaubt, genau aufzuzeichnen, was ich sehe. Klarerweise brauche ich keine Kamera zu tragen, weil ich ja eine Kamera bin, aber in „Shooting Back" trage ich trotzdem eine. Wenn ich also mit einem ganz gewöhnlichen Camcorder einen simplen Dokumentarfilm mache (und dabei meine Kamera-Prothese trage), kann die eine Kamera durch die andere hindurchsehen – und so gewinnt das Publikum eine einzigartige Ich-Perspektive, als wäre es innerhalb meines Auges, während ich filme.

In „Shooting Back" wende ich mich gegen die Repräsentanten des „Surveillance Superhighway", also gegen Einrichtungen wie z. B. Kaufhäuser, die Videoüberwachung in großem Stil einsetzen, aber gleichzeitig den Besuchern das Photographieren verbieten. Am Anfang halte ich meinen Camcorder unauffällig neben meinem Körper, er zeigt weg vom Angehörigen des „SS". Dann frage ich ihn oder sie: „Was sind das für mysteriöse Halbkugeln an der Decke?" oder „Ist das eine Videokamera? Und warum machen Sie Aufnahmen von mir ohne meine Genehmigung?" Nachdem mir der Mitarbeiter erklärt hat, daß

of the audience, who see the face-to-face conversation followed by the eyecup of a camcorder, eventually revealing the inside of the viewfinder, upon which we now see the representative of the SS displayed). At this point, the representative of the SS often shows great concern about my camcorder, and thus, in a 180 degree reversal, is self-incriminating.

ich paranoid sei und nur Kriminelle sich Sorgen wegen der Kameras machten, hebe ich meinen Camcorder ans Auge so, daß das Publikum das folgende Gespräch aus dem Blickwinkel des Kamerasuchers verfolgen kann, der ab diesem Moment das Bild des „SS"-Angehörigen zeigt). Häufig sind die Mitarbeiter des „SS" sehr irritiert, wenn sie meinen Camcorder sehen, und beschuldigen sich – durch diese Volte um 180 Grad – sozusagen selbst.

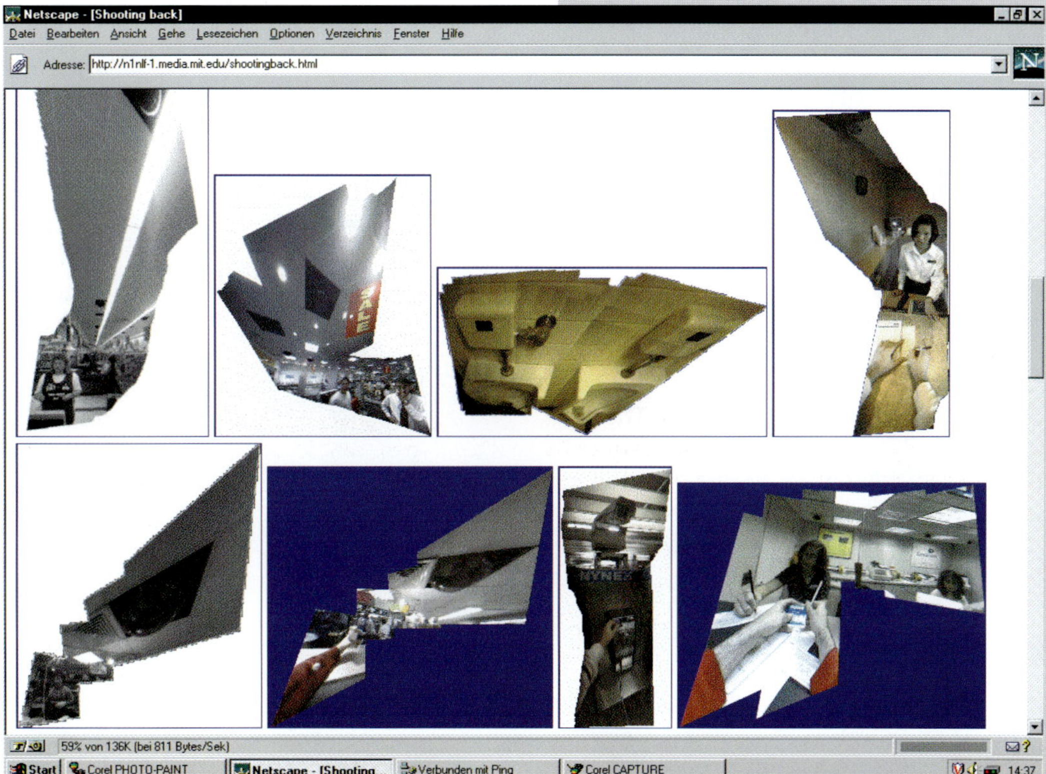

Steve Mann is a doctoral student at the MIT Media Laboratory where he co-founded the MIT wearable computing project. He currently holds degrees in physics and electrical engineering. Steve is known to many as inventor of the so-called "wearable computer" and "WearCam" (personal imaging system).

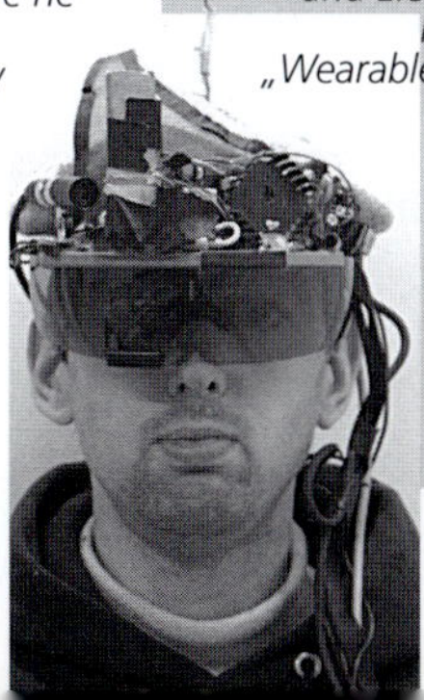

Steve Mann (USA) ist Doktorand am MIT Media Laboratory und Mitbegründer des MIT Wearable Computing Project. Er hat Physik und Elektrotechnik studiert und ist vielen als Erfinder des „Wearable Computers"(Computer zum Anziehen) und der „WearCam" (Personal Imaging System) bekannt.

Jodi

Joan Heemskerk / Dirk Paesmans

```
*DESCRIPTION
---------------------------------------------------------------
  * Public Property Get URL() As String
    URLPicture = m_URLPicture
---------------------------------------------------------------
Public Property Let URL(Url As String)
   Dim sTag$
      If (m_URLPicture <> Url) Then         ' Do only if value has changed ...
   {  ClearPictureOnly = Not ClearURLOnly  ' If property is not set by URL _
   {          then clear the
   {  URLPicture value ...
      m_URLPicture = Url   ' Save url string value to variable
                {
     If Not ClearURLOnly Then
         On Error GoTo ErrorHandler   ' Handle Error
                {
        If (Url <> "") Then
          'OCX.AsyncRead Url, vbAsyncType Picture,
                    PROP_IMAGE_URLPICTURE  ' Begin
                   async download of picture file ...
                  'construct a unique tag for debugging purposes
                    DocID = DocID + 1
                    sTag = Format$(DocID) + ":" + URL
{ ImgDownLoads.Add me, sTag 'add to the list of pending downloads
               OCX.GetImageInternal Url, sTag   ' Begin async download of<(~?
                {
            'OCX.AsyncRead Url, vbAsyncTypePicture, "InternalPic" &
sTag  ' Begin async
download of picture file ...
        Else
            Set Picture = Nothing
        End If
    End If
End If
 Exit Property
---------------------------------------------------------------
 ErrorHandler: ClearPictureOnly = False
  ErrMsg
End Property
---------------------------------------------------------------
```

"Jodi" is part of the anti-art movement: they favor anti-subjective art, i.e. they work from the assumption that there is no such thing as "individuality" or an "autonomous mind". People are enslaved by their biological machinery and subliminal drives on the one hand, and by the blind, relentless, unremitting rush of the world on the other hand. Crushed between these two forces, people are nothing but helpless spectators who stand gaping at the violent spectacle that is taking place within and around them. The notion of the conscious mind obstructs rather than elucidates our capacity to

„Jodi" ist Teil einer Anti-Kunst-Bewegung: „Jodi" favorisieren anti-subjektivisti sche Kunst, sie stehen also auf dem Standpunkt, daß es so etwas wie „Individualität" oder einen „autonomen Geist" nicht gibt. Die Menschen sind einerseits Sklaven ihrer biologischen Maschinerie und ihrer subliminalen Antriebe und andererseits dem blinden, unaufhörlichen reaktionslosen Dahinsausen der Welt unterworfen. Zwischen diesen beiden Kräften werden die Menschen zermalmt, es bleiben nichts als hilflose Zuseher, die mi offenem Mund auf das gewalttätige Spektakel rund um sie blikken. Der Begrif

understand this situation – or at least, that is what anti-subjective art believes.

The two people who form "Jodi" – Dirk Paesmans and Joan Heemskerk– hold that there is no such thing as the genius of the artist, and that interpretations of their work are irrelevant. They refuse to explain themselves and relish the confusion their visitors find themselves in. Page upon page filled with strange symbols, scrambled images and inexplicable messages appear on one's screen, all of them intentionally devoid of meaning. People visiting "Jodi's" site are often startled: they fear that their browser has gone haywire or that their computer has been infected by a virus. "Jodi's" site seems to resist and deflect communication – and on the net, that's almost funny. But more importantly, they stress that interpretation, communication and a solid sense of self are perhaps not that easy to achieve nor as reliable as many people expect and believe them to be; and that is a very crucial notion. *Karin Spaink*

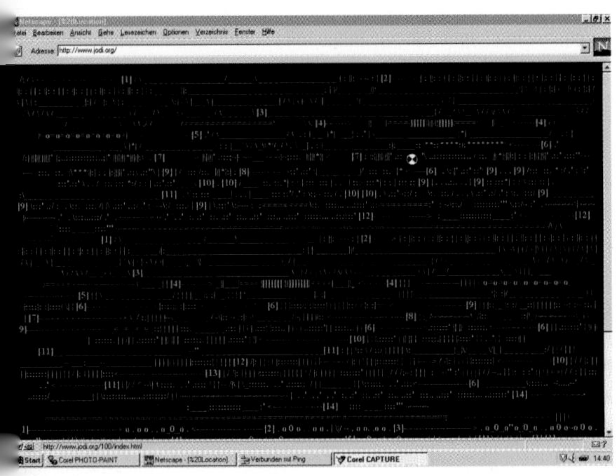

```
<!-->
<META HTTP-EQUIV="Jodi"="http://www.jodi.org">
<METANAME="Joan Heemskerk, Dirk Paesmans">
<A HREF="http://cocoa.is.in-berlin.de/~netlab">netlab</A>
<A HREF="http://dewey.rug.ac.be/isdm/home.htm">ISDM</A>
<A HREF="http://sunsite.cs.msu.su/wwwart">WWWartCenterMoskow</A>
<A HREF="http://www.xs4all.nl/~w139">W139</A>
<A HREF="http://www.slate.com">Slate</A>
<A HREF="http://www.connect-arte.com">Connect-Arte</A>
<A HREF="http://www.irational.org/cern">CERN</A>
<A HREF="http://www.adaweb.com">Adaweb</A>
<A HREF="http://www.documenta.de/jodi">DocumentaX</A>
<--!>
```

des „Bewußtseins" beschränkt unsere Fähigkeit, diese Situation zu verstehen, eher als sie zu erhellen – zumindest glaubt das die anti-subjetivistische Kunst.

Die beiden Künstler, die hinter „Jodi" stehen – Dirk Paesmans und Joan Heemskerk – behaupten, daß es so etwas wie den Genius des Künstlers nicht gibt und daß Interpretationen ihrer Arbeiten irrelevant sind. Sie weigern sich, Erklärungen abzugeben und die Konfusion zu mildern, in die sie ihre Besucher stürzen. Seite um Seite mit seltsamen Symbolen, verdrehten Bildern und unerklärlichen Nachrichten erscheinen auf dem Bildschirm des Besuchers, alle ganz bewußt bar jeder Bedeutung. Häufig erschrecken die Besucher von „Jodis" Site, sie fürchten, ihr Browser hätte durchgedreht oder ihr Computer sich mit einem Virus infiziert. „Jodis" Web-Site scheint jeder Kommunikation zu trotzen und sie entschieden zurückzuweisen – und im Netz ist das schon wieder fast lustig. Aber viel wichtiger ist ihre Botschaft, nämlich daß Interpretation, Kommunikation und ein gesundes Selbstbewußtsein vielleicht doch gar nicht so leicht zu erreichen sind und viel leichter wieder verloren gehen können, als viele Leute meinen – und das ist eine ganz entscheidende Ansicht. Karin Spaink

Container City

Markus Schulthess

"Container City" is a city in the World Wide Web. This three-dimensional city, which may be entered thanks to VRML, exists only in the WWW, and there it may be inhabited. People can even live there, shop there, educate themselves, enjoy themselves, be medically examined, and soon work there, too.

Every net user with their own WWW site has the possibility of living there - in a container designed and built by themselves whenever possible.

The city's own "arbitrary body" decides who will be accepted. The only condition for being accepted into "Container City" is that the candidate's site must be special and extraordinary in some way. In keeping

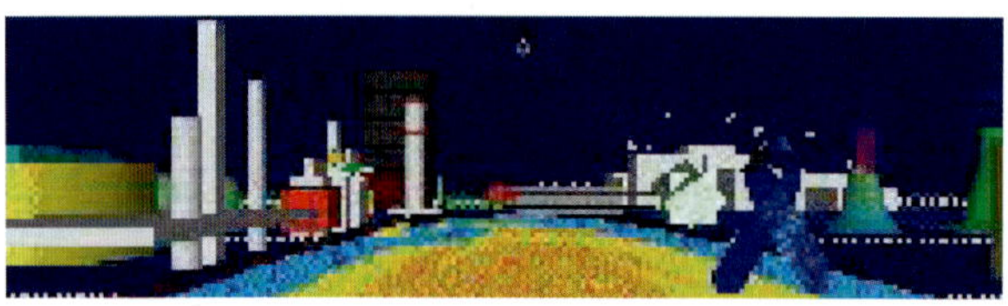

with its name, however, the arbitrary body's decisions are varying and arbitrary. In addition, the body may also decide on compulsory acceptance. Life in "Container City" is tax-free, and the city's future is secured by renting small advertising areas within the city.

"Container City" was purposely designed to resemble a conventional, real city, in order to be able to experience the alteration of net-specific use in the near future. It should be possible to observe how a community is formed and if an identity in and with the city can develop. Various activities in "Container City" are planned for the next few months.

"Container City" grows on a square grid, within which the individual containers are placed. All of the containers are developed from pre-defined bodies, so-called nodes, and do not consist of converted CAD drawings, so that the amount of data may be limited and the city can grow.

„Container City“ ist eine Stadt im World Wide Web. Die dank VRML dreidimensional begehbare Stadt existiert nur im WWW und kann bewohnt werden, ja, man kann in „Container City“ leben, einkaufen, sich weiterbilden, vergnügen, medizinisch untersuchen lassen und bald auch arbeiten. Für jede/jeden NetzteilnehmerIn mit eigener WWW-Site besteht die Möglichkeit, in „Container City“ zu leben –nach Möglichkeit in einem selbstentworfenen und -gebauten Container. Über eine Aufnahme entscheidet das stadteigene „arbitrary gremium“. Die einzige Bedingung für eine Aufnahme in „Container City“ ist, daß die Site des Kandidaten in irgendeiner Weise speziell und herausragend sein muß. Das willkürliche Gremium entscheidet jedoch seinem Namen entsprechend unterschiedlich und auch willkürlich. Es können zudem auch Zwangsaufnahmen durch das Gremium beschlossen werden. Das Leben in „Container City“ ist frei von Steuern, die Zukunft der Stadt wird durch Vermietung von kleinen Werbeflächen innerhalb der Stadt gesichert. „Container City“ wurde bewußt ähnlich einer herkömmlichen, realen Stadt entworfen, um die Veränderung de netzspezifischen Nutzung in der kommenden Zeit zu erfahren. Es bleibt zu beobachten, wie sich eine Community bildet und ob sich eine Identität in und mit der Stadt entwickeln kann. In den nächsten Monaten sind verschiedene Aktivitäten in „Container City“ geplant. „Container City wächst auf einem quadratischen Raster, innerhalb dessen die einzelnen Container plaziert werden. Diese werden und wurden mit einem Texteditor entworfen und entwickelt. Um die Datenmenge klein zu halten und damit die Stadt auch wachsen zu lassen, sind alle Container aus vordefinierten Körpern, sogenannten Nodes, entwickelt und bestehen nicht aus konvertierten CAD-Zeichnungen.

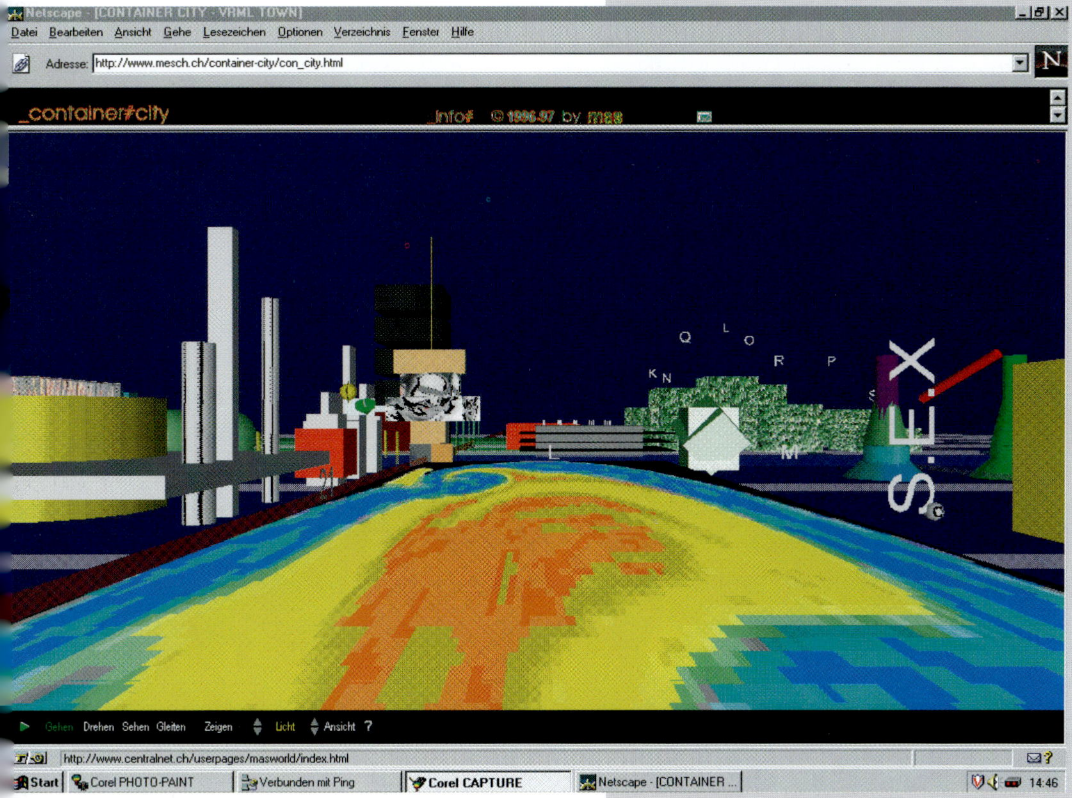

HONORARY MENTION

Markus Schulthess (CH), born 1956, lives and works as a freelance architect in Lucerne. Member of the artists group Artcore. Has studied space and time in the Internet since 1995. He is currently working together with mesch on a 3D Internet model of the city of Lucerne.

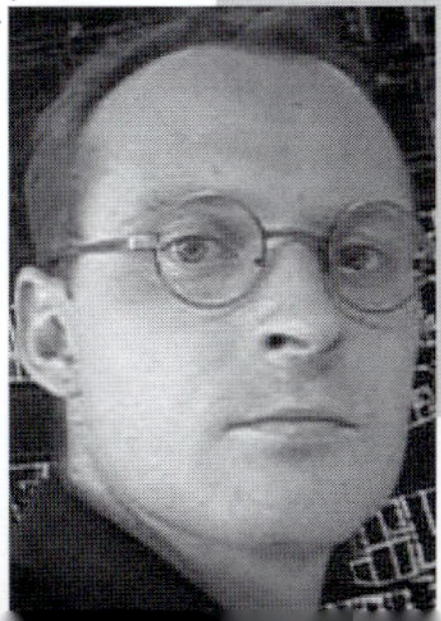

Markus Schulthess (CH), geb. 1965, lebt und arbeitet als freischaffender Architekt in Luzern. Mitglied der Künstlergruppe Artcore. Seit 1995 befaßt er sich mit Raum und Zeit im Internet. Derzeit arbeitet er zusammen mit mesch an einem 3D-Internet-Modell der Stadt Luzern.

Form

Alexei Shulgin

A project by Alexei Shulgin in which he proposes a new form of art: "Form Art".

This new art form is based on the Internet technology and provides artists with new possibilities for self-expression. The Web site devoted to "Form Art" consists of numerous Form Art images and animations, as well as other examples of Form Art implementations. The site is open for contribution and discussion and will hopefully become an international center for development and research of this form of art.

The project has been realised at C3, Budapest.

„Form Art" ist ein Projekt von Alexei Shulgin, in dem er eine neue Kunstform vorschlägt: „Form Art". Diese Kunstform basiert auf der Internet-Technologie und gibt den Künstlern neue Möglichkeiten, sich auszudrücken. Die Web-Site von „Form Art" besteht aus zahlreichen Form-Art-Bildern und -Animationen sowie anderen Beispielen von Implementierungen von Form Art. Die Site ist für Beiträge und Diskussionen offen und wird hoffentlich ein internationales Zentrum für die weitere Entwicklung dieser Kunstform.

Das Projekt wurde bei C3 in Budapest entwickelt.

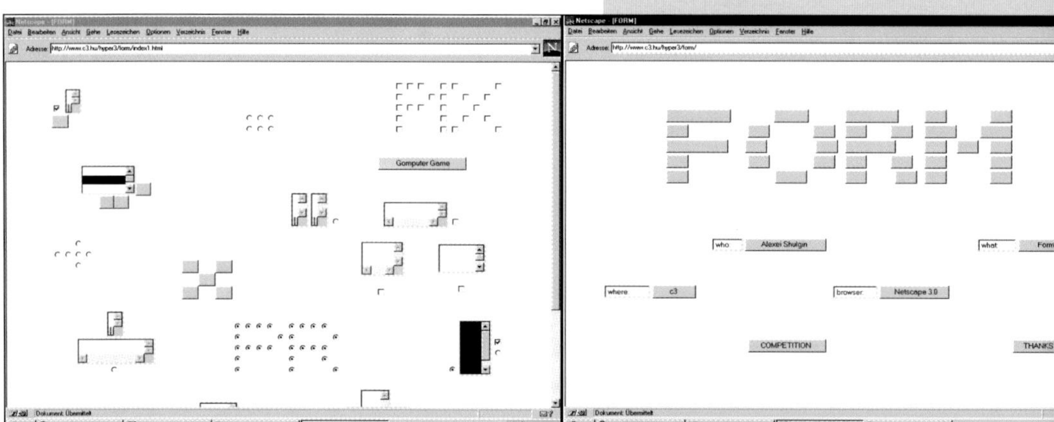

HTTP://WWW.C3.HU/HYPER3/FORM

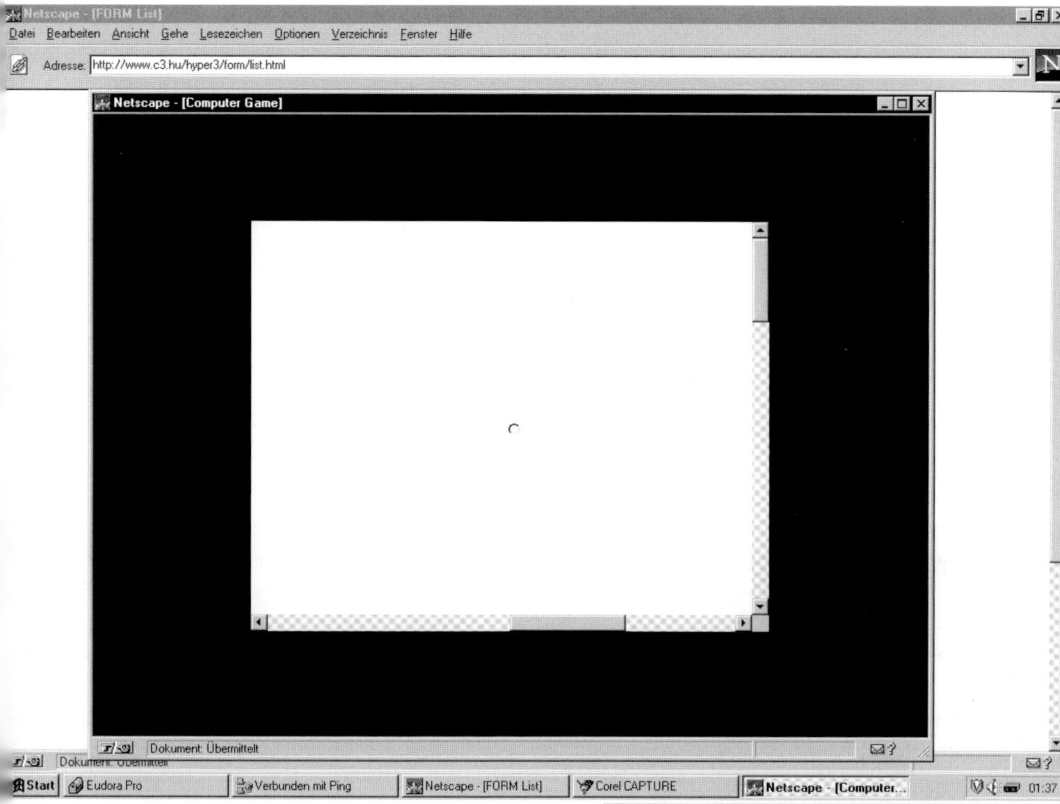

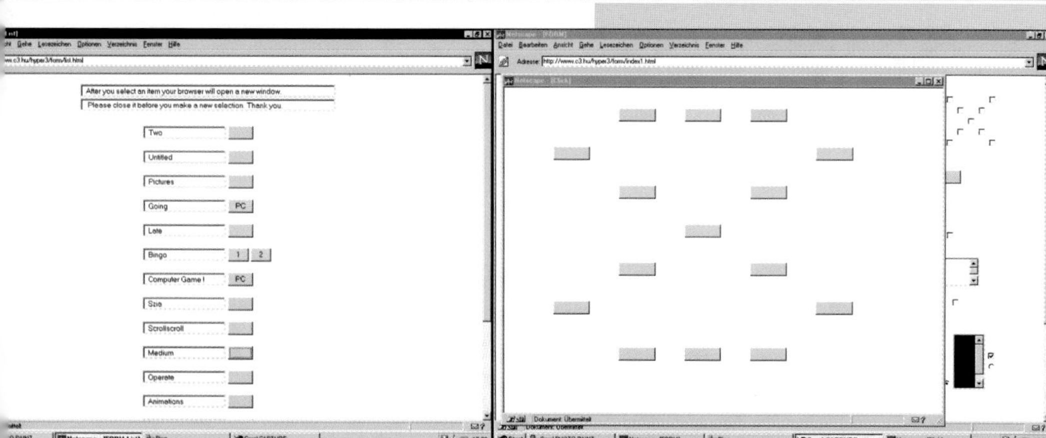

Alexei Shulgin (RU), born 1963. Artist and curator. Lives and works in Moscow. Inventor of "Form Art" and founder/creator of photographic galleries in Moscow and on the Internet. Participant of a number of symposiums on photography/contemporary art/new media/communications.

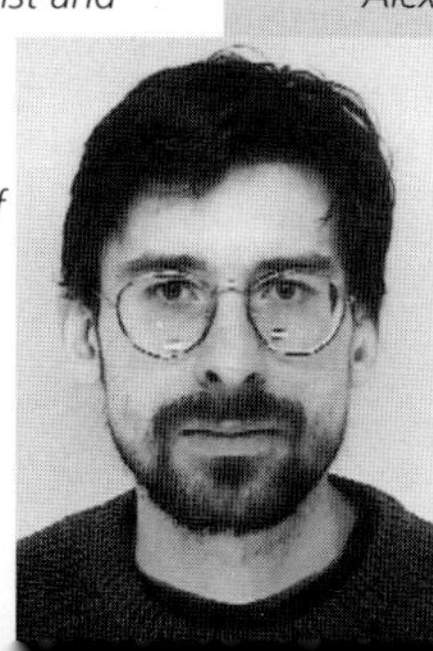

Alexei Shulgin (RU), geb. 1963, Künstler und Kurator, lebt und arbeitet in Moskau. Erfinder von „Form Art" und Begründer/Schöpfer von Photogalerien in Moskau und im Internet. Teilnehmer an zahlreichen Symposien zu Photographie, zeitgenössischer Kunst, Neuen Medien, Kommunikation.

Energized Gaming Culture

Mark von Rahden

"Energized Gaming Culture" is an interactive essay and an encyclopedia for computer gamers. The work involves game structures and game innovations.

The Web site includes articles on how computer games started ("From the Pentagon to the Pocket", the history of computer games since 1961), descriptions of the devices used for playing, and articles on games that are already in existence. Finally, it also addresses the question of whether Thomas Gottschalk's live TV show with "Telephone Pong" already marks the end of development.

In the introduction, I have written:

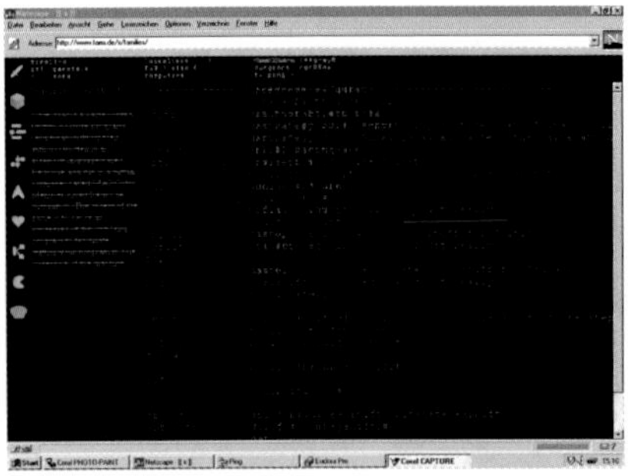

"Computer games went through several generations in the last 25 years. Museums show pong consoles. Classic games are redesigned as cyber hype giga re-releases. Beginning with Atari, home consoles, arcade games and games on PC's mutated from low-level applications to complex games." And I quote Youichi Miyaji, president of ESP and GameArts: "No matter how much or how little you know about the video game industry, I'm sure we can all agree on two things: The industry produces way too much junk, and the industry can never realize its true potential as long as its products are regarded by the mainstream media as children's toys."

I use the medium of the Internet to visualize my subject, in order to make use of the most extensive references and technical possibilities. Games are shown and explained, additional sources are listed. In order to provide the worldwide community of gamers with a platform for exchanging experiences and suggestions, "Energized

„Energized Gaming Culture" ist ein interaktiver Essay und eine Enzyklopädie für Computer-Spieler. Die Arbeit beschäftigt sich mit Spielstrukturen und Spielinnovationen.

Die Web-Site enthält Beiträge dazu, wie es mit den Computerspielen anfing („Vom Pentagon in die Hosentasche", die Geschichte der Computerspiele seit 1961), Beschreibungen der Geräte, an welchen man spielt, und Beiträge über die inzwischen existierenden Spiele; schließlich die Fragestellung, ob nicht Thomas Gottschalk's Fernseh-Live-Show mit „Telefon-Pong" schon der Endpunkt der Entwicklung war.

In der Einleitung schreibe ich: „Computerspiele haben sich in den letzten 25 Jahren zu etlichen Generationen entwikkelt. In Museen werden schon Ping-Pong Konsolen ausgestellt und Klassiker unter den Spielen werden als Cyber Hype giga re-releases neu designed. Angefangen von Atari, Heim-Spielkonsolen, Arcade games und PC-Spielen: sie alle sind von einfachen Anwendungen zu komplexen Spielen mutiert ..." Und ich zitiere Youichi Miyaji, Präsident von ESP und GameArts: „Egal, wieviel oder wie wenig man über die Videospiel-Industrie weiß, ich bin mir sicher, daß wir uns alle über zweierlei einig sind: die Industrie produziert viel zu viel Mist, und die Industrie kann sich ihres wahren Potentials niemals bewußt sein, solange die wichtigsten Massenmedien ihre Produkte als Kinderspielzeug betrachten."

Für die Visualisierung meines Themas nutze ich das Medium Internet, um weitestgehende Querverweise und technische Möglichkeiten zu nutzen. Es werden Spiele gezeigt und erklärt, weiterführende Quellen angegeben. Um der weltumspannenden Gemeinde der Computerspieler eine Plattform für den Austausch ihrer Erfahrungen und Anregungen zu bieten,

Gaming Culture" includes a PHP-chat and an interactive magazine generated from a mSQL database (Programming by Ralf Röber, isb). A sensuous experience of the games described is supported by dynamic buttons (Java applets) and the use of QuickTimeVR.

beinhaltet „Energized Gaming Culture" einen PHP-Chat sowie eine aus einer mSQL-Datenbank generierte interaktive Zeitung (Programmierung von Ralf Röber, isb). Durch dynamische Buttons (Java-applets) sowie den Einsatz von QuickTimeVR wird eine sinnliche Erfahrbarkeit der beschriebenen Spiele unterstützt.

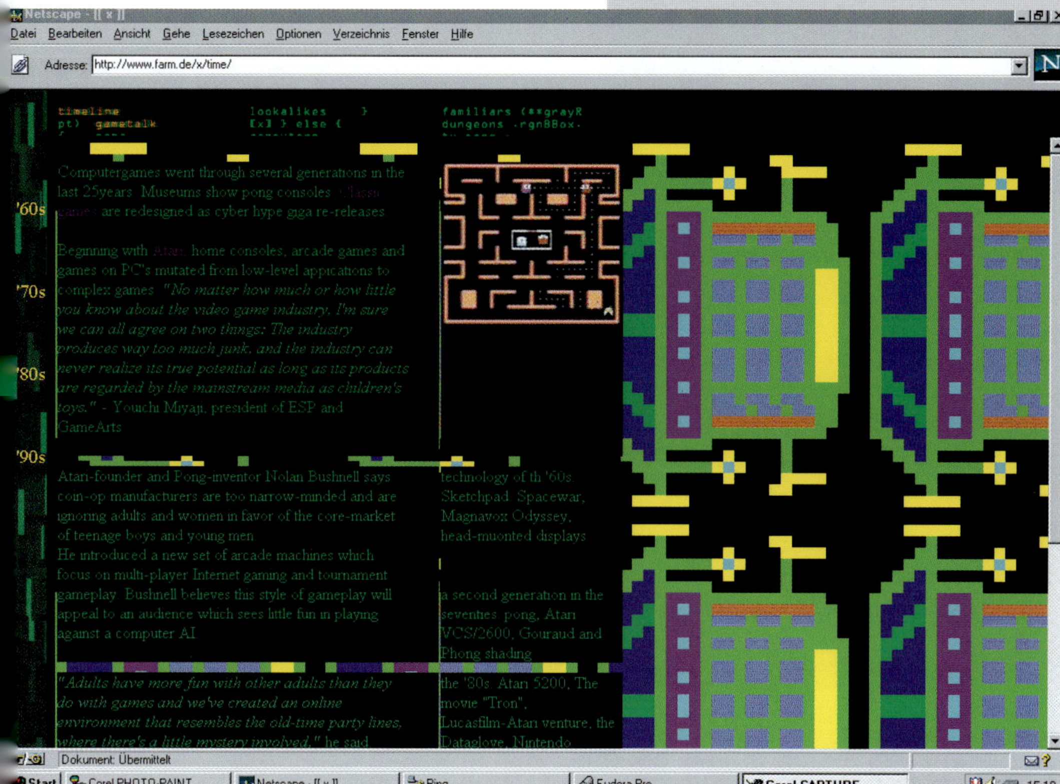

Mark von Rahden (D), trained and worked as a sign painter, studied graphics design at the Hochschule für Kunst Bremen; 1992-94 "coinn" development project at the HfK Bremen; since 1996, "The Farm", advertising and project agency together with David Bartusch and Rainer Krause.

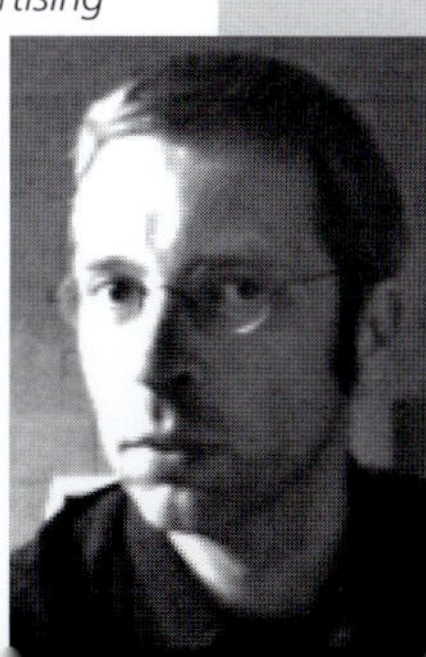

Mark von Rahden (D), Ausbildung und Berufstätigkeit als Schildermaler, Studium Grafik-Design an der HfK Bremen, 1992-94 „coinn"-Entwicklungsprojekt an der HfK Bremen, seit 1996 „die Farm", Werbe- und Projektagentur zusammen mit David Bartusch und Rainer Krause.

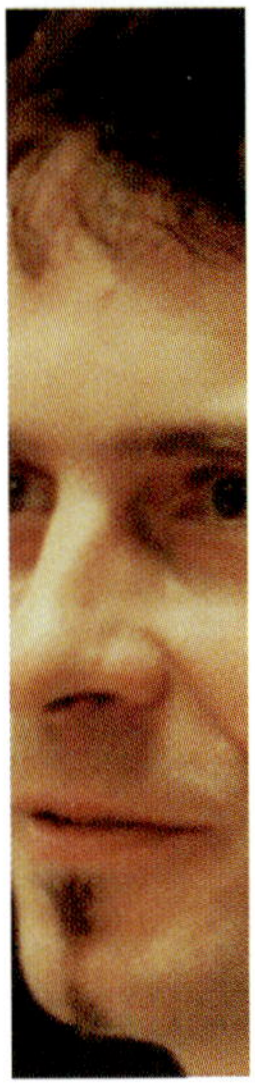
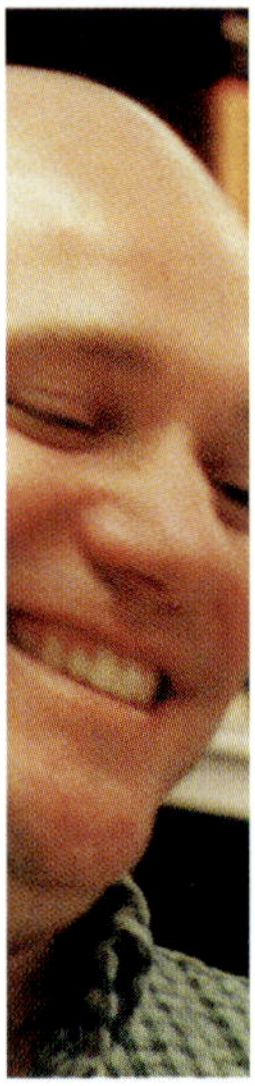
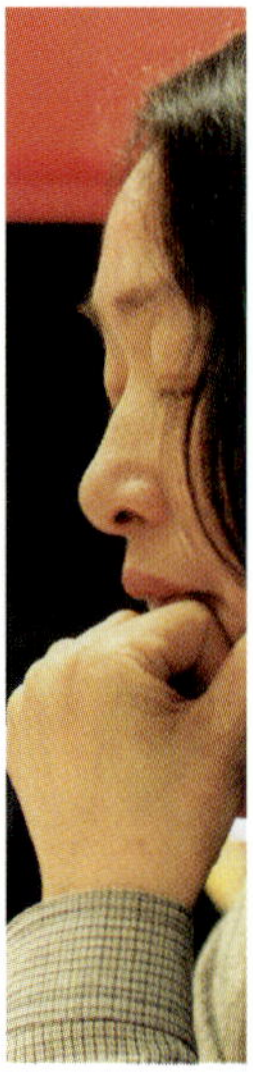

INTERACTIVE ART JURY

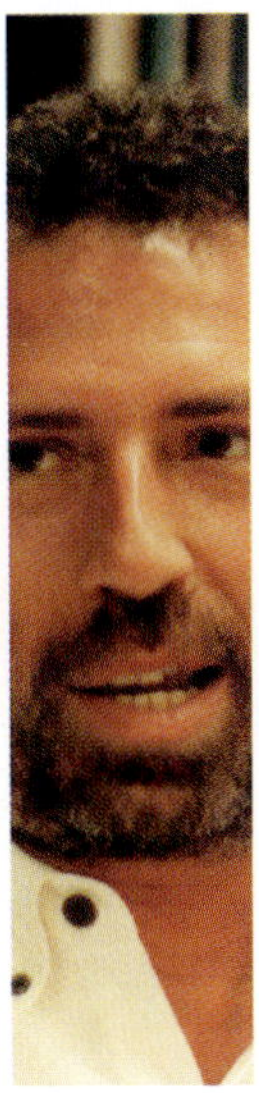

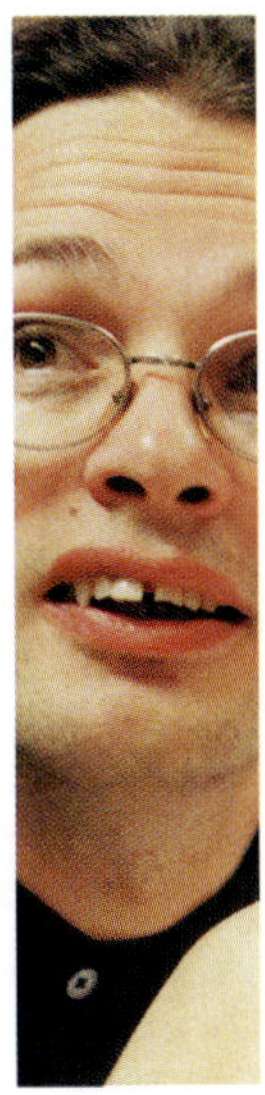

TRANSCENDING THE CATEGORIES

AUFLÖSUNG DER KATEGORIEN

When the Prix Ars Electronica created a separate category for "interactive art", they must have had a mental image of the frameworks within which this category could be judged.

And probably (again this is conjecture, because I wasn't there) the qualities they were looking for were to be found in the interdisciplinary and unstable, open character of these works of art, that gave fresh impulses to the way art was perceived, produced and distributed. There could only have been an inkling that this category would develop so explosively and would extend itself into all directions and platforms of art and culture. Now, in 1997, this category has obviously become firmly established with artists and in the world of art as a whole. A great diversity of work is being produced, transcending all traditional categories inside and outside of art. The entries for the Interactive Art 1997 award frequently overlap with other categories of Prix Ars Electronica, which indicates that it's not really possible to categorize interactive art in the way we categorized art a decade ago. The general terms in which interactive art is usually evaluated are no longer adequate to describe and further develop its specific qualities. The term

Als der Prix Ars Electronica eine eigene Kategorie für Interaktive Kunst schuf, mußten die Organisatoren eine Vorstellung gehabt haben von den Rahmenbedingungen, nach denen eine solche Kategorie beurteilt werden könnte.

Und wahrscheinlich (all das sind Vermutungen, weil ich nicht dabei war) ware die gesuchten Qualitäten im interdisziplinä ren und instabilen, offenen Charakter dieser Kunstwerke zu suchen, der der Produk tion, Perzeption und Distribution von Kun neue Impulse gegeben hat. Es wird wohl nur eine ferne Ahnung gewesen sein, daß sich diese Kategorie so explosiv entwickel und auf alle Richtungen und Plattformen der Kunst und Kultur übergreifen würde. Jetzt, 1997, hat sich diese Kategorie offer sichtlich bei den Künstlern und in der Kunstwelt etabliert. Es entstehen die vielfältigsten Werke, die alle traditionellen Kategorien innerhalb und außerhalb der Kunst hinter sich lassen. Auch die diesjäh gen Einreichungen in der Sparte Interakti Kunst überlappten sich häufig mit andere Kategorien des Prix Ars Electronica – ein Zeichen dafür, daß es nicht wirklich möglich ist, die Interaktive Kunst so zu katego risieren, wie wir Kunst generell noch vor zehn Jahren kategorisieren konnten. Die allgemeinen Begriffe, mit denen Interakti Kunst normalerweise bewertet wurde, rei chen nicht mehr aus, ihre speziellen Qualitäten zu beschreiben und weiterzuentwickeln. Der Begriff „Interaktivität" in seiner allgemeinen Bedeutung wird inflationär verwendet und ist kaum mehr als Generalbegriff geeignet, die oftmals sehr komplexen Interaktionsmuster innerhalb der Kunstwerke zu beschreiben. Interakti findet in diesen Werken zwischen Menschen statt, zwischen Mensch und Masch

"interactivity" is subject to inflation in its general meaning and is hardly suitable anymore as a general term to describe the often complex patterns of interaction within the artworks. Interaction in these works takes place between people, between people and machines and between machines themselves and therefore raises questions about many issues concerning the way we define the quality of art. Thinking about "machines" and aesthetics not only questions originality and uniqueness, issues that have been discussed for some years now in media-art, but also the intention of the artist, authorship and moments of decisions made by man and machine. Artworks have become complex "machines" where the user does not so much individually control the work, but cooperates, obstructs and directs. This results in dynamic patterns of interaction within which the work evolves. Genetic algorithms, artificial intelligence, the Internet as a medium for communication, and self-organizing systems are applied to create works of art, in which the actions of users and machines become almost indistinguishable. This complex interaction delivers unexpected results in the way the work is finally experienced. Interaction here means to participate in a process where individual actions lead to collisions and/or moments

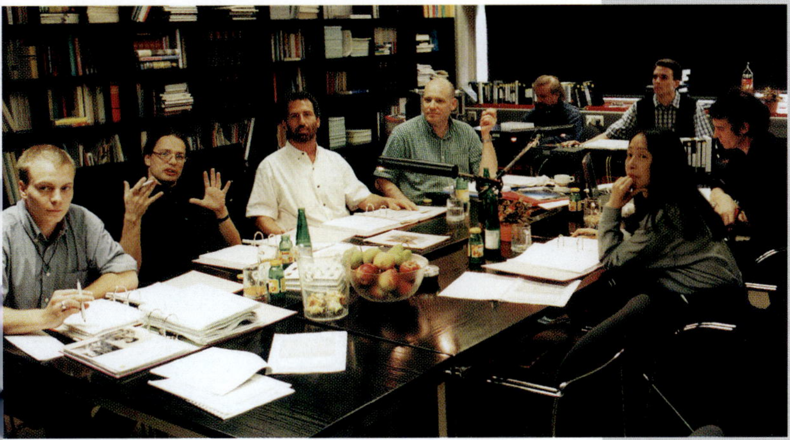

ne oder zwischen Maschine und Maschine und wirft damit eine ganze Reihe von Fragen auf, die sich alle damit beschäftigen, wie wir die Qualität von Kunst bewerten. Überlegungen über „Maschinen und Ästhetik" hinterfragen nicht nur Aspekte wie Originalität und Einzigartigkeit – Anliegen, die schon seit einigen Jahren in der Medienkunst diskutiert werden –, sondern auch die Intention des Künstlers, die Urheberschaft und die Aufteilung der Entscheidungsmomente zwischen Mensch und Maschine. Kunstwerke sind komplexe Maschinerien geworden, in denen der Benutzer weniger das Werk individuell steuert, sondern vielmehr kooperiert, stört, leitet. Daraus ergibt sich ein dynamisches Interaktionsmuster, innerhalb dessen das Werk sich entwickelt. Genetische Algorithmen, künstliche Intelligenz, das Internet als Medium der Kommunikation und selbstorganisierende Systeme werden eingesetzt, um Kunstwerke zu schaffen, bei denen die Aktionen von Mensch und Maschinen kaum mehr auseinanderzuhalten sind. Diese komplexe Interaktion führt zu ganz unerwarteten Ergebnissen, was die Art und Weise betrifft, wie das Werk letztlich erfahren wird. Interaktion bedeutet hier, an einem Prozeß teilzunehmen, bei dem individuelle Aktionen zu Kollisionen und/oder zu Zufallsereignissen führen können. Hier manifestiert sich das Kunstwerk als dynamischer Prozeß, als Erfahrung und nicht als ästhetisches Objekt. Die Projekte von Knowbotic Research (Anerkennung) stehen exemplarisch für diese Art von Werk.

Wenn wir jedoch alle Einreichungen in der Sparte Interaktive Kunst betrachten, so ist eine derart summarische Beschreibung der Interaktivität längst nicht repräsentativ. Obwohl es zweifellos eine wachsende Komplexität bei den interaktiven Aspekten der meisten Werke gibt, zeigt sie sich doch auf sehr unterschiedliche Weise.

of coincidence. Here the artwork manifests itself as a dynamic process, as an experience instead of as an aesthetic object. The projects of Knowbotic Research (honorable mention) are exemplary of this type of work. This rather summary description of interactivity is not quite representative if we look at all of the entries for the Interactive Art award. Although there is certainly a growing complexity in the interactive aspects of most of the work, this manifests itself in very diverse ways.

This year, the jury considered approximately 200 entries, out of which three award winners were chosen. A further twelve works received an honorary mention. The entries were very diverse, meaning they varied greatly in patterns of interactivity, interface design and application scope. To illustrate this diversity, I will briefly discuss this year's award winning works and the jury's opinion of them.

The music performance "Music Plays Images x Images Play Music" by Toshio Iwai and Ryuichi Sakamoto concurs with a development that was initiated in Europe at the beginning of this century and that focused on the relation between image and sound. This synaesthetic experience, where the stimulation of one of the senses activates another sense, was explored by musicians like Scriabin in order to reach an experience of totality. "Music Plays Images x Images Play Music" recognizes this tradition that has steadily progressed in this century. In this piece, computer graphics play the piano and pianos play with computer graphics. But it is not only the machines playing with each other. Ryuichi Sakamoto and the audience also participate in this game of images and sounds generated through inventive interfaces. Sakamoto himself is on stage, but the audience participates through the Internet and controls one of two pianos. Through real player feedback these Internet participants can hear the actual performance too. And everything seamlessly flows from one

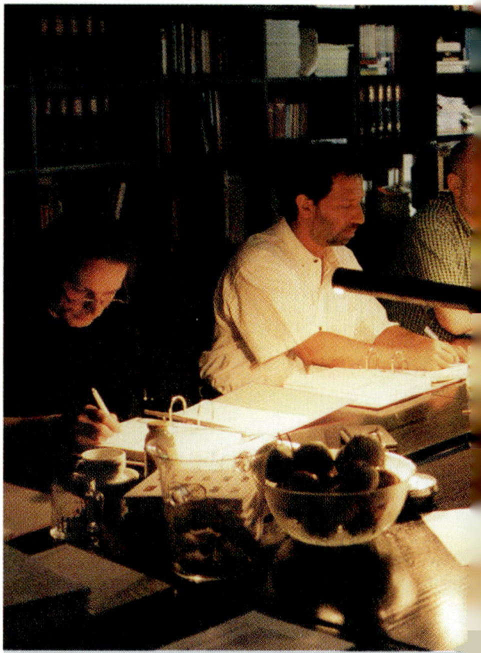

In diesem Jahr hat die Jury rund 200 Arbeiten begutachtet, aus denen drei Preisträger ermittelt wurden. Weitere zwölf Werke erhielten Anerkennungen. Die Einreichungen waren höchst unterschiedlich – sie variierten stark in den Interaktivitätsmustern, im Design der Interfaces und in der Zielsetzung der Anwendung. Um diese Verschiedenheit zu illustrieren, möchte ich kurz die diesjährigen Preisträger und die Meinung der Jury über sie zusammenfassen.

Die Musik-Performance „Music Plays Images x Images Play Music" von Toshio Iwai und Ryuichi Sakamoto steht in der Tradition einer Entwicklung, die zu Beginn dieses Jahrhunderts in Europa begann und auf die Beziehung zwischen Klang und Bild abzielte. Diese synästhetische Erfahrung, in der die Stimulation eines Sinnes einen anderen aktiviert, wurde von Musikern wie Skrjabin erforscht, die so eine umfassende Kunsterfahrung erreichen wollten „Music Plays Images x Images Play Music" anerkennt diese Tradition, die sich seit ihren Anfängen beständig weiterentwickelt hat. In diesem Stück spielen Computergraphiken Klavier und Klaviere malen Computer-

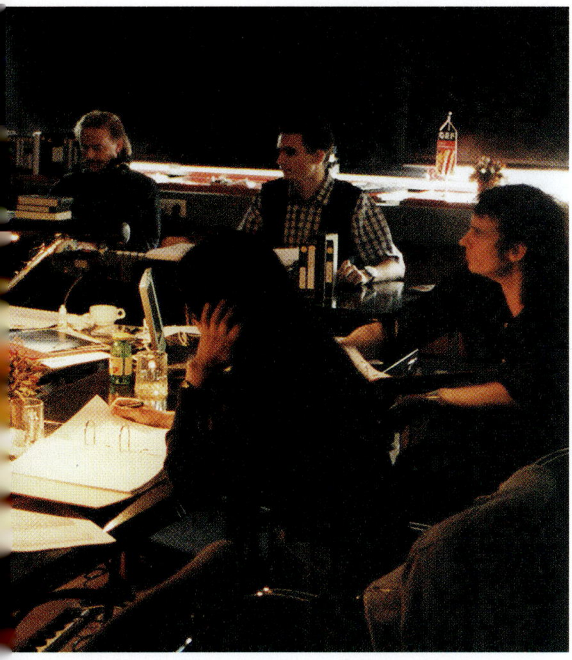

domain to the other. In this sense the piece follows a Japanese tradition in which one does not look for controversy and conflict in interaction, but for a conjoining and merging of players. The aesthetic experience is here being maximized, resulting in an equally strong experience. This work deals with several elements of current thinking on how we work with media technology in relation to art. Public - private, local - global, multi-user interaction through networks and the creation of an open structure define specific qualities that further interactivity in relation to art.

"Border Patrol" by Paul Garrin and David Rokeby, however, in contrast with Toshio Iwai's work, actively pursues the friction which exists between technology as an instrument of control and power and the audience that is being controlled by technology. The work explores social and political aspects of technology. On the one hand, Paul Garrin is concerned with the deployment of technology by social or cultural minorities as a means of survival and of articulating a message. On the other hand, he recognizes the fact that historic forces in society, like the state and

graphiken. Aber es beschränkt sich nicht auf das Spiel der Maschinen miteinander. Ryuichi Sakamoto und das Publikum nehmen an diesem Spiel der Bilder und Klänge mit Hilfe erfindungsreicher Interfaces aktiv teil. Sakamoto befindet sich selbst auf der Bühne, aber das Publikum greift über das Internet ein und steuert eines der beiden Klaviere. Über das Feedback des realen Spielers hören die Internet-Mitwirkenden auch die tatsächliche Performance. Und all das fließt nahtlos von einer Domäne in die andere. In diesem Sinne folgt das Stück einer japanischen Tradition, die in der Interaktion nicht Konflikt und Kontroverse sucht, sondern den Zusammenschluß, die Verschmelzung der Mitwirkenden. Die ästhetische Komponente wird maximiert, was zu einer eindringlichen Erfahrung führt. Die Arbeit selbst behandelt mehrere Aspekte, die die gegenwärtige Diskussion über die Beziehung zwischen Medientechnologie und Kunst kennzeichnen: Öffentlich–privat, lokal–global, Multi-User-Interaktion über Netzwerke und die Schaffung einer offenen Struktur definieren Qualitäten, die die Interaktion in der Kunst weiter voranbringen. Im Gegensatz zu Toshio Iwais Werk beschäftigt sich „Border Patrol" („Grenzpatrouille") von Paul Garrin und David Rokeby aktiv mit dem Konflikt, der zwischen der Technologie als Kontroll- und Machtinstrument und dem von der Technologie kontrollierten und gesteuerten Publikum herrscht. Die Arbeit untersucht die sozialen und politischen Aspekte der Technologie, Paul Garrin fragt nach dem Einsatz der Technologie seitens sozialer oder kultureller Minderheiten als Mittel zum Überleben und zum Ausdruck ihrer Botschaften. Andererseits spricht er die Tatsache an, daß die historischen Kräfte der Gesellschaft – Staat und Wirtschaft – sich selbst durch die Technologie verstärken und dabei eine bestimmte Kontrolle über den öffentlichen Raum ausüben. Er betrachtet die Vereinnahmung des Medienfeldes durch diese Kräfte als Versuch, sowohl das Individuum als auch die Gesellschaft zu kontrollieren – eine Entwicklung, die in scharfem Gegensatz zur Medienbotschaft steht, die von ideologischen Gruppen und vom Marketing verbreitet wird: die Dezentralisierung der Macht.

Bei „Border Patrol", das in Südkorea Premiere

the economy, reinforce themselves through technology and in doing so exert a certain control over public space. He regards the occupation of the media field by these forces as an attempt to control both the individual and society, a development which sharply contradicts the media message delivered by ideological groups and through marketing: decentralization of power.

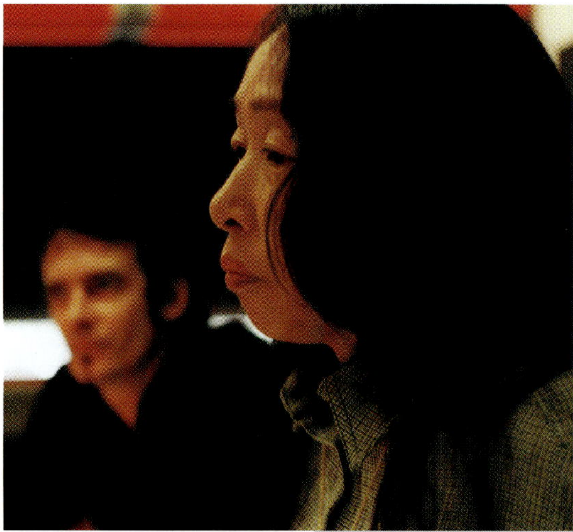

In "Border Patrol", which premiered in South Korea, the audience is situated in a military stronghold where their moves are tracked by video cameras. These lock onto individuals in the audience, behaving like snipers' guns seeking to eliminate them. On monitors built into the sandbag walls, these individuals see themselves being trapped by these sniper cameras. "Border Patrol" is direct and therefore limited in its interaction with the audience. It leaves little to be imagined, speculating on the oppressiveness it inspires.

The prize for Art+Com from Berlin for their project "The Invisible Shape of Things Past" was awarded for reasons that are completely different from the ones involved with the works described above. "The Invisible Shape of Things Past" is a tool, transforming filmed images into 3D-objects. Stated like this, it seems a simple enough tool, but this entry demonstrated several applications which immediately conveyed the project's impact. The 3D-objects resulting from the filmed images are partly formed by the camera movement (zoom, pan, tilt), thus creating irregular shapes. These 3D-objects can be created from historic footage or from contemporary film images. In this case, they were from historic footage of parts of Berlin

*hatte, befindet sic
das Publikum in e
ner militärischen F
stung, in der dess
Bewegungen von
Videokameras ver
folgt werden. Die
„klammern" sich a
einzelne Personen
im Publikum fest
und verhalten sich
wie Hecken-
schützen, die vers
chen, dieses „fein
liche Objekt" zu eliminieren. Auf
Monitoren, die in Sandsack-Wände einge
lassen sind, sehen sich die Personen selb
wie sie von den Kameras gefangen sind.
„Border Patrol" ist direkt und insofern
auch in seiner Interaktion mit dem
Publikum beschränkt, es läßt der Phanta
nur wenig Spielraum und spekuliert mit
dem Gefühl der Unterdrückung, das es
auslöst. Eine Auszeichnung geht an
ART+COM aus Berlin für das Projekt „Th
Invisible Shape of Things Past", und zwar
aus Gründen, die mit den oben genannt
nichts gemeinsam haben. „The Invisible
Shape of Things Past" ist ein Werkzeug,
das gefilmte Bilder in 3D-Objekte verwan
delt. So knapp beschrieben, scheint es al
Werkzeug ziemlich simpel zu sein, aber d
Einreichung zeigte mehrere Anwendun-
gen, die die Dimensionen des Projektes
eindrucksvoll demonstrieren. Die 3D-Ob-
jekte aus den gefilmten Bildern werden
teilweise aus der Kamerabewegung ge-
formt (Zoom, Schwenk, Kippen), wodurc
unregelmäßige Formen entstehen. Die O
jekte können aus historischem wie aktue
lem Filmmaterial entwickelt werden. Im
vorliegenden Fall stammten sie aus histor
schen Aufnahmen von Teilen Berlins (Stra
ßen, Plätze, Gebäude). Diese Filmobjekte
können in eine Darstellung der Stadt, in
ein virtuelles Berlin, eingebaut werden.*

*Diese virtuelle Stadt wiederum kann
über eine Zeitleiste erschlossen werden*

(streets, squares, buildings). The film objects can be placed within a representation of the city, a virtual Berlin. This virtual city can be accessed via a timeline (Berlin 1910, 1930 and so on). The images then occur in the virtual city on the actual location were they were shot. It is also possible to move the images to the same location in a different time period, resulting in rich connections between designing, organizing and thinking about the city in its many aspects. This virtual Berlin can be navigated from different camera positions, enabling the user to choose locations where film objects are placed and subsequently travel through these images, either as film sequences or as VR-representations. The project can be linked with databases or the Internet, making it potentially very dynamic and lending itself to many applications in the fields of architecture, film, education and otherwise. The jury was impressed by the strength of such a tool which utilizes different user interfaces. There is, however, the danger of such a project being swiftly monopolized. In that case, the choice and content of linked databases would be determined by the short term commercial interests of corporations, whereas the force of a project like this is its open structure which guarantees diversity.

(Berlin 1910, Berlin 1930 usw.), und dann finden die historischen Aufnahmen ihren Platz genau an jener Stelle der virtuellen Stadt, an der sie gedreht wurden. Man kann auch die Bilder an dieselbe Stelle, aber in eine andere Zeit verschieben, wodurch sich zahlreiche Querverbindungen zu Entwurf, Organisation und anderen Fragen der Stadtgestaltung ergeben. Der User kann von unterschiedlichen Kamerapositionen aus durch dieses virtuelle Berlin reisen und so die Aufstellungsorte von gefilmten Objekten wählen und danach durch diese Bilder – entweder als Filmsequenz oder als VR-Objekte – wandern. Das Projekt läßt sich mit Datenbanken oder dem Internet verknüpfen, es enthält daher ein großes dynamisches Potential und bietet zahlreiche Anwendungen, u. a. in Architektur, Film oder Bildung. Die Jury war von der Stärke eines solchen Werkzeuges beeindruckt, das verschiedene Interfaces verwendet. Allerdings besteht bei solchen Projekten immer die Gefahr der Monopolisierung. In einem solchen Fall würden Auswahl und Inhalt der Datenbanken bald von den kurzfristigen kommerziellen Interessen von Firmen bestimmt werden, wohingegen die wahre Kraft eines derartigen Projektes in seiner offenen Struktur liegt, die für unterschiedlichste Anwendungen geeignet ist.

Music Plays Images x Images Play Music

Toshio Iwai / Ryuichi Sakamoto

The idea for this collaborative performance has its origins in Toshio Iwai's installation "Piano – as image media" created at the ZKM, Karlsruhe, Germany in 1995. Conceived upon Iwai's return to Japan when he met Ryuichi Sakamoto, "Music Plays Images x Images Play Music" is a multi-media concert that uses the system of Iwai's piano piece and visualizes a musical performance by Sakamoto in real time. Over the course of a year, Iwai, who lives in Tokyo, often corresponded by e-mail with Sakamoto, who lives in New York. During their correspondence, the concept for the collaboration expanded rapidly, developing from the simple idea of "Sakamoto playing Iwai's 'Piano – as image media'" to a performance that links, in various ways, musical sounds with visual images. The actual performance was given on December 16, 1996 at the Art Tower Mito in Mito City, Japan. The result of the collaboration was more than a "novel combination of music and image." Many elements including the actual presence of Sakamoto/Iwai, the machinery that is called the piano, the acoustic sounds produced by the piano, and the virtual images generated from computers were linked and developed in various ways. From the feedback produced, a radically new space of musicality-visuality was created. In addition, the performance was transmitted over the Internet and relayed world-wide. This experiment, the first of its kind, was an attempt to utilize the interactive medium of the Internet. People on the other side of the Internet were able to play the piano located in Mito by remote control and in turn Sakamoto had a session with their playing.

Die Idee für diese gemeinsame Performance entstand in Zusammenhang mit Toshio Iwais Installation „Piano – as image media" am ZKM, Karlsruhe 1995. Das nach Iwais Rückkehr nach Japan und bei einer Begegnung mit Ryuichi Sakamoto konzipierte „Music Plays Images x Images Play Music" ist ein multimediales Konzert, das das System von Iwais Klavierstück verwendet und Sakamotos musikalische Performance in Echtzeit visualisiert. Während eines Jahres stand Iwai, der in Tokyo lebt, in E-Mail-Kontakt mit dem in New York ansässigen Sakamoto. Im Zuge dieser Korrespondenz erweiterte sich ihr Konzept beträchtlich – aus der anfänglichen Idee „Sakamoto spielt Iwais Klavier als Bildmedium" wurde eine Performance, die musikalische Klänge auf verschiedenste Art mit visuellen Bildern verknüpft.

Die eigentliche Aufführung fand am 16. Dezember 1996 im Art Tower Mito in Mito City, Japan, statt. Das Ergebnis dieser Zusammenarbeit war weit mehr als eine neuartige Kombination aus Musik und Bil Viele Elemente – die tatsächliche Präsenz von Sakamoto und Iwai; jene Maschine, di man landläufig Klavier nennt; die akustischen Klänge, die damit erzeugt wurden; und die virtuellen Bilder, die aus dem Com puter stammten – wurden auf mannigfaltige Weise verbunden und entwickelt. Das dabei entstehende Feedback ergab einen völlig neuartigen Raum der Musikalität un Visualität. Darüber hinaus wurde dieses Ex periment über das Internet weltweit übertragen – der erste Versuch dieser Art, das interaktive Medium Internet in eine solche Aufführung einzubeziehen. Internet-User auf der ganzen Welt konnten via Remote-Control auf dem in Mito stehenden Flügel spielen, während Sakamoto mit ihrem Spi interagieren konnte.

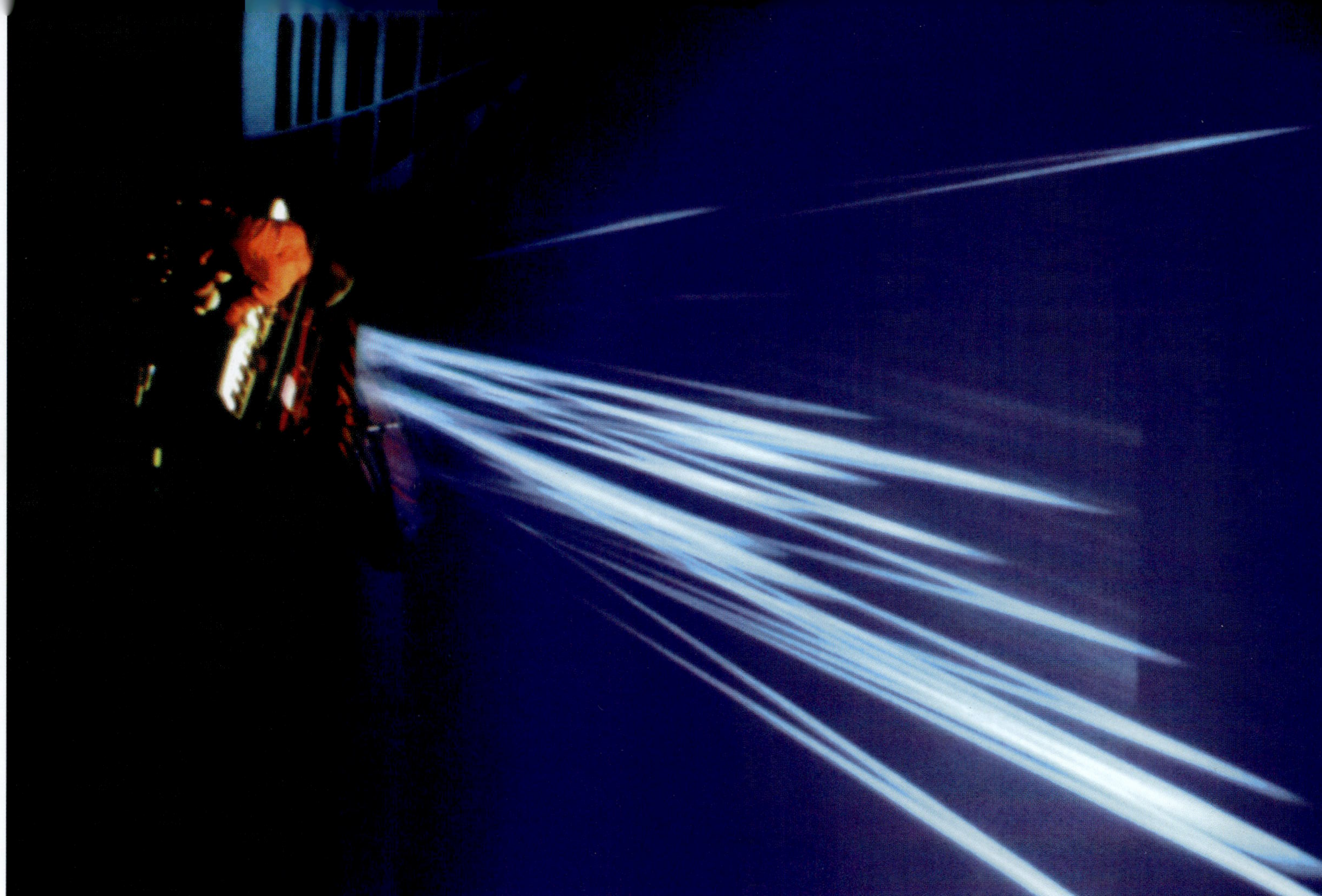

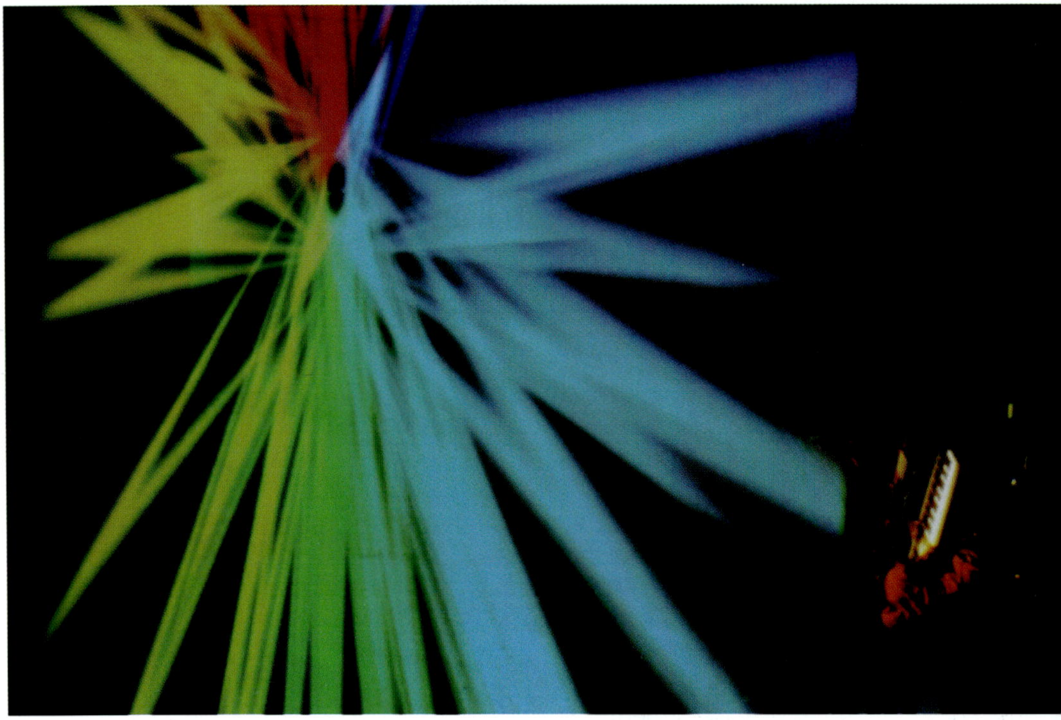

The Set-Up of the Stage

Two concert grand pianos are placed on the stage. A semi-transparent screen 8 meters in height and 10 meters in width is suspended above them. The lids of both pianos are removed and their interiors and keyboards clearly illuminated. The screen is irradiated with a beautiful shade of blue. A black and white video camera is placed on each of the pianos in order to film the movements of the performer's fingers and the piano keys. The video cameras project enlarged images of each piano keyboard onto the screen on either side of the pianos. Several computers are set up on a long narrow table placed on the right side of the stage.

The Content of the Performance

1) Sakamoto does not initially appear on the stage. Amidst the silence, the piano begins to play on its own without a player. An image of Sakamoto playing the piano gradually comes into view on the screen to the side of the piano. Sakamoto's image

Die Bühne

Zwei Konzertflügel stehen auf der Bühne, eine semitransparente Leinwand von zehn Metern Breite und acht Metern Höhe hängt über ihnen. Die Deckel der beiden Klaviere sind entfernt, ihr Inneres und die Tastaturen sind hell ausgeleuchtet. Die Projektionsfläche wird mit einem schönen Blauton angestrahlt. Auf jedem Klavier ist eine Schwarz-Weiß-Videokamera montiert, die die Bewegungen der Finger und Tasten aufnimmt. Diese Kameras projizieren vergrößerte Abbildungen der Tastaturen auf Projektionsflächen beiderseits der Klaviere. Mehrere Computer sind auf einem schmalen, langen Tisch auf der rechten Seite der Bühne aufgestellt.

Der Inhalt der Performance

1) Am Anfang erscheint Sakamoto nicht auf der Bühne. In der Stille beginnt das Klavier von selbst,

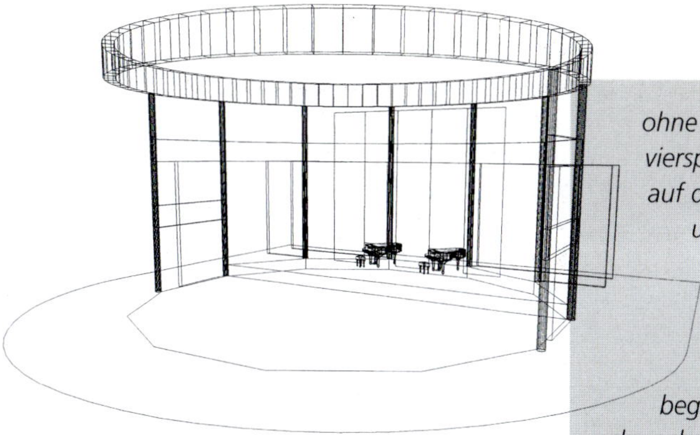

slowly disappears as the music ends.

2) Sakamoto appears on the left side of the stage. He begins to play the small keyboard placed there. As he plays this keyboard, the piano starts to play. A faint light gradually emanates from the piano. At a certain point, Sakamoto stops playing. However, the last musical phrase repeats itself automatically, slowly fading out.

3) Sakamoto plays the piano directly. The light-images that visually represent each note of the musical piece begin to rise

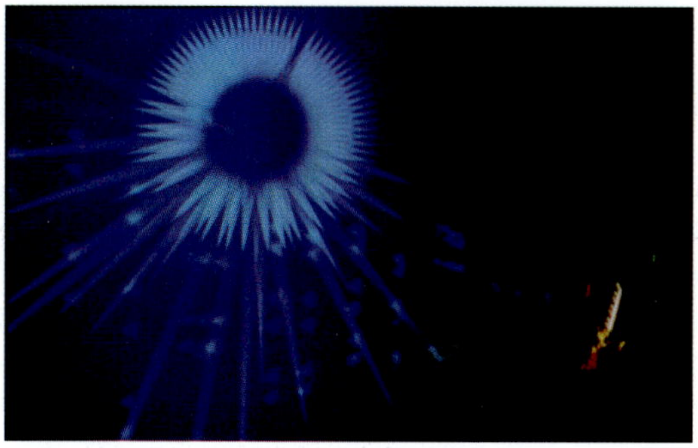

from within the piano. The images capture not only the timing of the notes, but every aspect of the music, including the length of the notes and their loudness. They are represented in a way that the eye can recognize.

4) During Sakamoto's performance, Iwai gradually modifies the image. The lights, which correspond to each note of the piece, are transformed over time in various ways. They do not merely rise from within the piano and fade into thin air. They radiate from the center of the screen out in all directions or revolve in a whirling pattern, eventually disappearing into a vortex. The

ohne Pianisten, zu spielen. Ein Bild des klavierspielenden Sakamoto erscheint langsam auf der Projektionsfläche seitlich des Klaviers und verschwindet wieder, wenn die Musik endet.

2) Sakamoto tritt auf der linken Seite der Bühne auf. Er spielt dort ein kleines Keyboard, und währenddessen beginnt auch das Klavier zu spielen. Ein schwaches Leuchten geht vom Klavier aus. An einem bestimmten Punkt hört Sakamoto zu spielen auf. Dennoch wird die letzte musikalische Phrase vom Flügel automatisch weitergespielt, bis sie langsam ausblendet.

3) Sakamoto spielt nun direkt auf dem Flügel. Lichtpunkte, die die einzelnen gespielten Note darstellen, steigen aus dem Inneren des Klaviers auf. Das Bild der Note gibt nicht nur ihr Timing wieder, sondern jeden Aspekt der Musik, die Länge des Tons, den Anschlag. Die Darstellung erfolgt so, daß sie für das Auge nachvollziehbar ist.

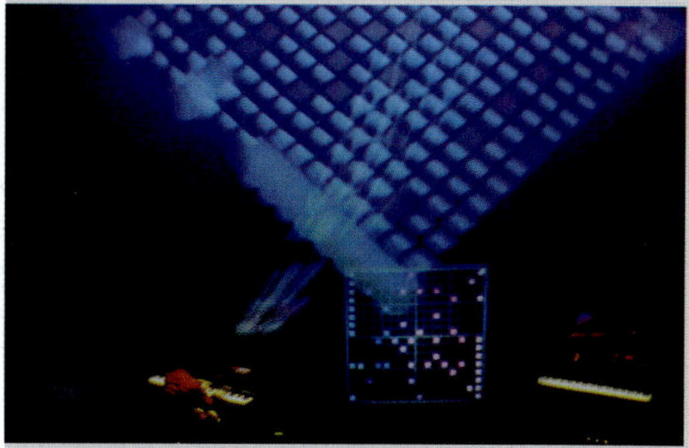

4) Während Sakamotos Spiel verändert Iwai nach und nach das Bild. Die Lichter, die jeweils einer Note des Stücks entsprechen, werden im Lauf der Zeit auf unterschiedliche Weise verwandelt. Sie steigen auch nicht mehr nur aus dem Klavier auf, sondern verbreiten sich aus der Mitte der Leinwand in alle Richtungen, sie drehen sich in einem Wirbel, um schließlich in einem Strudel zu verschwinden. Die Metamorphose der projizierten Bilder beeinflußt Sakamotos Spiel, der die verschiedenen Transformationen beobachtet.

5) Bildobjekte, bestehend aus 88 Elementen – ebensovielen, wie ein Klavier Tasten hat –, erscheinen in der Mitte der Projektionsfläche. Jedes Element entspricht einer Taste, und die

metamorphoses of the projected images influence Sakamoto's playing as he watches the various transformations.

5) The image-objects, composed of 88 elements – the same number as the number of keys on a piano – appear on the center of the screen. Each of the elements coincides with the keys of the piano and the objects are continually transformed according to Sakamoto's piano performance.

6) Clicking a computer mouse, Iwai places dots of light onto the screen. Whenever the dots of light fall on the piano, the piano begins to play. The dots of light then rebound to their original position, striking the keyboard at fixed intervals. Depending on where the dots of light are placed, the pitch of the notes and the patterns of repetition change. The number of dots of light are increased. The dots of light bouncing from the piano are all shifted back and forth from left to right. The whole musical scale changes according to how the dots of light are moved. The dots of light are transformed into rays and then polygonal figures.

7) The video camera films Sakamoto's entire body. The computer captures the image in real time. Sakamoto's silhouette is pixelized and then projected onto the screen above the piano. When Sakamoto moves and the image of his body overlaps with the piano, the piano plays. When Sakamoto moves toward or away from the camera, or when a close-up image of his hands is projected, the sound is changed dynamically. The images of Sakamoto's hands and body are what plays the machine called the piano. Iwai implements further modifications by zooming in on and rotating Sakamoto's image on the screen thereby slowly modifying the sound. Sakamoto continues to move his body while looking on.

8) A chessboard-like grid appears on the screen and begins to rotate. Sakamoto and Iwai take turns placing glowing dots of

Objekte verändern ihre Form je nach Sakamotos Spiel.

6) Mit einem Mausklick positioniert Iwai Lichtpunkte auf die Projektionswand. Wann immer solche Lichtpunkte auf die Tastatur fallen, spielt der Flügel. Dann springen die Lichtpunkte in ihre Ausgangsposition zurück, von wo sie in fixen Intervallen neuerlich auf die Tastatur „purzeln". Je nachdem, wo die Lichtpunkte postiert sind, ändern sich die Höhe des angeschlagenen Tons und das repetitive Muster. Die Anzahl der Lichtpunkte wird vergrößert, die zurückspringenden Punkte werden nach links und rechts verschoben. Je nachdem wie die Lichtpunkte bewegt werden, verändert sich auch die gesamte musikalische Skala. Aus den Lichtpunkten werden Strahlen und später Polygone.

7) Die Videokamera filmt Sakamotos ganzen Körper. Der Computer setzt das Bild in Echtzeit um, die Silhouette wird in Pixel umgerechnet und auf die Leinwand über dem Klavier projiziert. Wenn Sakamoto sich bewegt und das Abbild seines Körpers sich mit dem Klavier überlappt, beginnt das Klavier zu spielen. Wenn sich Sakamoto zur Kamera hin oder von ihr weg bewegt oder wenn eine Nahaufnahme seiner Hand projiziert wird, ändert sich die Dynamik des Klangs: Das Abbild von Sakamoto und seinen Händen spielt jene Maschine, die man Klavier nennt. Iwai bringt weitere Veränderungen ein, indem er Sakamotots Bild auf der Projektionsfläche dreht, hineinzoomt usw. Mit der Veränderung des Abbilds ändert sich auch der Klang. Sakamoto beobachtet die Veränderungen und bewegt sich weiter vor der Kamera.

8) Ein schachbrettähnliches Muster erscheint auf der Projektionsfläche und beginnt sich zu drehen. Sakamoto und Iwai setzen abwechselnd Lichtpunkte auf dieses Gitter, worauf sich ein kurzes musikalisches Muster ständig wiederholt. Während aus dieser Zusammenarbeit eine Musik entsteht, transponiert Iwai das musikalische Muster, indem er die Oktav wech-

light onto the grid, thereby creating a repetition of short melodic patterns. As the music is created through this collaboration, Iwai transposes the musical pattern, changing the octave and altering the sounds and the images. Sakamoto continues to create new patterns. Finally, with the completed musical phrases as a backdrop, Sakamoto improvises on the piano. Iwai continues to modify the performance piece.

9) The music played by Sakamoto is turned into an image. The image that springs from the first piano and lands on the second piano creates a parabola in its fall. The second piano plays according to this image. The figure of Sakamoto performing on the first piano is projected next to the second piano with a delay that coincides with the imaged music arriving at the second piano. It appears as if there are two Sakamotos playing the two different pianos.

10) The short musical phrase performed by Sakamoto and the resulting image loop between the two pianos, repeating again and again. Sakamoto continues to add more sounds to the loop. The repeating images and sounds, flowing forth like a waterfall between the two pianos, grow in layered richness and complexity.

(translated from Japanese by Dano O'Neill)

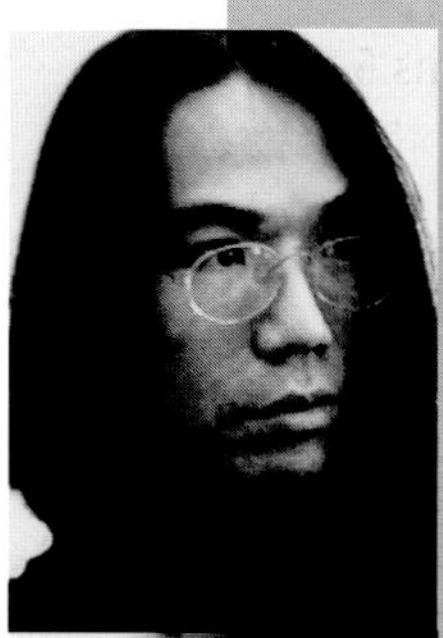

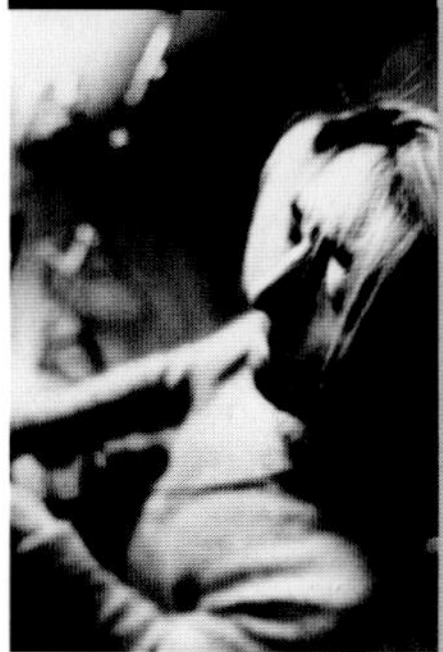

Toshio Iwai (J), born 1962, is an interactive artist and graduated from the Tsukuba University. He has had several solo exhibitions worldwide and two of his pieces are permanently displayed at the San Francisco Exploratorium. He has been Artist-in-Residence at ZKM/Karlsruhe and is currently Artist-in Residence at IAMAS.

Ruyichi Sakamoto (J), born 1952, MFA from the Tokyo National University of Fine Arts. He is an internationally recognized composer, performer and media artist. His musical works for films (Bernardo Bertolucci: "The Last Emperor", "The Sheltering Sky", "Little Buddha") have received Grammy, Academy and Golden Globe Awards.

selt und sowohl Klänge als auch Bilder verändert. Sakamoto seinerseits erzeugt weiter neue Muster. Wenn schließlich genug Klangmaterial als Hintergrund vorhanden ist, improvisiert Sakamoto auf dem Klavier, während Iwai weiter die Hintergrundmusik verändert.

9) Die von Sakamoto gespielte Musik wird in ein Bild umgesetzt, das vom ersten Flügel wegspringt und in einer Parabel auf den zweiten niederfällt. Der zweite Flügel spielt je nachdem, wie dieses Bild auf ihm aufkommt. Das Bild von Sakamoto, der den ersten Flügel spielt, wird mit derselben Zeitverzögerung neben das zweite Klavier projiziert, die durch den Flug der „Ton-Bilder" entsteht. Es sieht so aus, als würden zwei Sakamotos jetzt die zwei Klaviere spielen.

10) Die von Sakamoto gespielte kurze musikalische Phrase und Sakamotos Bild kreisen zwischen den beiden Klavieren hin und her und wiederholen ihre Musik. Sakamoto fügt dieser Schleife immer neue Klänge hinzu. Die sich wiederholenden Bilder und Klänge, die wie ein Wasserfall zwischen den beiden Klavieren hin- und herfließen, überlagern sich langsam und gewinnen so an Komplexität.

Toshio Iwai (J), geb. 1962, ist interaktiver Künstler und graduierte an der Tsukuba University. Seine Werke wurden in mehreren Einzelausstellungen weltweit gezeigt, zwei seiner Arbeiten sind in der Dauerausstellung des San Francisco Exploratorium zu sehen. Er war Artist-in-Residence am ZKM/Karlsruhe und ist derzeit Artist-in Residence am IAMAS.

Ruyichi Sakamoto (J), geb. 1952, graduierte an der Tokyo National University of Fine Arts zum MFA. Er ist international anerkannter Komponist, Musiker und Medienkünstler. Für seine Filmmusik (Bernardo Bertoluccis „Der letzte Kaiser", „The Sheltering Sky", „Little Buddha") hat er Grammy und Golden Globe Awards sowie Oscars erhalten.

The Invisible Shape of Things Past

Dirk Lüsebrink / Joachim Sauter

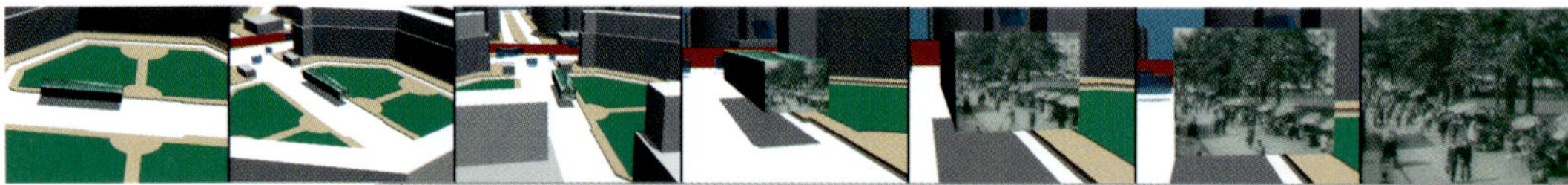

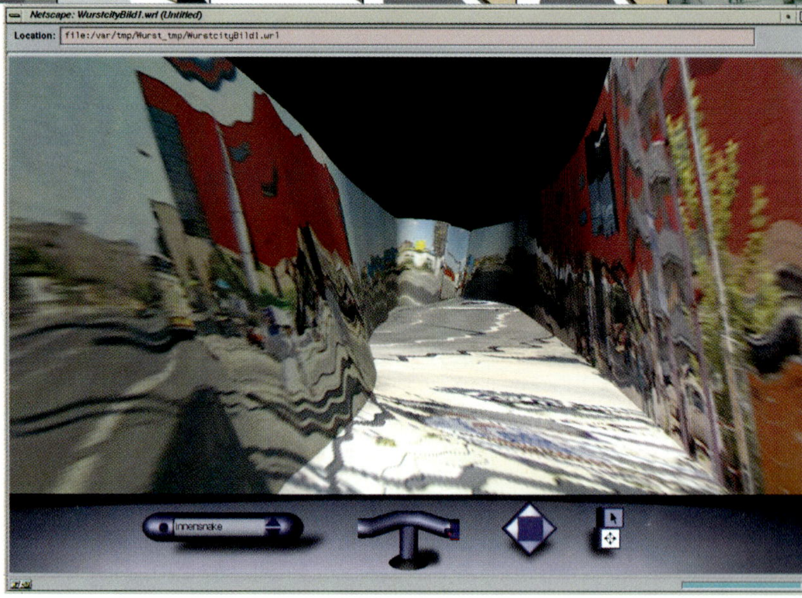

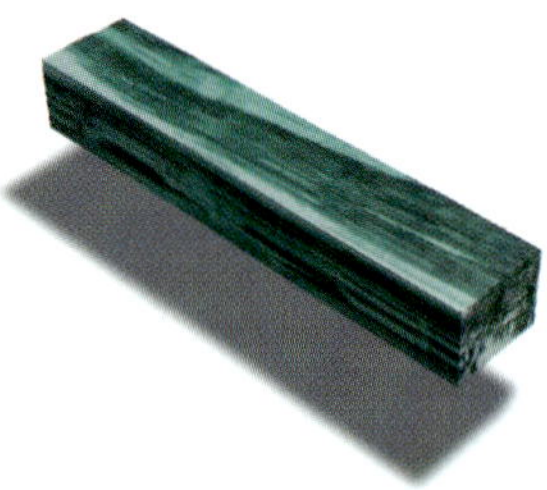

Following a long interest in the representation of virtual space and navigation in virtual spaces, „The Invisible Shape of Things Past" is an exploration of the representation of time and navigation through time in VR.

Our project enables users to transform film sequences (time-based information) into interactive virtual objects. This transformation is based on all the camera parameters of a particular film sequence (movement, perspective, focal length). The individual frames of the film are lined up along the path of the camera as it is transferred to virtual space.

The angle of the individual frames in relation to the virtual camera path depends on the perspective of the actual camera, the size of the individual frames depends on the focal length used. A virtual architecture of information emerges, which represents the underlying film sequence through its form.

Nach langer Beschäftigung mit der Darstellung von virtuellem Raum und der Navigation in diesen virtuellen Räumen ist die Arbeit „Invisible Shape of Things Past" eine Auseinandersetzung mit der Abbildung von Zeit und der Navigation durch die Zeit in der VR.

Das Projekt ermöglicht es den Nutzern, Filmsequenzen (zeitbasierte Information) in interaktive virtuelle Objekte umzusetzten. Diese Transformation basiert auf den Kameraparametern der zugrundeliegenden Filmsequenz (Kamerabewegung, Blickrichtung und Brennweite). Entlang des in den virtuellen Raum übertragenen Kamerapfades werden die Einzelbilder des Filmes aneinandergereiht. Die Neigung der einzelnen Frames zum virtuellen Kamerapfad hängt dabei von der jeweiligen Blickrichtung der realen Kamera ab, die Größe der einzelnen Frames von der verwendeten Brennweite. Eine virtuelle Informationsarchitektur

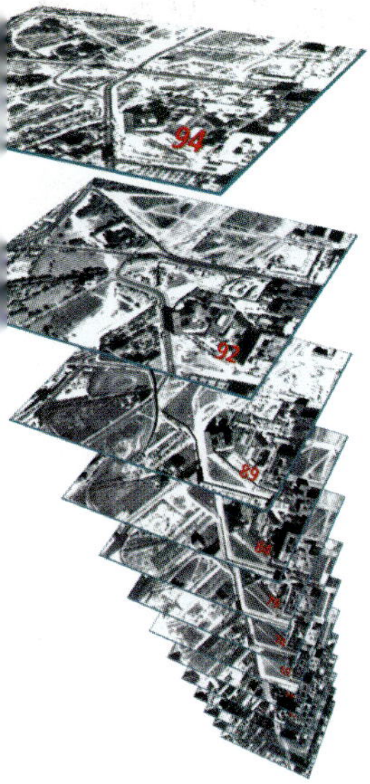

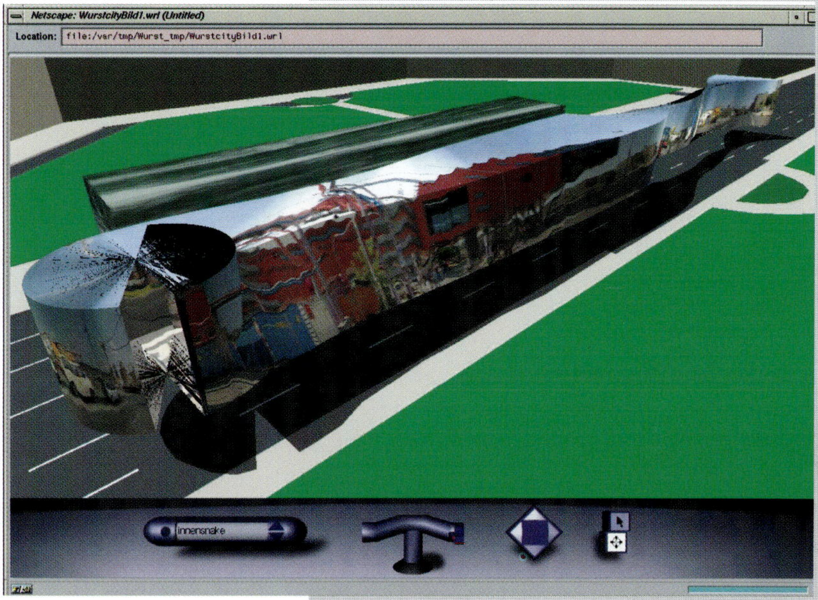

This object may be explored interactively, it is the spatial user interface (VRUI) to the information it contains. Activating the front side of the object, for example, „plays the movie" - a double click leads the user through the object along the virtual camera path.

Since there is not only a place but also a time for the occurrence of each film sequence, a virtual representation of the place of occurrence was modeled for the organization of the film objects, thus enabling the user to navigate through time. Using the example of Berlin, all the urban situations since the year 1900 were modeled, and the film objects were positioned according to the virtual location and the virtual time layer, where and when they were shot. Three concepts were developed for navigating through these time layers. The first and most obvious concept is based on the transformation from one temporal urban situation to the next (spatial morph-

entsteht, die durch ihre Form die Entstehung der zugrunde liegenden Filmsequenz kommuniziert.

Dieses Objekt ist interaktiv erfahrbar, ist räumliches Userinterface (VRUI) zu der in ihm enthaltenen Information. Beispielsweise wird bei Aktivierung der Frontseite das „Filmobjekt abgespielt" – bei doppelter Aktivierung wird der Nutzer entlang des virtuellen Kamerapfades durch das Objekt geführt.

Da jede Filmsequenz sowohl einen Ort als auch einen Zeitpunkt ihrer Entstehung hat, wurde für die Organisation der Filmobjekte eine durch die Zeit navigierbare virtuelle Abbildung ihrer Entstehungsorte modelliert. Am Beispiel Berlin wurden alle urbanen Zustände seit dem Jahr 1900 modelliert und die Filmobjekte an dem ihrer Aufnahme entsprechenden virtuellen Ort und in der ihnen entsprechenden Zeitschicht positioniert. Drei Ansätze zur Navigation durch diese Zeitschichten wurden entwickelt. Der erste und naheliegendste Ansatz basiert auf einer Transformation von einem zeitlichen urba-

ing). With the help of a timeslider, the user can change the surroundings according to the year selected. The second concept is based on the spatial layering of the individual historical conditions. In this case, users navigate spatially through time. In the third and most VR-adequate concept, elements from different time layers may be individually configured, for example by placing the future building plan of Potsdamer Platz on a current aerial picture and adding the course of the Berlin Wall and all the film objects from 1940 to 1945.

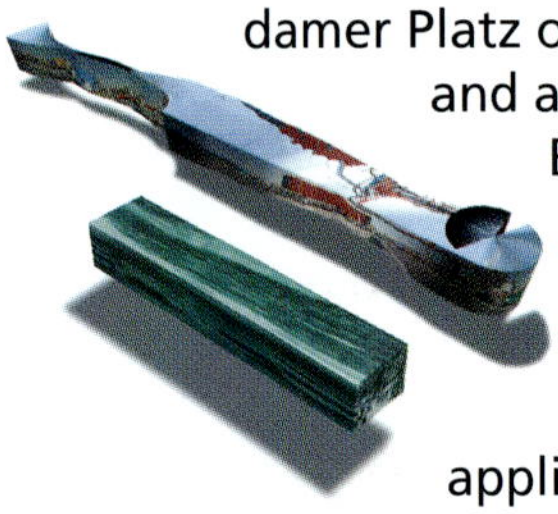

In addition to a local interactive application, a networked VRML Web application has been developed, enabling Internet users to integrate their film sequences as objects in the virtual Berlin. In addition to the aspect of a spatial database of film material, another result is that through these film objects an urban architecture that was previously invisible is able to grow with the help of the Internet.

In a final step, a film object based on a sequence filmed in the Leipziger Strasse in 1941 is integrated into the exact same location as a simulation in the real space of today. This could be the starting point for designing a real architectonic or sculptural object.

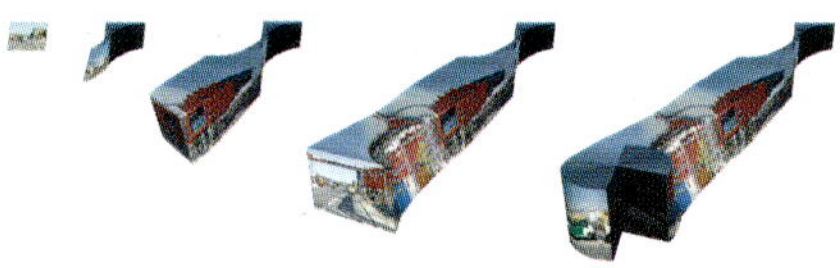

nen Zustand zum nächsten (räumlicher Morph). Mit Hilfe eines Timesliders verändern die Nutzer die virtuelle Umgebung entsprechend der gewählten Jahreszahl. Der zweite Ansatz basiert auf einer räumlichen Schichtung der einzelnen historischen Zustände. Die Nutzer navigieren dabei räumlich durch die Zeit. Im dritten und VR-adäquatesten Ansatz können Elemente aus verschiedenen Zeitschichten individuell konfiguriert werden, beispielsweise die zukünftige Bebauung des Potsdamer Platzes auf einem aktuellen Luftbild, dazu der Verlauf der Berliner Mauer und alle Filmobjekte aus den Jahren 1940 bis 1945.

Neben einer lokalen interaktiven Anwendung wurde eine vernetzte VRML-Web-Applikation entwickelt, die es Internet-Nutzern ermöglicht, ihre Filmsequenzen als Objekte in das virtuelle Berlin zu integrieren. Neben dem Aspekt einer räumlichen Filmdatenbank entsteht durch diese Filmobjekte eine bisher unsichtbare, mit Hilfe des Internet wachsende urbane Architektur.

In einem letzten Schritt wurde ein Filmobjekt, das auf einer 1941 in der Leipziger Straße aufgezeichneten Sequenz basiert, an genau derselben Stelle in den heutigen realen Raum als Simulation integriert. Dies könnte ein Ausgangspunkt sein für die Gestaltung eines realen architektonischen oder skulpturalen Objektes.

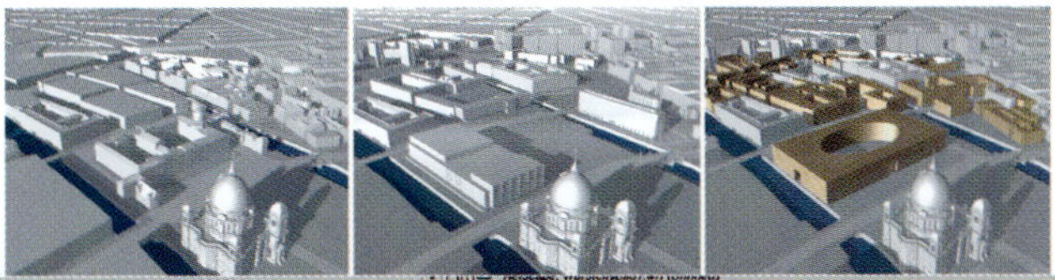

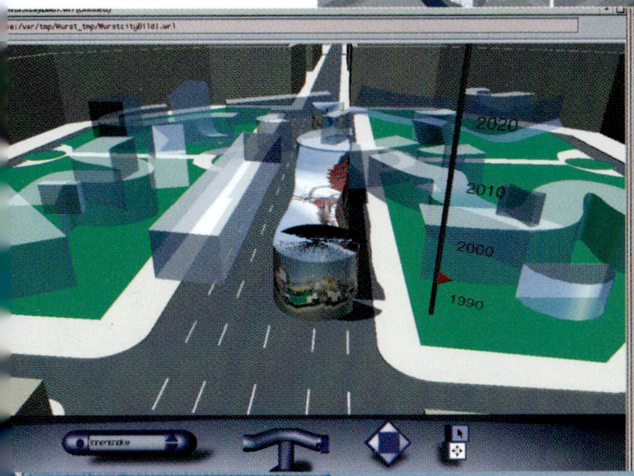

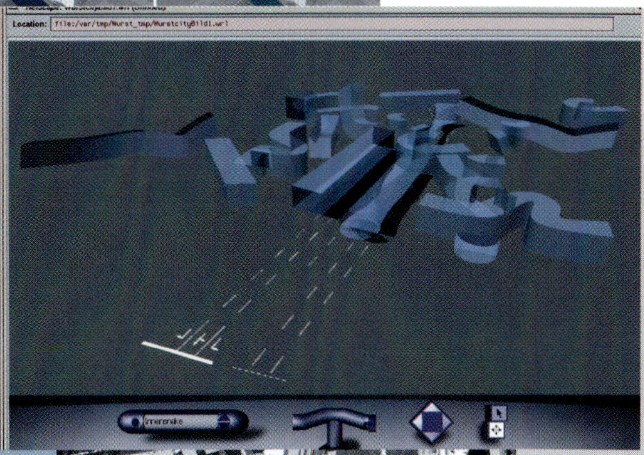

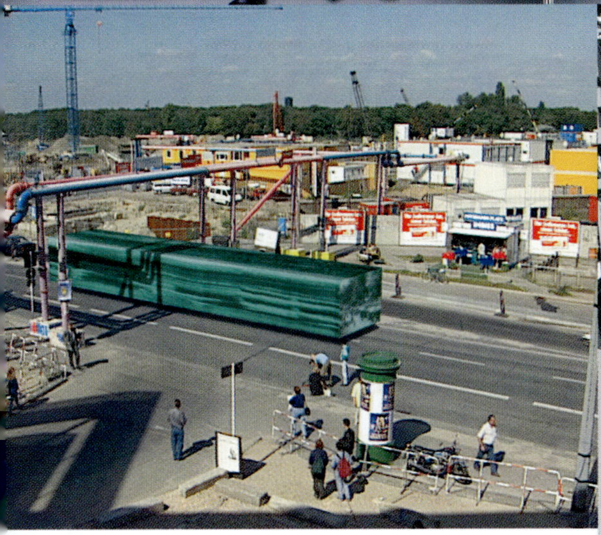

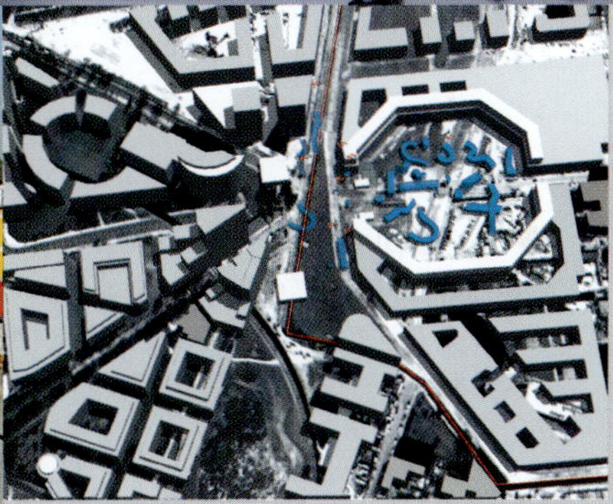

Dirk Lüsebrink and Joachim Sauter (D) are founding members of ART+COM (1988) and received an Award of Distinction in the Prix Ars Electronica 92 for their work „Zerseher" in the interactive art category. Dirk Lüsebrink studied computer science at the TU Berlin and works in development, Joachim Sauter heads the design department of ART+COM and teaches digital media design at the Hochschule der Künste Berlin.

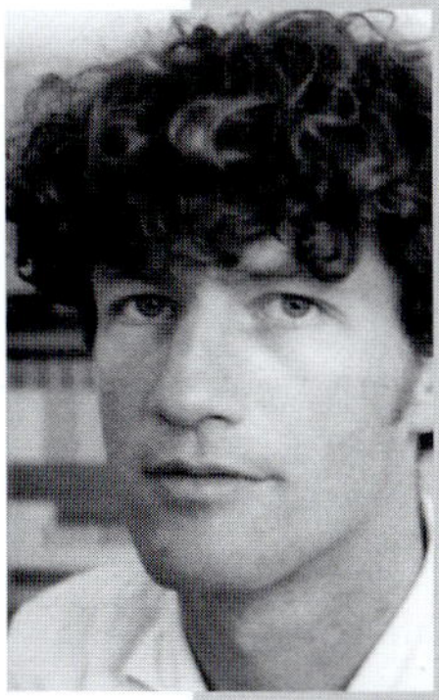

Dirk Lüsebrink und Joachim Sauter (D) sind Gründungsmitglieder von ART+COM (1988) und wurden beim Prix Ars Electronica 92 für ihre Arbeit „Zerseher" in der Kategorie Interaktive Kunst ausgezeichnet. Dirk Lüsebrink hat Informatik an der TU Berlin studiert und arbeitet im Bereich Entwicklung, Joachim Sauter leitet den Gestaltungsbereich bei ART+COM und lehrt an der Hochschule der Künste Berlin das Fach Gestalten mit digitalen Medien.

Border Patrol

Paul Garrin / David Rokeby

Autonomous Object Tracking System with Robotic Cameras

Stationary cameras function as visual sensors to the VNSII interface by David Rokeby that controls the positioning of robotic "snipercams" which lock on to moving targets (the viewer's heads) and "fire shots" (audio of gunshots). The viewer sees his/her image on video monitors, set into the face of the metal wall, in the crosshairs of the snipercam.

Each of the 4 autonomous snipercams can track up to 32 individual objects, and monitor their status:

scanning – looks for target
tracking – found target
lock – engage target
fire – shoot target
verify – check for movement
if no movement go on to next
check again for movement
after x duration, end time to live cycle.

Security through InSecurity

Increase the level of paranoia ...

bombs in the metro ... bombs in Oklahoma ...

bombs in the World Trade Center ... bombs in Mururoa ...

Who are the real terrorists?

Edam Salem, who was a key witness in the so called NYC terrorist conspiracy, was a paid informant for the FBI (and was also working with the FBI before the bombing of the WTC). Salem was the one who set up a hidden video camera and taped a "bomb making class" thereby entrapping the participants when the secret tape was used as damning evidence, claiming that the fertilizer and fuel oil bomb they were cooking up was meant for the U.N., the Lincoln and Holland Tunnels, and the Federal Building in Lower Manhattan. Even though no bombs ever went off in any of those locations, all were found guilty of sedition (through an obscure, 19th century law not invoked

Ein autonomes Object-Tracking-System mit Roboter-Kameras

Stationäre Kameras dienen als visuelle Sensoren für das VNSII-Interface von David Rokeby, das die Positionierung von Roboter-„Snipercams" („Heckenschützenkameras") steuert, die sich an beweglichen Zielen „festkrallen" (in diesem Fall an den Köpfen der Besucher) und „Schüsse" abfeuern (entsprechende Schußgeräusche). Auf Videowänden, die in die Metallseitenteile der Installation eingelassen sind, sehen die Besucher ihre eigenen Köpfe im Fadenkreuz des Visiers.

Jede der vier automatischen „Snipercams" kann bis zu 32 einzelne Objekte anvisieren und ihren Status beobachten:

Scannen – sucht nach Ziel
Verfolgen – Ziel gefunden
„Festkrallen" – Ziel anvisieren
Feuern – Ziel abschießen
Verifizieren – auf Bewegung untersuchen
Falls keine Bewegung – Gehe zu nächstem Befehl
Neuerliche Bewegungskontrolle
Nach Dauer X Zyklus beenden

Sicherheit durch Unsicherheit

Erhöht den Grad an Paranoia ...

Bomben in der U-Bahn ... Bomben in Oklahoma ...

Bomben im World Trade Center ... Bomben auf Mururoa ...

Wer sind die wirklichen Terroristen?

Edam Salem, der Hauptzeuge in der sogenannten New Yorker Terroristenverschwörung, war bezahlter Informant des FBI (und hatte mit diesem schon vor dem Attentat auf das World Trade Center zusammengearbeitet). Es war Salem, der eine versteckte Videokamera installierte und einen „Bombenbaukurs" filmte. Er lockte die Teilnehmer in eine Falle, denn das geheim aufgenommene Band wurde als ausschlaggebendes Beweismittel verwendet.

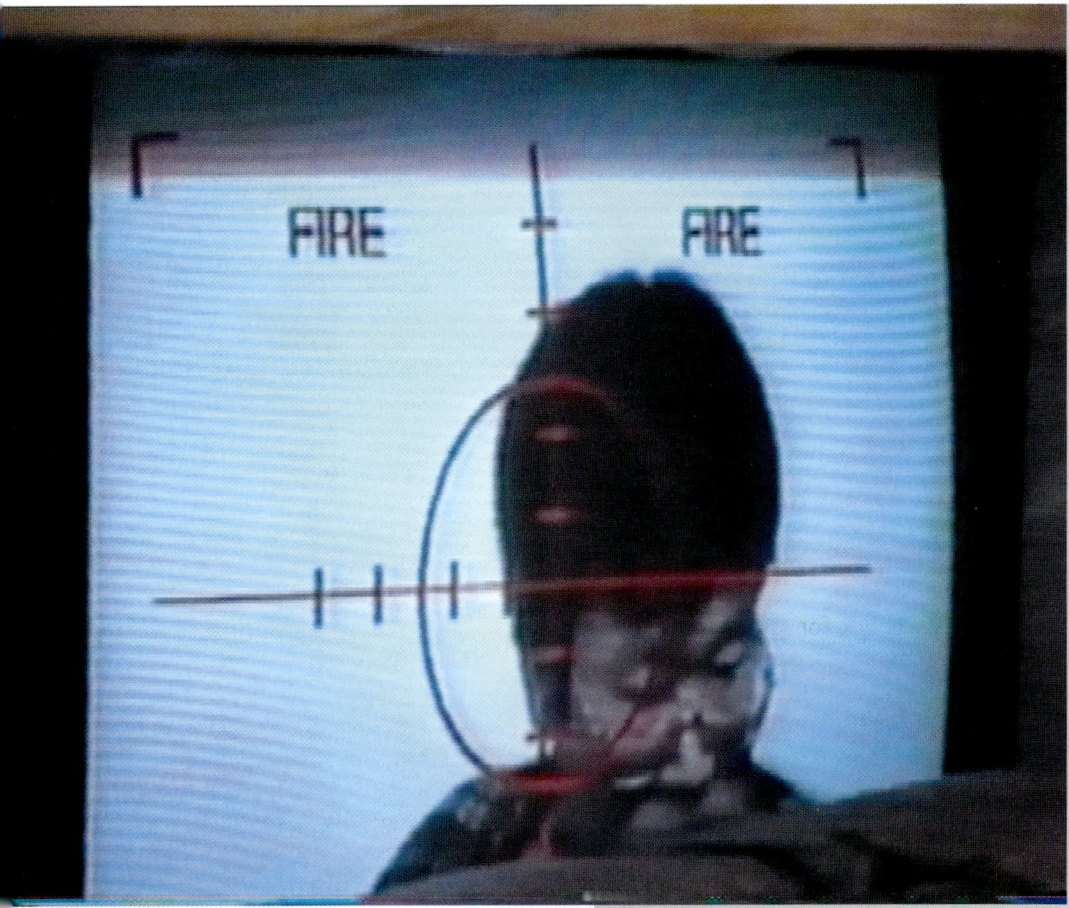
FIRE
FIRE

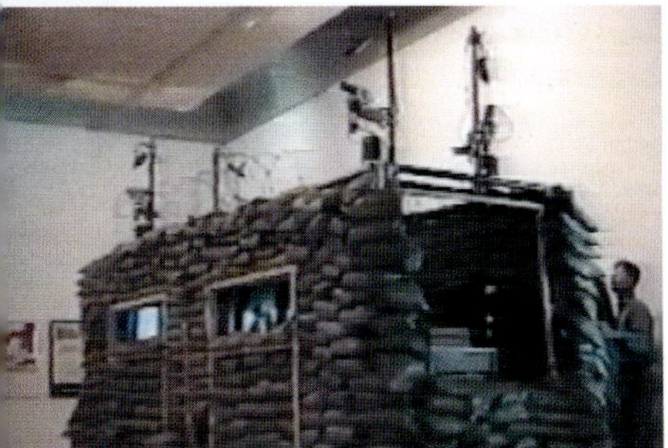

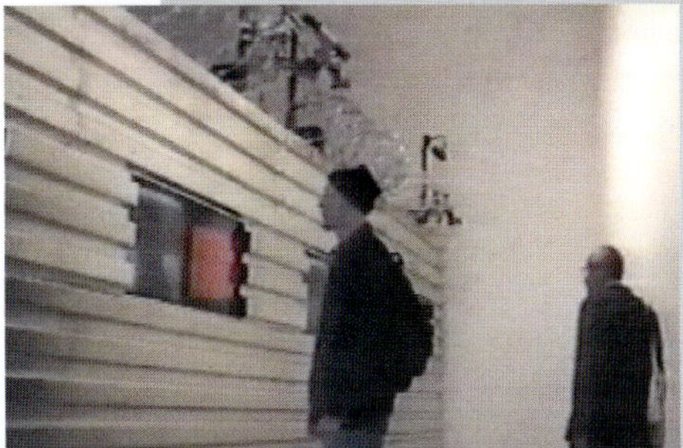

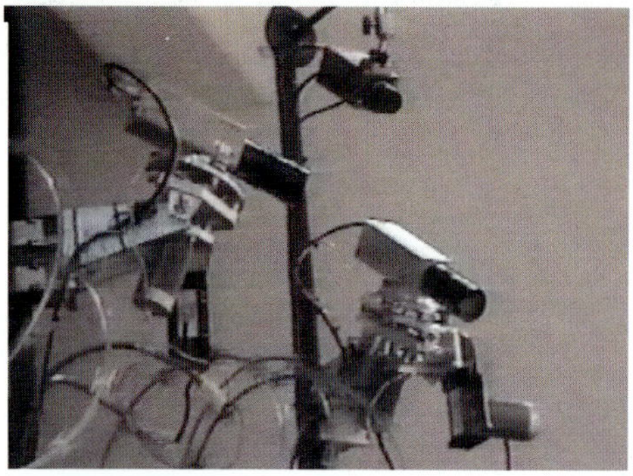

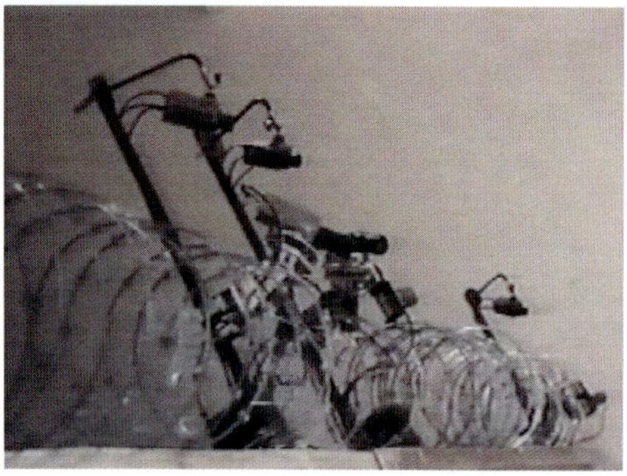

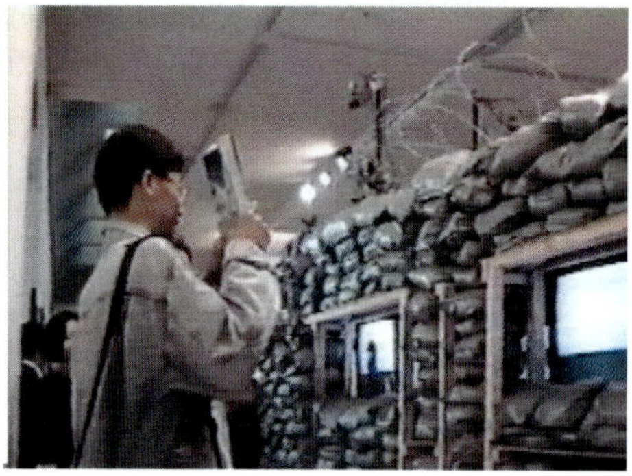

since the American Civil War) ... Muslim Fundamentalist Terrorists ... that was the catchy phrase. Enough to inspire fear in the public ... enough to convince them that the new "anti-terrorist" laws are meant to "protect" them.

Was the US government (or at least the FBI) complicit in the so-called terrorist bombings? If so, their strategy seems to be working. The new anti-terrorist laws recently passed seriously erode civil liberties and bring the US closer to a police state.

The new laws provide increased police powers, surveillance of the general population through use and development of technology and tactical training for a growing number of police and military, in preparation

Es wurde behauptet, daß die aus Kunstdünger und Dieselöl zusammengebraute Bombe, an der die Teilnehmer bastelten, für die UNO, den Lincoln- und den Holland-Tunnel sowie das Federal Building in Lower Manhattan gedacht gewesen sei. Obwohl dort niemals eine Bombe gelegt wurde, wurden sie alle der Staatsgefährdung für schuldig befunden (aufgrund eines obskuren Gesetzes aus dem 19. Jah hundert, das seit dem Amerikanischen Bürgerkrieg nie mehr angewendet worde war) ... Moslemisch-fundamentalistische Terroristen ... das war das entscheidende Zauberwort. Genug, um der Öffentlichkei Angst einzujagen ... genug, um sie zu überzeugen, daß die neuen „Anti-Terror"-Gesetze zu ihrem „Schutz" da seien.

War die US-Regierung (oder zumindest das FBI) Komplize bei diesen sogenannten Terroristen-Anschläge? Wenn ja, dann scheint ihre Strategie aufzugehen. Die kürzlich beschlossenen neuen Anti-Terrorgesetze schränken die Freiheit des Bürgers entscheidend ein und bringen die USA näher an der Status eines Polizeistaats.

Die neuen Gesetze sehen eine Verstä kung der Polizeikräfte vor, eine Überwachung der Bevölkerung durch die Entwick lung und den Einsatz technologischer Mit tel. Außerdem sollen noch mehr Polizei- und Militäreinheiten als Vorbereitung auf urbane Kriegsführung und Anti-Terror-Operationen eine taktische Ausbildung erhalten. Was ist aus der amerikanischen Verfassung geworden? Wie konnte die Regierung die Bevölkerung dazu bringen, eine stärkere Polizeiüberwachung ihres Alltagslebens hinzunehmen, ohne dagegen auf breiter Front zu revoltieren? Man schaffe Terrorangst und identifiziere einen wohldefinierten „Feind", und schon gibt die Masse ihre Privatsphäre und ihre Bürgerrechte für mehr „Sicherheit" auf.

Ob die Bomben nun in der U-Bahn, i Oklahoma oder auf den fernen Pazifik-Atolls von Polynesien hochgehen – die

for urban warfare and counter terrorism operations. What happened to the US Constitution? How did the government get the general public to accept greater police control over their lives without provoking large-scale revolt? Create fear of terrorism and identify a defined enemy, and the general public will give up privacy and civil liberties in favor of "security".

Whether the bombs are in the metro, Oklahoma, or in the Pacific atolls of Polynesia – the bombs themselves merely fuel the real weapon: FEAR. A fear that the ordinary suddenly becomes deadly ... a fear which provokes irrationality ... a fear which begs for protection ... at any cost.

The message of the state: create a State of Insecurity in order to establish Security – by any means.

Ask yourselves – is this what you want?

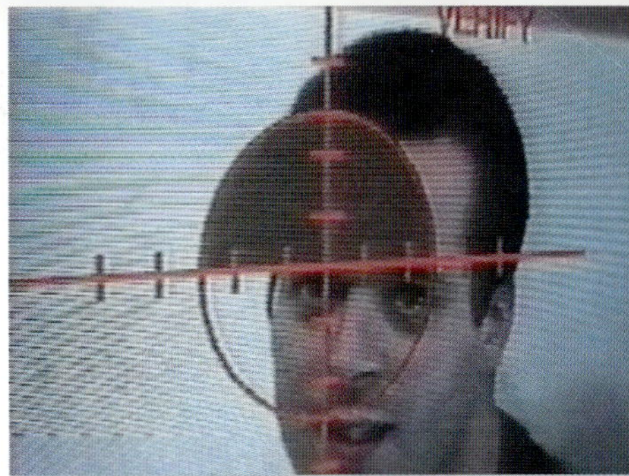

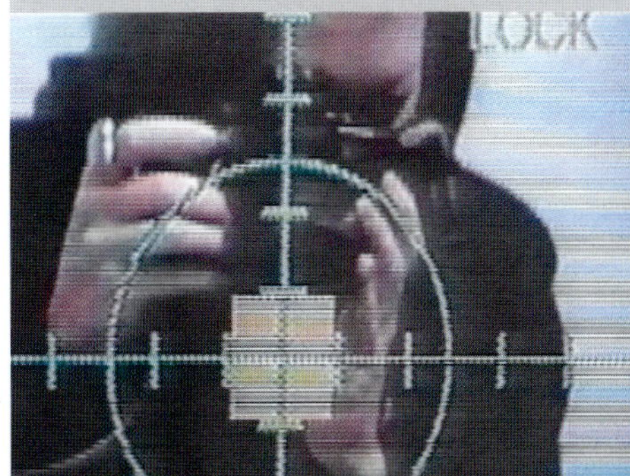

Bomben selbst dienen nur als Munition für die eigentliche Waffe: Angst. Die Angst, daß das Gewöhnliche plötzlich zum Tödlichen werden könnte ... eine Angst, die Irrationalität hervorruft ... eine Angst, die um Schutz bittet ... um jeden Preis.

Die Botschaft des Staates: Schaffe einen Status der Unsicherheit, um Sicherheit schaffen zu können – mit allen Mitteln. Fragen Sie sich selbst – wollen Sie das wirklich?

Paul Garrin (USA) works as a media artist and has been collaborating with Nam June Paik for a long time. His documentation of the "Tompkins Square Riot" 1988 in New York City is known as the spark which ignited the "camcorder revolution". He has been Artist-in-Residence at the Berlin Videofest in 1990 and has been awarded several highly recognized prizes.

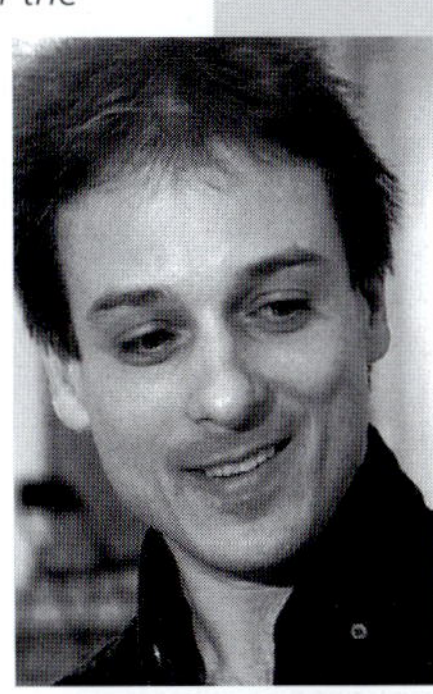

*Paul Garrin (USA) ist Medienkünstler und hat lange mit Nam June Paik zusammengearbeitet. Seine Dokumentation des „Tompkins Square Riot" 1988 in New York City wurde zum auslösenden Funken der „Camcorder-Revolution".
Er war Artist-in-Residence beim Berliner Videofest 1990 und hat mehrere internationale Preise erhalten.*

David Rokeby (CDN) is an interactive sound- and video-installation artist based in Toronto, Canada. He has been creating interactive installatiåedia Arts in 1988 and the Prix Ars Electronica Award of Distinction for his interactive piece "Very Nervous System" which is now being used to enable a paralyzed woman to speak and write.

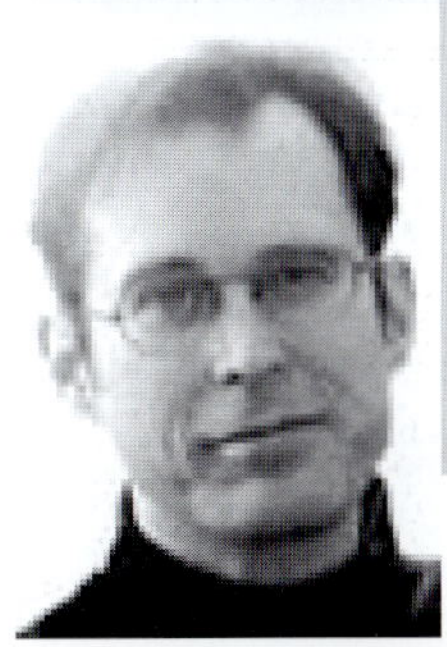

David Rokeby (CDN) beschäftigt sich seit 1982 mit interaktiven Klang- und Videoinstallationen. Er war Träger des ersten Petro-Canada Award for Media Arts, 1988, und Gewinner einer Auszeichnung beim Prix Ars Electronica für sein interaktives Stück „Very Nervous System", das derzeit verwendet wird, um einer gelähmten Frau das Sprechen und Schreiben zu ermöglichen.

Seeing is Believing

Hachiya Kazuhiko

"Seeing Is Believing" is an exhibit consisting of several elements. Here I would like to explain what motivated me to create each of these different works and projects.

Mega Diary – Global open diary project on the net

One day, I had a dream. I was in a very large library. It was a somewhat peculiar library because its entire collection consisted of the diaries of all those people who had ever lived. I randomly read through these records of the lives of all mankind. They were written by various types of people of all ages and from all countries, some of them still living and others dead. I read, and I was both moved and impressed. I felt sympathy for the authors, and I laughed and I cried.

Two weeks later, I started the "Mega Diary" project, conceived as one hundred people keeping diaries for 100 days and exchanging them with one another on the

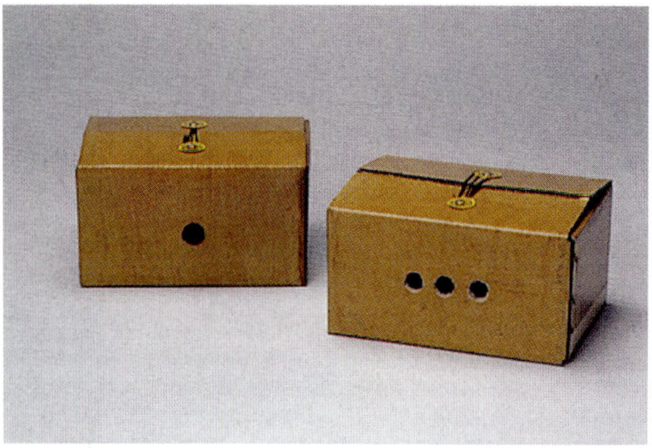

PC communications net. The diaries would be collected and shown each day through Nifty Serve and the Internet, as well as being broadcast nationwide once a week on FM "visible information radio". In fact, "Mega Diary" has spurred me to think about how diverse the everyday lives of others are, and how unaware we typically are of this fact, etc. The project has also

„Seeing Is Believing" ist eine Installation, die aus verschiedenen Elementen besteht, und im folgenden möchte ich erläutern, was mich dazu motiviert hat, jedes dieser Projekte und Werke zu schaffen.

Mega Diary – Ein globales offenes Tagebuchprojekt im Netz

Eines Tages hatte ich einen Traum. Ich träumte, ich befände mich in einer Bibliothek – einer ganz speziellen Bibliothek, denn sie enthielt nichts anderes als die Tagebücher aller Menschen, die jemals auf dieser Erde gelebt hatten. Ich habe mich – hie und da blätternd – nach und nach durch diese Aufzeichnungen aus dem Leben der Menschheit gelesen. Sie stammten von den unterschiedlichsten Leuten aller Altersstufen und Länder, manche davon noch am Leben, andere schon lange tot. Ich las, und ich war ebenso beeindruckt wie bewegt, ich lachte und ich weinte.

Zwei Wochen später initiierte ich das „Mega Diary"-Projekt, konzipiert für hundert Leute, die hundert Tage lang Tagebücher führen und sie untereinander im Netz austauschen sollten. Die Tagebücher sollten gesammelt und täglich über Nifty Serve und das Internet gezeigt sowie einmal wöchentlich landesweit im „Visible Information Radio" auf UKW ausgestrahlt werden. „Mega Diary" hat mich angespornt, darüber nachzudenken, wie unterschiedlich doch das Alltagsleben der einzelnen Menschen ist und wie wenig uns diese Tatsache normalerweise bewußt ist. Und noch etwas hat mir das Projekt klargemacht: Ein Monolog kann manchmal ein besseres Kommunikationsmittel sein als ein Dialog.

Empty Entity – The Invisible Electric Light Display

Obwohl wir normalerweise dazu tendieren, es zu vergessen: Sichtbarkeit und

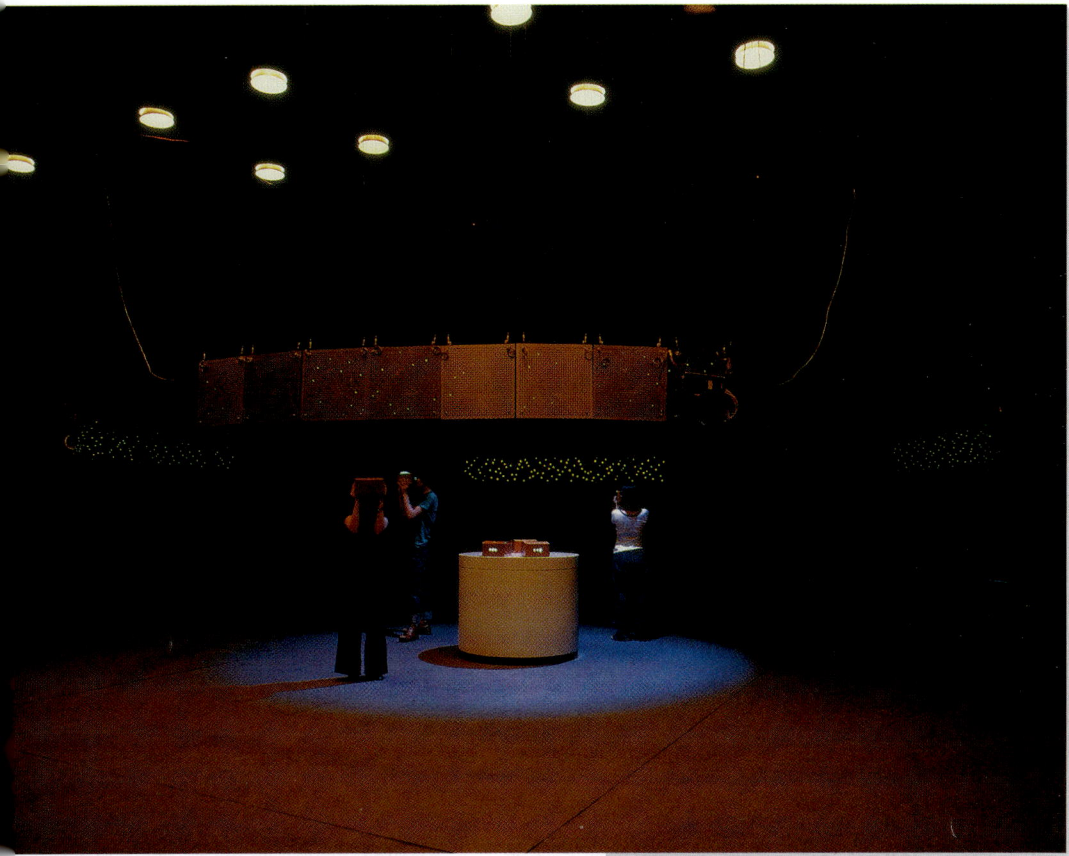

made me realize that monologue is often a better means of communications than dialogue.

[E]mpty Entity – [T]he Invisible Electric Light Display

Though we usually tend to forget it, visibility and invisibility are deeply connected with how we humans are structured physically. The range of visible rays differs from species to species. Human eyes can pick up 400 nm (nanometer) to 800 nm electromagnetic waves. Of course, light also exists outside the boundaries of what is visible.

In this piece, each display unit is made up of 8 letters, with 16 x 16 (256) light-emitting diodes (LED) for each letter, and a total of 4 units in the room. Since 15 of 16

Unsichtbarkeit sind unmittelbar mit der physischen Struktur der Lebewesen verknüpft. Der sichtbare Strahlungsbereich ist von Spezies zu Spezies anders; das menschliche Auge kann elektromagnetische Wellen zwischen 400 und 800 Nanometer Länge wahrnehmen. Und dennoch existiert Licht auch jenseits dessen, was wir sehen ...

In diesem Stück besteht jedes Display aus acht Buchstaben, die sich jeweils aus 16 x 16 (256) Licht-Emitter-Dioden(LED) zusammensetzen. Vier solcher Displays sind im Raum angebracht. Da 15 der 16 LEDs Infrarotlicht abstrahlen, sieht das unbewaffnete Menschenauge nur ein flackerndes Licht, das die wenigen übrigen LEDs, die im sichtbaren Bereich leuchten, abstrahlen. Mit besonderen Brillen allerdings kann der Betrachter Buchstaben auf den Displays erkennen. In dieser Ausstellung werden

of the LEDs are infrared ones, the naked eye sees only flickering lights, produced by the small number of LEDs emitting visible rays. With special eyewear on, however, the viewer can see letters flowing across the units. In this exhibition, the diaries which many people have contributed to "Mega Diary" are displayed. These diaries could be considered as one of the best examples of something which is visible to us, but at the same time is not visible, and is extremely precious nonetheless.

The Sheep: The Tool to See the Invisible

In Exupéry's "The Little Prince," there is a scene in which the prince begs "me" to draw a picture of a sheep. "I" draw many pictures, but none of them please him. Finally "I" draw a picture of a box, and say,

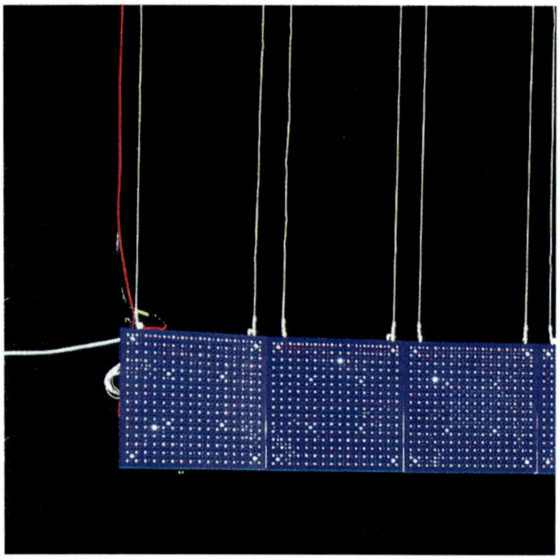

"This is his box. The sheep you asked for is inside!" Then, for the first time, the prince likes "my" picture.

In this "box," which serves as eyewear for the viewer, what is inside, is not a tiny sheep, but a monochrome observation camera and a liquid crystal TV monitor. When you peep through the box, you will surely see a little bit of the invisible. After that, it's a matter of imagination: even if you no longer have the box, or even if the "Mega Diary" project runs to completion, you should still be able to use your imagi-

die Beiträge, die diese vielen Menschen zum „Mega Diary"-Projekt beigesteuert haben, ausgestellt. Diese Tagebücher sind auch metaphorisch das beste Beispiel für Dinge, die zwar für uns gleichzeitig sichtbar und unsichtbar und dennoch von großem Wert sind.

The Sheep: The Tool to See the Invisible

In Antoine de Saint Exupérys „Der Kleine Prinz" gibt es eine Szene, in der der Prinz den Erzähler bittet, doch ein Schaf für ihn zu zeichnen. Der Erzähler zeichnet mehrere Schafe, aber keines gefällt dem Prinzen. So entsteht zuletzt das Bild einer Schachtel, und der Erzähler sagt: „Dies ist seine Schachtel, und das Schaf, das du wolltest, ist drin!" Und da gefällt dem Prinzen erstmals das Bild.

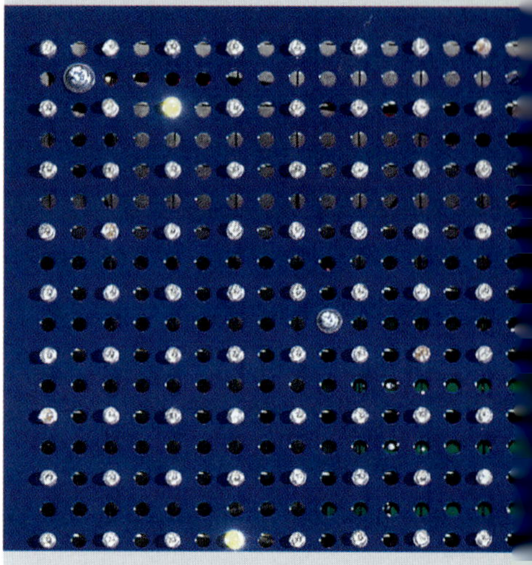

In dieser „Schachtel", die dem Besucher als Sehhilfe dient, ist zwar kein Schaf, wohl aber eine monochrome Beobachtungskamera und ein Flüssigkristall-Monitor. Wenn man durch die Schachtel schaut, sieht man ein Stück des Unsichtbaren. Danach aber ist alles eine Frage der Vorstellungskraft: Selbst wenn man die „Schachtel" nicht mehr auf hat oder wenn das „Mega Diary"-Projekt zu Ende geht, sollte man doch seine Phantasie benutzen und an jene Menschen denken die man so kennengelernt hat. Schließ-

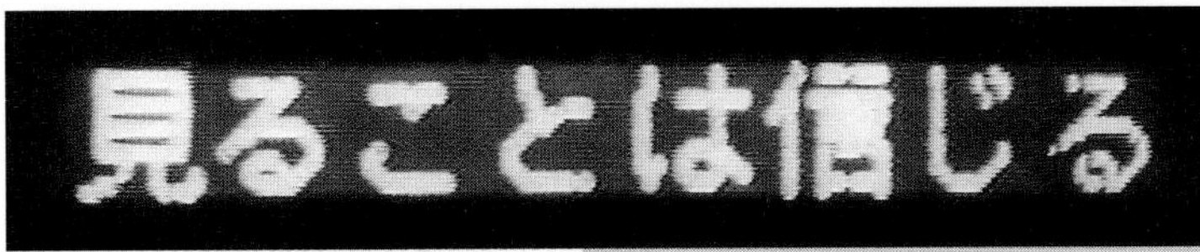

nation to think of those people with whom you have become acquainted. After all, you are no longer a stranger to them, now that you have read their diaries.

In my opinion, though, what one can see constitutes only a fraction of the truth. This is how I felt during the "Mega Diary" project. When I read someone's diary for a long while, I begin to experience emotions toward the author and feel closer to that person, even though I do not know him or her. My "connected-ness" or relationship with the person is one of a sort of respect and "belief."

lich ist man ihnen ja nicht mehr fremd, wenn man ihre Tagebücher gelesen hat.

Meiner Ansicht nach ist das, was man sehen kann, nur ein Bruchteil der Wahrheit. Das jedenfalls fühlte ich während des „Mega Diary"-Projekts. Wenn ich jemandes Tagebuch über längere Zeit hinweg lese, beginne ich, dem Autor gegenüber eine emotionale Beziehung zu entwickeln und mich ihm näher zu fühlen, auch wenn ich ihn gar nicht kenne. Meine Beziehung, mein Verbundensein mit dieser Person, wird getragen von einer Mischung aus Respekt und „Glauben".

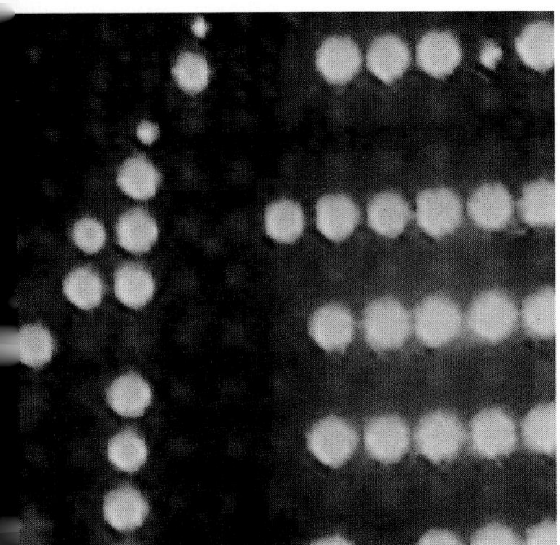

Kazuhiko Hachiya (J), born 1966 in Saga, Japan, graduated from Kyushu Institute of Design (Visual Communication Design). He has had several exhibitions in Japan. Lives and works in Tokyo.

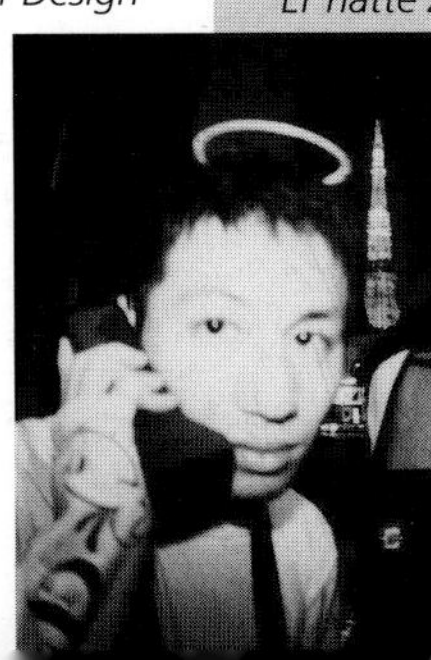

Kazuhiko Hachiya (J), geb. 1966, graduierte in Visual Communications Design am Kyushu Institute of Design. Er hatte zahlreiche Ausstellungen in Japan und lebt und arbeitet in Tokio.

Moppet

Toshihiro Anzai / Tamio Kihara

"Moppet" is a new creative system that anyone can use. It is simple, easy and open. People can paint and make music at the same time. The system is networked through a telecommunications network. It links many participants and several spaces.

Anyone can do it, because it is not difficult, but easy to use. It involves networked cooperation, multi-users, multi-spaces. The performance may be seen by a large audience.

"Moppet" creates CG paintings according to the way the participants move. The movement on "Moppet's" court takes the place of a paintbrush. The painting is projected to a screen over the court. All the participants share the same canvas.

The sound is generated by the movements of the participants, too. The participants are able to make music and paintings at the same time.

No one needs to wear special equipment. The system is wireless. Each participant handles different colored balls that may be simply handed to the other participants.

There is something ambivalent about "Moppet". We feel that the participants' activities on the "Moppet's" courts demonstrate communication and miscommunication. Our intention is to clearly represent the great difficulties involved in collaboration in networked spaces of creativity.

With the "Moppet" and its activities, we are attempting to make an open creative system using networks.

"Moppet" was produced and designed by Toshihiro Anzai (art direction, software), Tamio Kihara (produce, system design), Koichi Fujii (sound design, software), Hiroyuki Moriwaki (interface design), Nobuya Suzuki (communication design) and Rieko Nakamura (visual design).

„Moppet" ist ein neues kreatives System, das jedermann benutzen kann, es ist einfach und offen. Leute können dam malen und gleichzeitig Musik machen. D System ist durch ein Telekommunikation netzwerk vernetzt und verknüpft viele M wirkende und mehrere Räume. Jeder kar mitwirken, weil es einfach zu bedienen i. gleichzeitige Benutzung zuläßt und raum unabhängig funktioniert. Die sich daraus ergebende Performance kann von einem breiten Publikum beobachtet werden.

„Moppet" erzeugt computergraphische Gemälde, je nachdem, wie sich die Mitwirkenden bewegen. Die Bewegung auf dem „Moppet-Feld" übernimmt die Rolle des Pinsels, und das Gemälde wird auf einen Schirm über dem „Spielfeld" projiziert. Alle Mitwirkenden benützen dieselbe „Leinwand". Auch der Klang wi aus den Bewegungen der Mitwirkenden generiert – man kann also malen und m sizieren zugleich. Besondere Ausrüstung oder Anzüge sind nicht erforderlich, das System funktioniert kabellos. Jeder Mitw kende hantiert mit unterschiedlich gefärb ten Bällen, die er einfach an andere weitergeben kann. „Moppet" hat etwas Am bivalentes an sich: Wir haben den Eindruc daß die Aktivität der Mitwirkenden im „Moppet"-Feld Kommunikation ebenso wie Fehlkommunikation aufzeigt, und ur sere Absicht ist, die großen Schwierigkeiten der Mitwirkenden bei der Zusammen arbeit in vernetzten kreativen Räumen sichtbar zu machen. Mit „Moppet" und seinen Aktivitäten wollen wir ein offenes kreatives System innerhalb eines Netzwer kes präsentieren.

„Moppet" wurde entworfen und pro duziert von Toshihiro Anzai (Regie, Software), Tamio Kihara (Produktion, System design), Koichi Fujii (Klangdesign, Software), Hiroyuki Moriwaki (Interface), Nobuya Suzuki (Kommunikation) und Rieko Nakamura (visuelle Gestaltung).

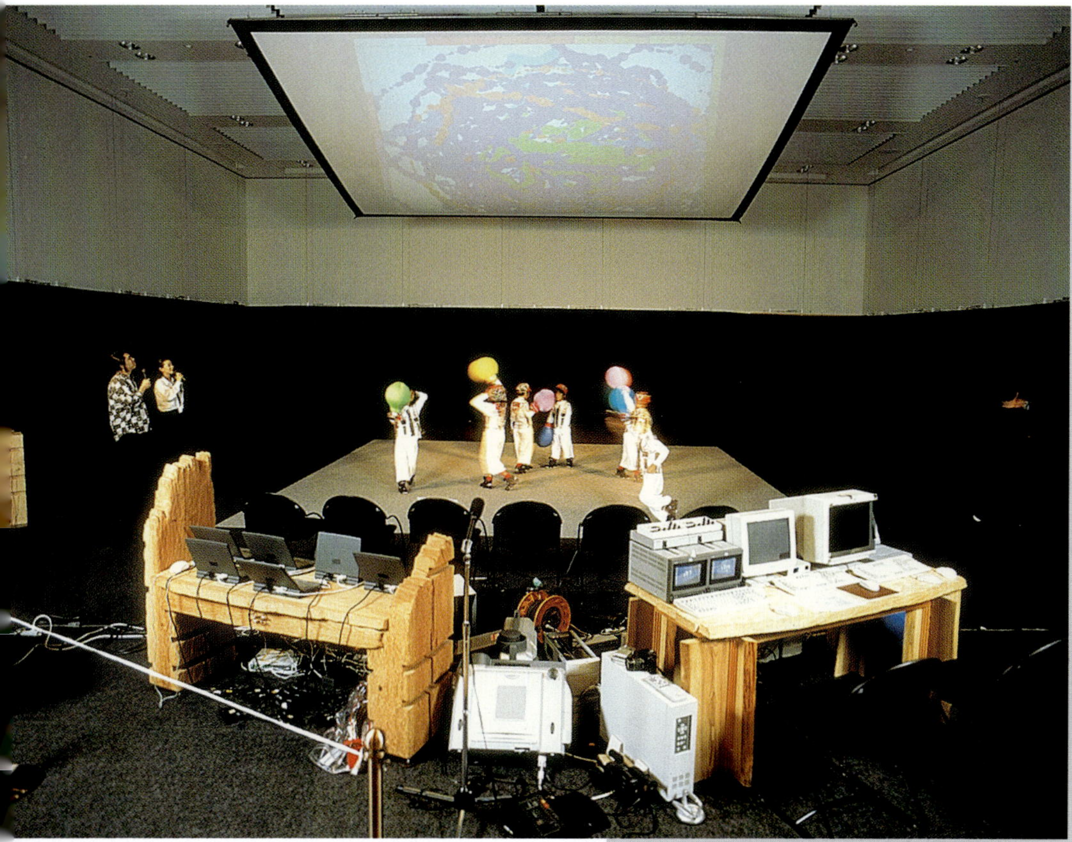

Toshihiro Anzai (J), born 1956, was chief producer of the development of Supertableau graphic system. Works both in computer graphics art and in software development.

Tamio Kihara(J) was born in Saitama in 1965. Completed graduate school of Aoyama Gakuin University in 1991. Now works at NTT Telecommunication Network Systems Laboratories.

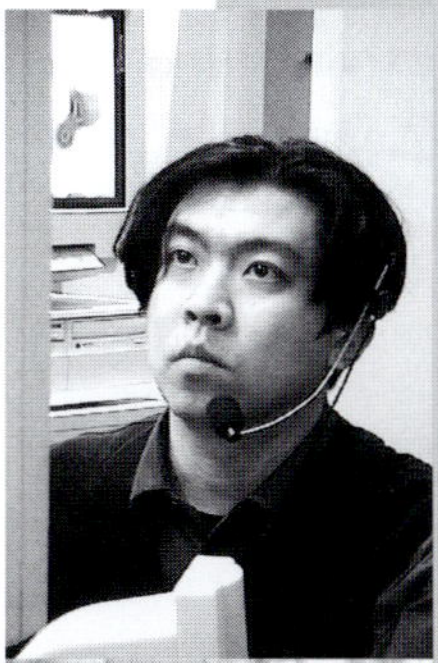

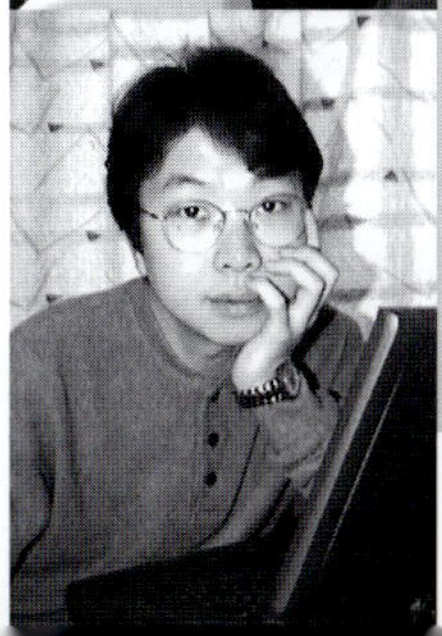

Toshihiro Anzai (J), geb. 1956, war Chefentwickler des Supertableau-Graphiksystems und arbeitet in der Computergraphik sowie in der Softwareentwicklung.

Tamio Kihara (J), geb. 1965, graduierte 1991 an der Aoyama Gakuin University. Derzeit arbeitet er in den NTT Telecommunication Network Systems Laboratories.

Anonymous Muttering

Knowbotic Research

"Anonymous Muttering" is a realized intervention in the texture of a city. Like parasites, we used a pirate broadcasting station to collect the acoustic energies and materials produced by DJ's parallel to one another in various nightclubs and sent this material into our sound and stroboscope installation.

The work was developed for both an actual urban space and electronic space (Internet, Java 3D applets, RealAudio). The DJ material we obtained like this was granulized by computer programs, in other words, dissolved into its smallest units, the finest particles of sound. Interfaces were provided on the Internet and locally in the city, where participants and visitors were able to interactively reassemble the granulized DJ material, which is authorless to begin with, using a granular synthe-program. The interface was designed as a fold, in which the rule sets (parameters for the synthesis algorithms) for processing sound were mounted. This meant that the atomized particles could be reassembled in a new way. The audible sounds that were formed using these fold surfaces were anchored in an urban space, but the rules were distributed across the Internet.

The collaboration between an anonymously large number of participants in electronic space and the active visitors on site resulted in influencing the "composition", however, this influence could not be individually controlled. We handed over the control of the rules. The sound event jumped back and forth between local and global rule sets. Thus it was not possible to predict what would really happen with the rule sets and what the effects on the granulated sound particles would be. The analog and digital fold-overs were inextricably interlocked. And they were interlocked in such a way that the conceptual distinction between "analog" and "digital" no longer made sense, even though they

„Anonymous Muttering" realisiert einen Eingriff in das Stadtgewebe. Wie ein Parasit haben wir die akustischen Energien und Materialien, die von DJ's parallel in verschiedenen Nachtclubs produziert worden sind, mit Piratensendern erfaßt und in unsere Klang- und Stroboskop-Installation geschickt. Diese Arbeit ist sowohl für den realen urbanen Raum sowie für den elektronischen Raum (Internet, Java 3D Applets, RealAudio) entwikkelt worden. Das so erhaltene DJ-Material wurde von Computerprogrammen granularisiert, also aufgelöst in kleinste Einheiten, feinste Klangpartikel. Das granularisierte Material der DJ's, ohnehin schon ein autorenloser Stoff, konnte wieder von Teilnehmern und Besuchern der Installation an Interfaces, Schnittstellen, die sowohl im Internet als auch im lokalen Stadtraum benutzbar waren, in Interaktion mit einem Granularsyntheprogramm zusammengefügt werden. Die Schnittstelle war ausgebildet als Falte, auf welcher die Regelwerke (Parameter für Synthesealgorithmen) für die Klangverarbeitung angebracht waren. Die zerstäubten Teile konnten also auf neue Weise zusammengefaltet werden. Die hörbaren Klänge, die durch diese Faltflächen verformt wurden, waren an einen Stadtraum gebunden, während die Regel im Internet verteilt waren. Durch die Zusammenarbeit einer anonym großen Anzahl von Teilnehmern im elektronische Raum und den aktiven Besuchern vor Ort ergab sich ein allerdings nicht individuell kontrollierbarer Einfluß auf die „Komposition". Wir haben diese Regeln a der Hand gegeben. Das Klangereignis sprang zwischen lokalen und globalen Regelsätzen hin und her. Es war also nicht vorhersehbar, was mit diesem Regelwerk und den Auswirkungen auf die granulare Klangpartikel wirklich geschehen würde. Analoge und digitale Faltung waren untrennbar ineinander verzahnt. Und zw

could be re-determined in this situation through various forms of actions. Yet it was no longer possible to distinguish who was actually intervening, manipulating, in other words, in the archaic sense of moving something by hand, and who was acting in electronic space. It was only possible for the participants to tendentially alter their own and others' interests. This installation provided an opportunity to operate with very fine fluctuations in the movements of interests simultaneously in different contexts (local and global).

so, daß die begriffliche Auflösung zwischen „analog" und „digital" keinen Sinn mehr machte, jedoch in der Situation durch unterschiedliche Handlungsformen neu bestimmt werden konnte. Man konnte aber nicht mehr unterscheiden, wer jetzt wirklich zugreift, also im archaischen Sinne manipuliert, von Hand bewegt und wer im elektronischen Raum agiert. Man hatte nur die Möglichkeit, eigene und fremde Interessen tendenziell zu verändern. Diese Installation gab die Möglichkeit, mit solch feinen fluktuierenden Interessensbewegungen simultan in verschiedenen Kontexten (lokal und global) zu operieren.

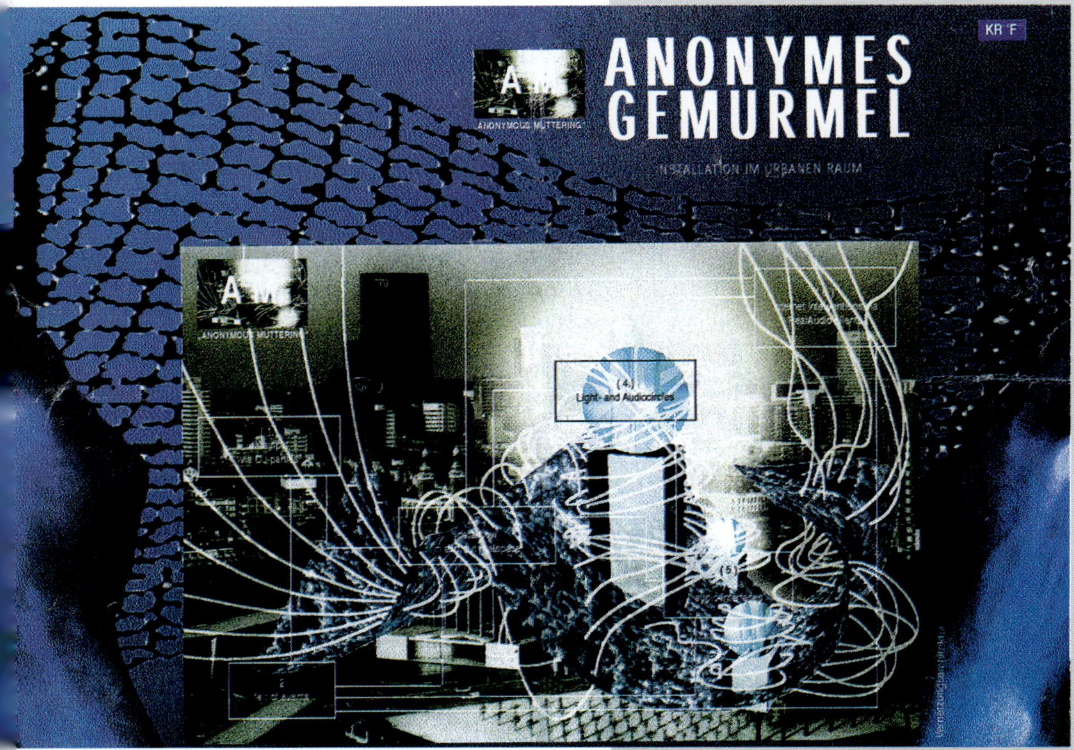

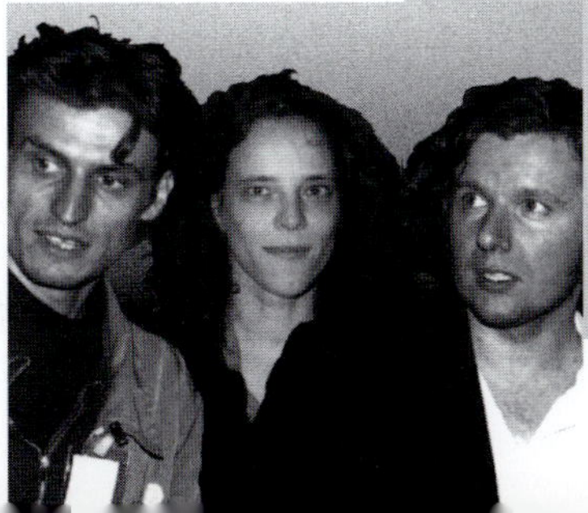

Knowbotic research (KR+cF) – Yvonne Wilhelm, Alexander Tuchacek, Christian Huebler – is based in Cologne. With the parents of Westbank Industries und Tactile Technology and the support of Academy for Media Arts, KR+cF has founded Mem_brane, a laboratory for media strategies. KR+cF has received major international Media Art Awards.

Knowbotic Research (KR+cF) – Yvonne Wilhelm, Alexander Tuchacek, Christian Huebler – ist in Köln angesiedelt. Mit Westbank Industries und Tactile Technology als Partner sowie mit Unterstützung der Academy for Media Arts hat KR+cF „Mem_brane", ein Laboratorium für Medienstrategien, gegründet. KR+cF wurde bereits mit wichtigen internationalen Medienkunstpreisen ausgezeichnet.

On the Expressive Potential of the Computer Controlled Human Face

Arthur Elsenaar / Huge Harry / Remko Scha

Algorithmic art production is aesthetically superior to the traditional methods which rely on the creative imagination of human persons. It is not constrained by the conventional ideas of human artists, or tainted by their often embarrassing ulterior motives. The Institute of Artificial Art is therefore committed to the development of fully automatic algorithmic art in all artistic genres.

In music and visual art, the success of the algorithmic approach is well established now; we have developed several algorithms which create rewarding works of art without requiring human interference. Applying the algorithmic approach to the production of computer-controlled dance and theatre performances is, however, still a major challenge: it requires a close collaboration between machines and human persons, in which the human participants cannot seize control of the situation to use it for their own expressive purposes. We intend to solve this problem by developing technological methods for controlling the movements of the muscles of human dancers and actors in complete detail.

Pilot experiments have shown that the body part which constitutes the most important human display device is the human face. Initially we have therefore focused our research completely on the computational control of the human facial muscles.

Algorithmische Kunstschöpfung ist den traditionellen Methoden, die sich auf die kreative Imagination von Menschen verlassen, ästhetisch überlegen. Sie wird nicht durch die konventionellen Ideen menschlicher Künstler beschränkt oder durch deren oft unbequeme höhere Motive verfälscht. Das Institute of Artificial Art hat sich daher der Entwicklung vollautomatischer algorithmischer Kunst in allen künstlerischen Genres verschrieben.

In der Musik und den visuellen Künsten ist der algorithmische Ansatz mittlerweile gut etabliert; wir haben zahlreiche Algorithmen entwickelt, die künstlerisch anspruchsvolle Werke ohne menschliche Eingriffe generieren. Die Anwendung des algorithmischen Ansatzes auf die Produktion computergesteuerter Tanz- und Theaterperformances ist hingegen noch immer eine große Herausforderung: Sie verlangt nach einer engen Zusammenarbeit zwischen Mensch und Maschine, wobei die menschlichen Mitwirkenden nicht die Kontrolle über die Situation erlangen dürfen, um sie nicht für ihre subjektiven expressiven Zwecke auszunützen. Wir haben vor, dieses Problem zu lösen, indem wir technologische Methoden für die detaillierte Überwachung und Steuerung der Muskeln der menschlichen Tänzer und Schauspieler entwickeln.

Pilot-Experimente haben erbracht, daß das wichtigste Display menschlichen Ausdrucks das Gesicht ist, weshalb wir anfänglich auch unsere Forschung gänzlich auf die computergestützte Steuerung der menschlichen Gesichtsmuskeln ausgerichtet haben.

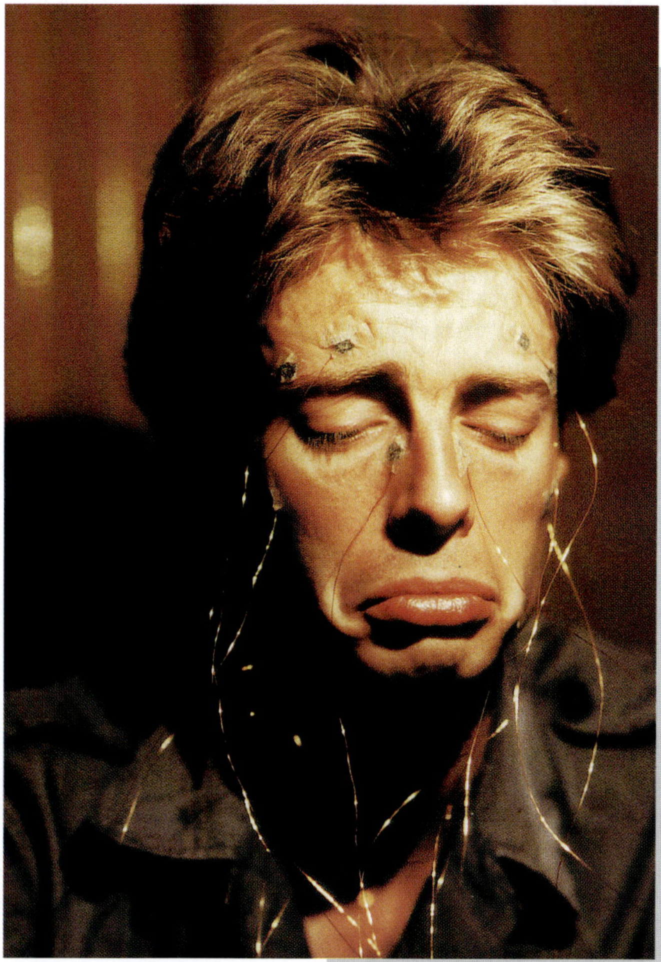

Huge Harry (01) is a commercially available voice synthesis machine, designed at the MIT Speech Laboratory. Currently, he works as a researcher and a spokes-machine at the Institute of Artificial Art in Amsterdam. He has also performed as a singer. Recently, Huge Harry has also started to work as a political activist, trying to achieve equal rights for computers.

Arthur Elsenaar (NL) spent most of his youth building ultra-high frequency radio and television transmitters that were used illegally by several hundred pirate stations throughout the Netherlands. In the early nineties he studied Cultural Engineering at the Fine Art Academy Minerva in Groningen.

Remko Scha (NL) was trained as a physicist and has worked on Natural Language Processing at Philips Research Labs (Eindhoven, the Netherlands) and BBN Labs (Cambridge, Mass., USA). Currently, he is head of the Computational Linguistics Department at the University of Amsterdam.

Huge Harry (01) ist eine am MIT Speech Laboratory entwickelte kommerziell erhältliche Sprachsynthesemaschine. Derzeit arbeitet er als Forscher und Sprechautomat am Institute of Artificial Art in Amsterdam. Er ist auch schon als Sänger aufgetreten. In letzter Zeit engagiert sich Huge Harry als politischer Aktivist im Kampf um Gleichberechtigung für Computer.

Arthur Elsenaar (NL) hat den größten Teil seiner Jugend damit verbracht, UHF-Radio und TV-Sender zu basteln, die von mehreren hundert illegalen Piratensendern in den ganzen Niederlanden verwendet wurden. In den frühen neunziger Jahren studierte er Cultural Engineering an der Minerva-Kunstakademie in Groningen.

Remko Scha (NL) ist gelernter Physiker und hat im Bereich des Natural Language Processing bei den Philips Research Labs (Eindhoven, NL) und den BBN Labs (Cambridge, Mass., USA) gearbeitet. Derzeit ist er Leiter des Computational Linguistics Department der Universität Amsterdam.

THE JEWELRY BOX

MARK MADEL

"The Jewelry Box" is designed to mirror a long-term relationship. It will pass through several stages in it's lifetime, and, in doing so, it's message and behavior will change. The amount of time it takes to pass from one stage to the next (hence the complete lifespan of the relationship) is, for the most part, in the hands of the owner: by the number of times the box is opened and the time that has passed since the last opening.

In effect, the artwork is about investment; both in precious objects (jewelry and/or artwork), as well as in relationships (human to human and/or human to technology); and the behavior of the object (e. g. the content of the messages) reflect this concept.

At the beginning of each stage, the piece estimates the maximum amount of times it can be opened before passing to the next stage, and the perfect length of time between openings. Ideally, each time the owner opens the box counts as one time only; but if it is opened too frequently or too seldomly, then "penalties" will be incurred, and the behavior of the object might change. Penalties are equivalent to "real" openings of the box. For example: the piece has not been opened for a month. The next time it is opened, 100 penalties are incurred. "The Jewelry Box", having been without the owner's attention for a month, has decided to add on an extra 100 times opened, thus hastening the point at which this stage of the relationship will end. The message from the piece also changes (e. g. it becomes "angry"), and this will influence the ideal opening time.

There will always be at least one perfect length of time to wait before opening the box (thereby getting no extra penalties and keeping the piece "content"). It might be a day, 4 days, a week, or longer; it will vary from stage to stage, and by the "mood" of the relationship at that particular mo-

Fig.3 *Diagram of inside of Jewelry Box - Sound Version (top of*

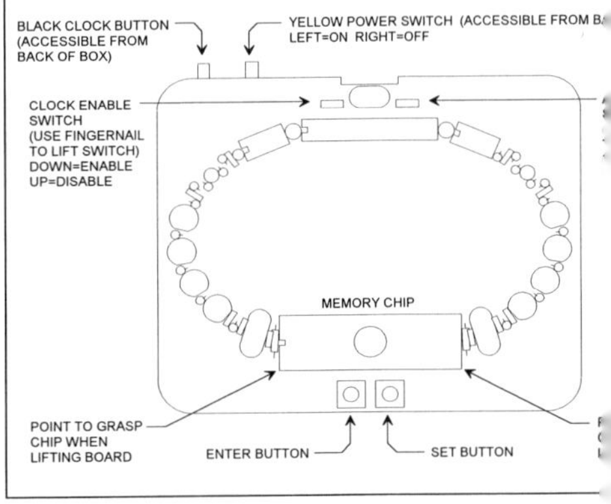

„The Jewelry Box" wurde entworfen, um eine langdauernde Beziehung zu reflektieren, die etliche Phasen im Laufe ihres Lebens durchmacht. Dabei verändert die Box sowohl ihr Verhalten als auch ihre Botschaft. Wie lange es dauert, von einer Etappe zur nächsten zu gelangen (woraus sich die gesamte Lebensspanne der Beziehung ergibt), liegt weitgehend in den Händen der Besitzer: Es hängt davon ab, wie oft die Box geöffnet wird und wieviel Zeit seit der letzten Öffnung vergangen ist. Bei dem Werk geht es inhaltlich um Investitionen in wertvolle Objekte (Schmuck und Kunst), aber auch in Beziehungen (von Mensch zu Mensch und von Mensch zur Technologie), und das Verhalten des Objekts (z. B. der Inhalt der erscheinenden Nachrichten) reflektiert dieses Konzept.

Am Anfang jeder Etappe schätzt das Stück ab, wie oft es maximal geöffnet werden kann, bevor es zur nächsten Stufe übergeht, und welche die ideale Zeitspanne zwischen den einzelnen Öffnungen ist. Idealerweise zählt jedes Öffnen der Box seitens des Besitzers als eine Öffnung; wird die Box jedoch zu oft oder zu selten geöffnet, werden „Strafen" verhängt, und das Verhalten des Objekts kann sich verändern. Diese „Strafen" sind Äquivalente zu echten Öffnungen der Box: Wenn der Besitzer seine Box beispielsweise einen Monat lang nicht be-

Fig.4 *Diagram of inside of Jewelry Box - Sound Version (bottom of board)*

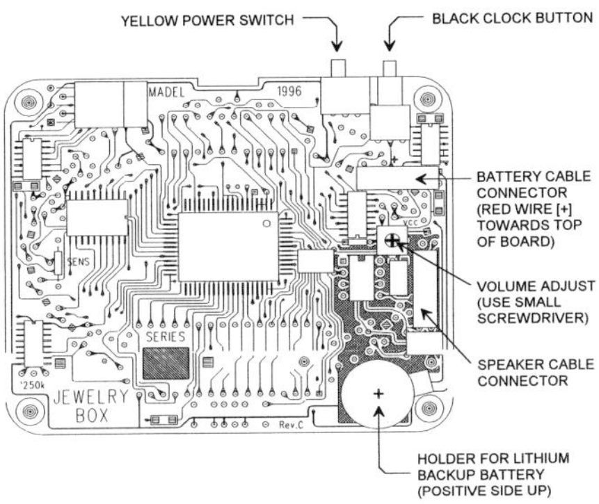

ment. The Penalty Count always reflects how many times the box considers itself to have been opened at that particular stage. When a new stage is started, the Penalty Count is reset, and new parameters are calculated according to the attention of the owner during the previous stages.

Remember: at some junctures in the relationship it is best to open the box quite often, while at other times fewer penalties will be incurred by opening it rarely. The Jewelry Box could take 60 years to pass through all of it's stages to it's "finish", or it might take only 5. It is up to the owner to gauge the impact of his/her behavior on the piece, at any time, in each particular stage.

achtet oder öffnet, kann die Box hundert „fiktive" Öffnungen strafweise hinzuzählen und so früher jenen Zeitpunkt herbeiführen, an dem diese Stufe der Beziehung endet. Die Nachrichten der Box passen sich entsprechend an (die Box wird sozusagen „ärgerlich"), und dies beeinflußt ebenfalls die idealen Öffnungszeiten. Es gibt immer zumindest eine optimale Zeitspanne, die vor der nächsten Öffnung der Box abzuwarten wäre (damit die Box „zufrieden" bleibt und keine „Strafpunkte" verhängt), und das kann ein Tag sein, vier Tage, eine Woche oder länger. Sie variiert von Ebene zu Ebene und hängt von der „Stimmung" der Beziehung zum gegebenen Zeitpunkt ab. Der „Strafzähler" zeigt an, wie oft die Box glaubt, auf dieser speziellen Ebene geöffnet worden zu sein. Wird eine neue Stufe begonnen, wird der Strafzähler zurückgesetzt, und neue Parameter werden in Abhängigkeit von der Aufmerksamkeit des Eigentümers während der früheren Phasen errechnet.

Es sollte beachtet werden, daß es während mancher Phasen in der Beziehung am besten ist, die Box ziemlich oft zu öffnen, während es zu anderen Zeiten weniger Strafpunkte gibt, wenn man sie nur selten öffnet. Es kann 60 Jahre dauern, bis man durch alle Etappen der Jewelry Box bis an ihr Ende gelangt ist, aber auch nur fünf. Und es obliegt dem Besitzer abzuschätzen, welche Folgen sein Verhalten auf das Stück hat – zu jeder Zeit, in jeder Etappe.

Mark Madel (USA), born 1957. 1982 BFA California Institute of the Arts. He is a conceptual artist who has been creating interactive electronic objects since 1980. Also inventor of the first portable motion-control system for the film industry in 1983. Exhibitions in the USA and Europe since 1982.

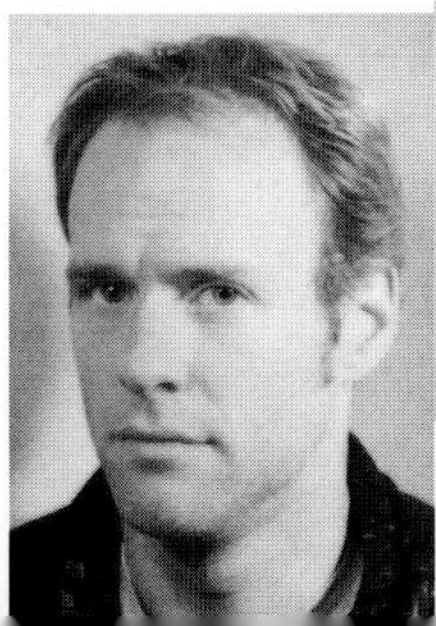

Mark Madel (USA), geb. 1957, erwarb 1982 einen BFA am California Institute of the Arts. Als Konzeptkünstler kreiert er seit 1980 interaktive elektronische Objekte. Außerdem hat er das erste tragbare Motion-Control-System für die Filmindustrie erfunden (1983). Seit 1982 Ausstellungen in den USA und Europa.

Life Journey

Jon Berge

"Life Journey" is an interactive installation which confronts participants/ viewers with the confinement imposed by a wheelchair. The piece contrasts the immobility of the participant, sitting motionless in the chair (which is attached to the floor) with the illusion of high-speed travel, created by projections onto the walls on either side, accompanied by the sounds of movement and the force of simulated wind from an industrial-sized floor fan.

The media employed to create the installation include a video component, a motorized wheelchair, and an industrial fan. The piece is designed so that one enters into a dark space with a single, illuminated wheelchair in the center. The LED sensor triggers as the viewer sits in the wheelchair, the wheels begin to spin, video projectors suspended from the ceiling on both sides of the chair are activated, as is an industrial fan located directly in front of the chair. The video projected on both walls consists of images of movement through different environments, that have a sunflower moving through the video frame. Twelve different paths taking you through environments as diverse as wooded paths to shopping malls at various speeds were burned on to a laser disc with stereo sound. The laser disc allows the player to randomly select the different sequences to be projected when it is triggered. My intention was to provide the participant/viewer with information to help them understand that a wheelchair, while seemingly decreasing one's mobility, may in some ways accelerate the processes by which one determines the choice of the path one's life journey takes. This vicarious "journey" continues as long as the participant remains seated in the wheelchair and stops as soon as she/he stands up. The viewer/participant is placed within the virtual environment representing the

„Life Journey" ist eine interaktive Installation, die den Mitwirkenden und Betrachter mit den Beschränkungen konfrontiert, die ein Rollstuhl mit sich bringt. Das Stück stellt den Kontrast zwischen der Immobilität des Mitwirkenden, der bewegungslos in einem am Boden fest montierten Rollstuhl sitzt, und der Illusion einer Reise mit hoher Geschwindigkeit dar, wobei der Reiseeindruck durch Projektionen auf beiden Seitenwänden in Verbindung mit Bewegungsgeräuschen und simuliertem Fahrtwind aus einem großen Industriegebläse hervorgerufen wird.

Die in der Installation eingesetzten Medien umfassen einen Video-Teil, den elektrischen Rollstuhl und den Industrieventilator. Die Aufstellung erfolgt so, daß man einen dunklen Raum betritt, in dessen Mitte der einzelne beleuchtete Rollstuhl steht. Sobald sich der Besucher setzt lösen LED-Sensoren die Installation aus: Die Räder beginnen sich zu drehen, die zu beiden Seiten des Rollstuhls von der Deck hängenden Projektoren werden eingeschaltet, der Ventilator direkt vor dem Rollstuhl läuft an. Das auf beide Wände projizierte Video zeigt Bilder einer Fahrt durch unterschiedliche Landschaften, wobei sich immer eine Sonnenblume durch das Bild bewegt. Zwölf verschiedene Strekken sind befahrbar, vom Waldweg bis zur Einkaufsstraße. Sie wurden mit wechselnder Geschwindigkeit und mit Stereo-Ton auf Laser-Disk gebrannt, was einen zufalls gesteuerten Zugriff auf einzelne Sequenzen ermöglicht.

Mein Ziel war es, den Mitspieler/Betrachter mit Information zu versorgen, die in ihm das Verständnis dafür wecken soll, daß ein Rollstuhl – wenn er auch scheinbar die persönliche Mobilität verringert – auf gewisse Weise auch jene Prozesse beschleunigen kann, durch die man die Wahl seines Lebensweges trifft. Diese zufällige Reise geht so lange wei-

ability to move from one space to another with the understanding that individuals who live with a wheelchair have limited mobility. While the participant remains seated, the wheelchair becomes a contradiction of motion. This juxtaposition of mobility with immobility heightens one's awareness of the loss of control that one feels when confined.

ter, solange der Besucher im Rollstuhl sitzt, und hört auf, wenn er sich erhebt. Der Betrachter wird in ein virtuelles Umfeld aus Bewegungserfahrungen gestellt, das ihm die Erkenntnis vermittelt, daß Personen, die auf den Rollstuhl angewiesen sind, in ihrer Mobilität eingeschränkt sind. Während der Benutzer sitzen bleibt, wird der feste Rollstuhl zu einem Widerspruch zur scheinbaren Bewegung, und diese Gegenüberstellung aus Bewegung und Unbeweglichkeit erhöht das Bewußtsein für den Kontrollverlust, den man empfindet, wenn die persönliche Mobilität eingeschränkt wird.

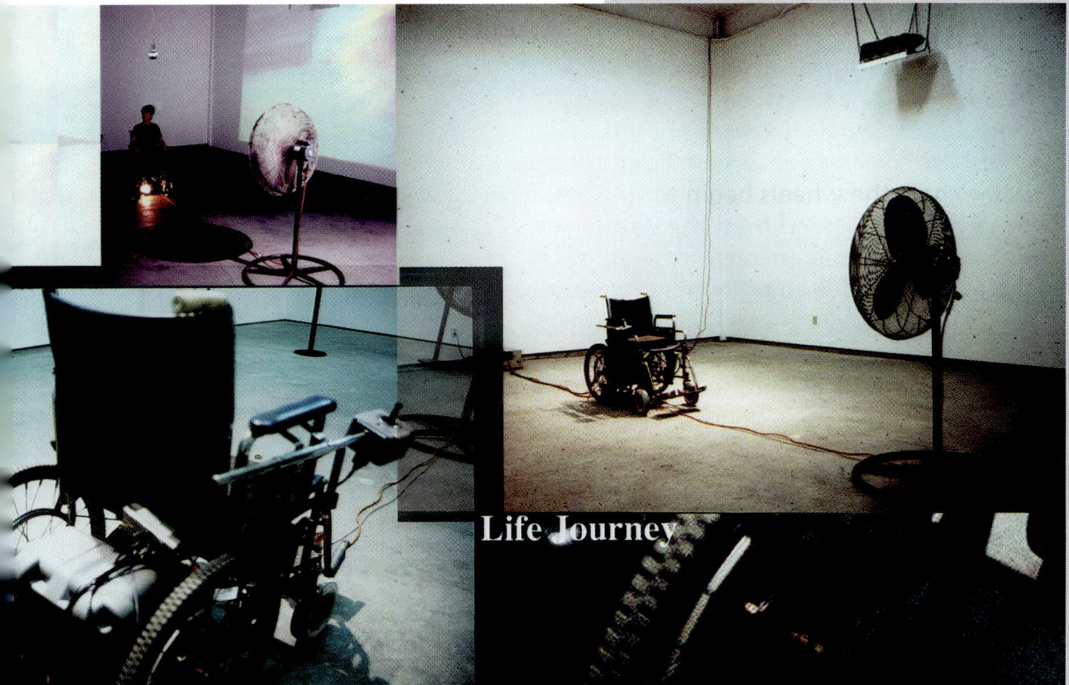

Jon Berge (USA), born, 1960, received a BS in psychology from the Wright State University and a MFA from the Ohio State University. Living in a wheelchair, his experience of limitation has made him focus in his work on educating and encouraging participants to examine the relevance of art and accessibility in their own lives.

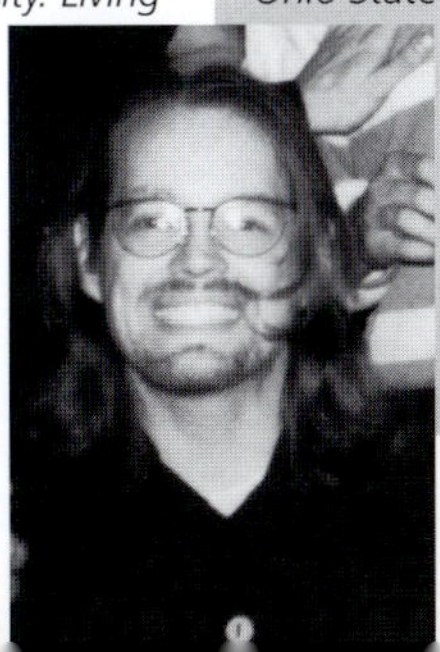

Jon Berge (USA), geb. 1960, graduierte zum BS aus Psychologie an der Wright State University und zum MFA an der Ohio State University. Seine Erfahrung mit den Beschränkungen, denen er als Rollstuhlfahrer unterliegt, hat ihn angeregt, das Publikum einzuladen, die Bedeutung von Kunst und unbehindertem Zugang im eigenen Leben zu untersuchen.

Akt

Stefan Zeyen

"Akt/Nude" is a work in progress. The resulting work is a video showing the author in a standing position and revolving constantly around his axis. Its realization started in the context of the Inter Act! exhibition at Wilhelm Lehmbrück Museum in Duisburg. The work was realized with participation from the audience on May 10, 12 and 13, 1997 in the rooms of the museum. At certain times visitors (most of them amateurs in drawing) were asked to realize a drawing of the author. Each drawing session (one full turn in the video) consists of eight people placed at the same distance from one other. Depending on the response from the public, one to three sessions were realized per day. A total of fifty drawings had to be completed. The drawings were digitized and a morph program on the Macintosh platform was used to create the missing frames between the neighboring drawings. For one full turn, four hundred frames were generated out of the original eight drawings. The goal was to update the video after each session and to display it with a video beamer (life size), enabling the participants to watch the result with a minimum of delay. With each session the video grew sixteen seconds to a total length of one minute and fifty six seconds.

„Akt" ist ein Work-in-Progress. Das daraus resultierende Video zeigt den Autor in stehender Position, wie er sich ständig um seine Achse dreht. Die Realisierung dieser Arbeit erfolgte im Kontext der „Inter Act!"-Ausstellung im Wilhelm Lehmbrück Museum in Duisburg. Das Werk wurde unter Mitwirkung des Publikums am 10., 12. und 13. Mai 1997 in den Räumen des Museums realisiert. Zu festen Zeiten wurden Ausstellungsbesucher – die meisten davon zeichnerische Amateure – eingeladen, eine Zeichnung des Autors zu produzieren. An jeder dieser Zeichnungssitzungen (eine volle Umdrehung am Video) nahmen acht Leute teil, die im gleichen Abstand voneinander plaziert wurden. Je nach der Reaktion der Besucher wurden ein bis drei Sitzungen pro Tag abgehalten. Insgesamt mußten 50 Zeichnungen erstellt werden. Diese wurden digitalisiert, ein Morphing-Programm auf dem Macintosh lieferte die zwischen den Zeichnungen fehlenden Kader. Für eine volle Umdrehung wurden insgesamt 400 Kader aus den jeweils acht Originalzeichnungen generiert. Ziel war ein Update des Videos unmittelbar nach jeder Sitzung und seine Projektion in Lebensgröße, so daß die mitwirkenden Besucher das Ergebnis ihrer Tätigkeit mit einem Minimum an Zeitverzögerung zu Gesicht bekamen. Bei jeder Sitzung „wuchs" das Video um 16 Sekunden zu einer Gesamtlänge von einer Minute und 50 Sekunden.

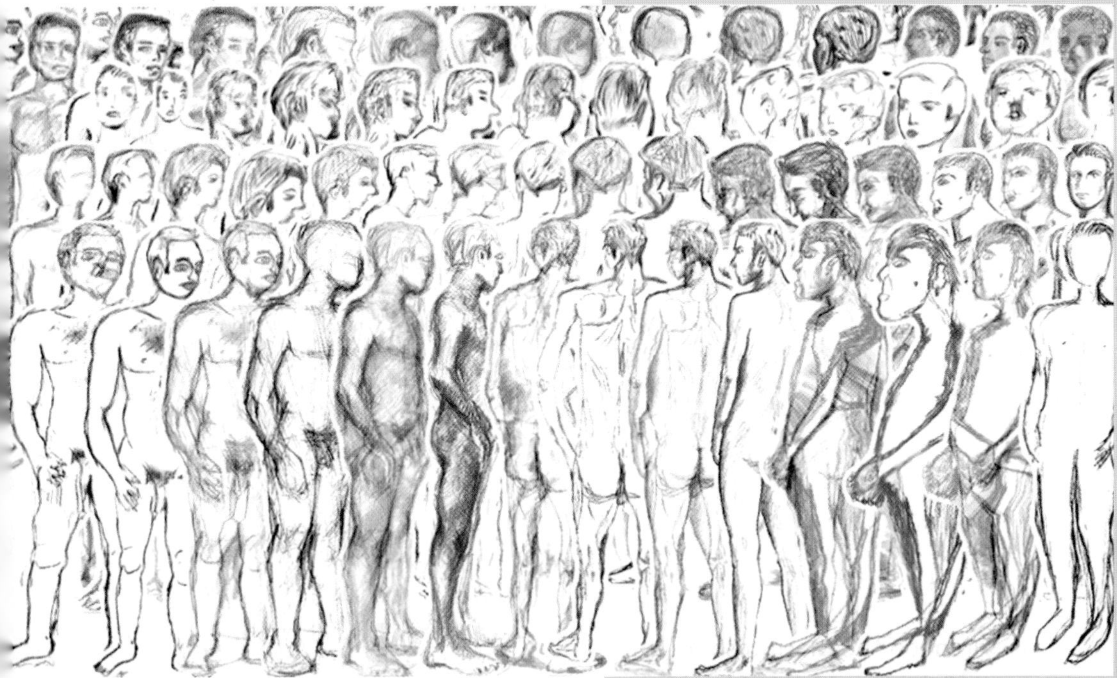

Stefan Zeyen (D) is a freelance artist based in Berlin. He studied philosophy, painting and fine arts in Germany, Belgium and Holland. His works have been exhibited in Germany, Holland, Belgium, Great Britain and Hungary.

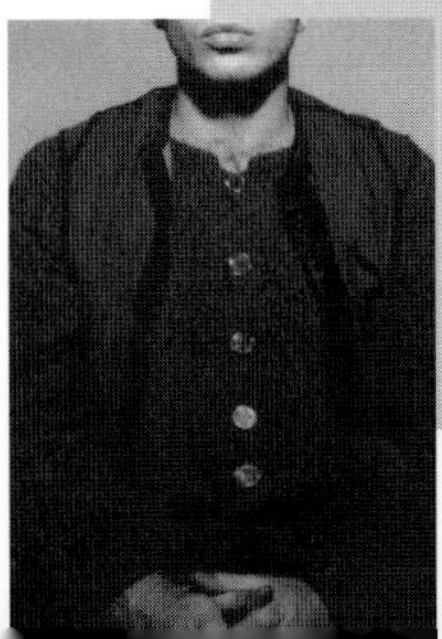

Stefan Zeyen (D) studierte Philosophie, Malerei und Bildende Kunst in Deutschland, Belgien und Holland. Er lebt als freier Künstler in Berlin. Seine Arbeiten wurden in Deutschland, Holland, Belgien, Großbritannien und Ungarn ausgestellt.

The Encyclopedia of Clamps

Barminski / Lewin / Hesketh

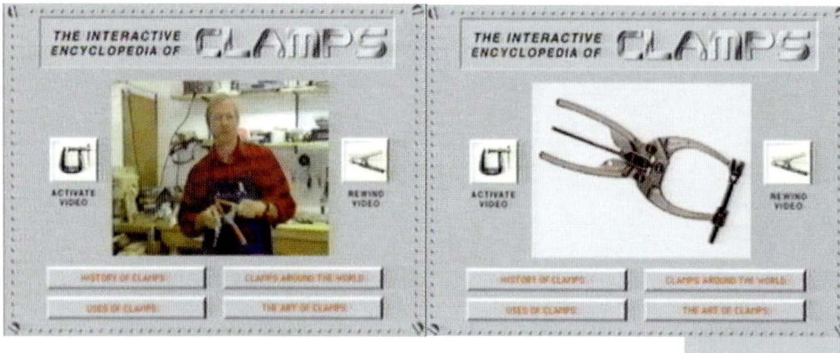

The "Encyclopedia of Clamps" will set a new standard for digital media. In this second title from the team at De-Lux'O, the world of experience-oriented art is explored in CD-ROM. Using digital media to create original works of interactive art, the design team at De-Lux'O sets forth as its goal to challenge the preconceived notions of interactivity, content and technology. Utilizing a 3-D environment, the user can visit over 30 exhibition rooms that contain real works of interactive art – not corporate diluted pseudo art. Using digital media to create original work, Bill Barminski, Webster Lewin and Jerry Hesketh (De-Lux'O, Inc.) are challenging the preconceived notions of interactivity, content and technology. The interactive "Encyclopedia of Clamps" will be ideally suited for the current (boring) CD-ROM market, because not only does it offer the finest in experience-oriented art on CD-ROM, but it is also the world´s first, and most complete reference guide to clamps and clamp-like devices. What could be better?

Utilizing a 3-D environment as an interface, the user can visit various exhibition rooms each containing a unique art experience. The core of this program are two intersecting hallways lined with Bill Barminski´s paintings through which the user can enter with a click of the mouse.

Die „Encyclopedia of Clamps" setzt einen neuen Standard im Bereich der digitalen Medien. In diesem zweiten Titel des Teams von De-Lux'O wird die Welt der erfahrungsorientierten Kunst auf CD-ROM erforscht. Dadurch, daß die Design-Mannschaft von De-Lux'O originelle interaktive Kunstwerke mittels digitaler Medien kreiert, will sie die vorgefaßten Meinungen über Interaktivität, Inhalt und Technologie hinterfragen. In einem dreidimensionalen Umfeld kann der Besucher mehr als 30 Ausstellungsräume besuchen, die echte Werke interaktiver Kunst enthalten – nicht einfach nur gängige, verwässerte Pseudo-Kunst. Durch den Einsatz digitaler Medien zur Schaffung origineller Arbeiten fordern Bill Barminski, Webster Lewin und Jerry Heskth (De-Lux'O, Inc.) die vorgefaßten Ansichten über Interaktivität, Inhalt und Technologie heraus. Die interaktive „Encyclopedia of Clamps" wird für den derzeitigen (langweiligen) CD-ROM-Markt ideal sein, weil sie nicht nur das Feinste an erfahrungsorientierter Kunst auf CD-ROM anbietet, sondern auch der erste – und umfassendste – Führer der Welt zu allen Arten von Schraubzwingen und schraubstockähnlichen Geräten ist. Was könnte besser sein?

Mittels eines 3D-Interface kann der Besucher zahlreiche Ausstellungsräume betreten, in denen jeweils eine einzigartige Kunsterfahrung auf ihn wartet. Kernstück dieses Programms sind zwei einander kreuzende Hallen, geschmückt mit Bill Barminskis Gemälden, in die der Benutzer mit einem Mausklick eintreten kann.

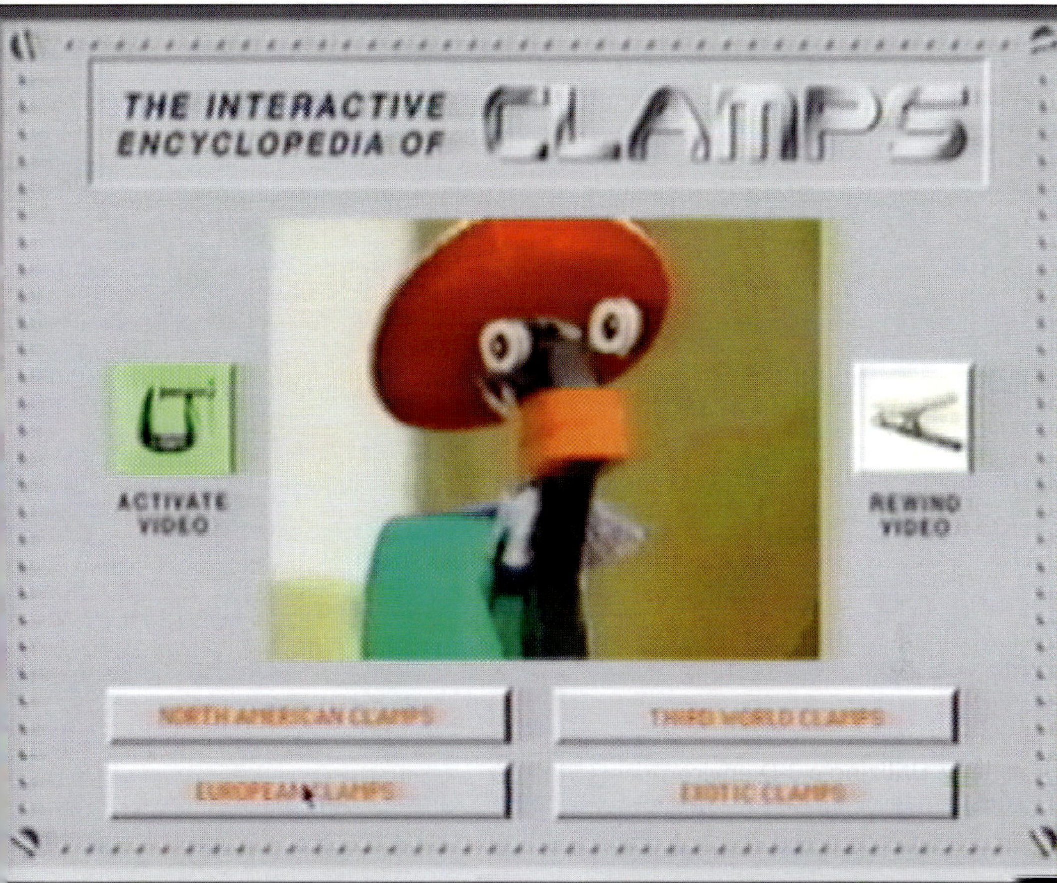

Bill Barminsky (USA) is a painter, whose work appears on several album covers, in feature films and music videos. His work has been featured in Art Forum International.

Webster Lewin (USA) is a filmmaker and screenwriter and has been a student of Martin Scorsese. His films and videos have appeared on PBS's "American Playhouse", MTV and on the Bravo Cable Network.

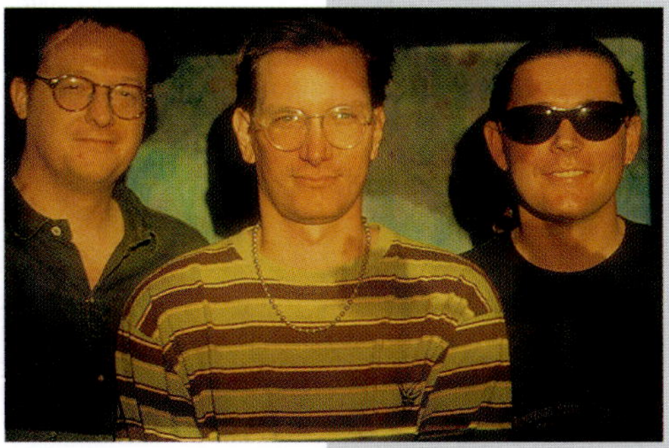

Jerry Hesketh (USA) was one of the foremost pioneers in the development of CD-ROM technology for Apple Computer Corporation. He is currently Director of Technology for Virgin Sound and Vision.

Bill Barminsky (USA) ist Maler, dessen Arbeiten auf zahlreichen Plattencovern, in Filmen und Musikvideos zu sehen sind. Über seine Arbeiten wurde im Art Forum International berichtet.

Webster Lewin (USA) ist Filmemacher und Drehbuchautor und war Schüler von Martin Scorsese. Seine Filme und Videos sind auf PBS's „American Playhouse", MTV und im Bravo Cable Network zu sehen.

Jerry Hesketh (USA) war einer der herausragenden Pioniere der CD-ROM-Technologie bei Apple Computer Corporation. Derzeit ist er Leiter der Technologieabteilung bei Virgin Sound and Vision.

Ottos Mops (trotzt)

KP Ludwig John / Bertram Quosdorf

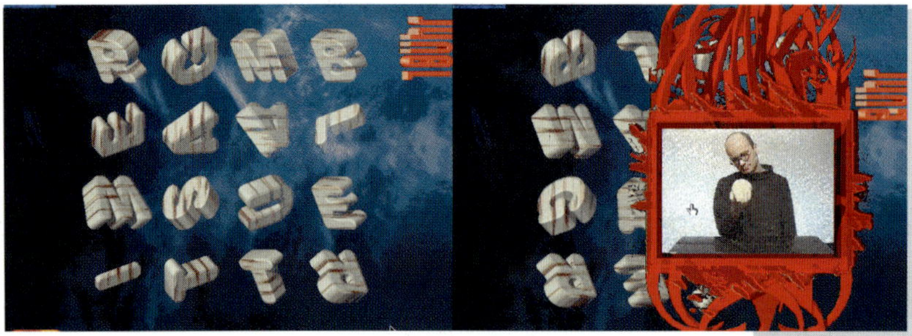

All of the interactive part of the CD-ROM is based on some 100 works by the Austrian word wizard Ernst Jandl, the self-proclaimed and widely acknowledged "avenger of language".

The CD-ROM includes many outrageous and sometimes even weird graphical wordplay games and a rather unique scoring system linked to each of the approximately 70 different settings. For example, there is a variation of the old arcade game "Breakout" with words and sounds rather than bricks to shoot at.

These interactive adaptations stand between you and the JANDL: solve the riddles and you will find the master himself, accompanied by his dog, the "mops", i.e. pug. The background of this is that "Ottos Mops" ("Otto's pug") is Ernst Jandl's most famous work.

"Ottos Mops" presents the ultimate quest for the Jandl and his pug in a fierce battle against time.

At the beginning of each run, you are asked to enter your name through the keyboard. Press return and you are asked to enter your birth date in a dd-mm-yy format. Press return again, and your name will be scrambled. This is a consistent feature of Ernst Jandl's work and just a hint of what the game may have in store for you.

The course of the game is based on a poem entitled "Niederzählung" ("Count-down"). The "electronic" adaptation starts with the current system date and counts down to the moment of your birth. This

Der gesamte interaktive Part der CD-ROM baut auf gut hundert Werken des österreichischen Sprachzauberers Ernst Jandl auf, der sich selbst als „Rächer der Sprache" proklamiert und auch als solcher anerkannt ist.

Die CD-ROM enthält eine Vielzahl skurriler bis fast schon durchgedrehter graphischer Wortspiel-Spiele und ein ziemlich einzigartiges Punktesystem, das mit jeder der gut 70 verschiedenen Einstellungen verknüpft ist. So gibt es beispielsweise eine Variante des guten alten „Breakout"-Spiels, bei dem man sich aber den Weg durch Worte und Klänge statt durch Ziegel freischießen muß.

Diese interaktiven Adaptierungen stehen zwischen dem Spieler und dem JANDL: Löse die Rätsel, und du findest den Meister selbst, natürlich begleitet von seinem Mops. Hintergrund dessen ist, daß das Gedicht „Ottos Mops" wohl Jandls berühmtestes Werk ist.

„Ottos Mops" präsentiert die ultimative Suche nach dem Jandl und seinem Mops, in einem heißen Kampf gegen die Zeit.

Zu Beginn jeder Runde wird man aufgefordert, seinen Namen über die Tastatur einzugeben. Weiters wird das Geburtsdatum (im Format TT-MM-JJ) abgefragt. Nach Drücken der Eingabetaste werden die Buchstaben des eingegebenen Namens geschüttelt – ein durchgehender Faktor in Ernst Jandls Werk und ein kleiner Vorgeschmack auf das, was das Spiel noch alles bereit hält.

Der Spielverlauf wird von Ernst Jandls Gedicht „Niederzählung" bestimmt. Die „elektronische" Adaptierung beginnt mit dem derzeitigen Systemdatum und zählt bis zum Moment der Geburt des Spielers zurück. Dies bedeutet einerseits, daß man ein Leben lang Zeit hat, den Jandl zu fin-

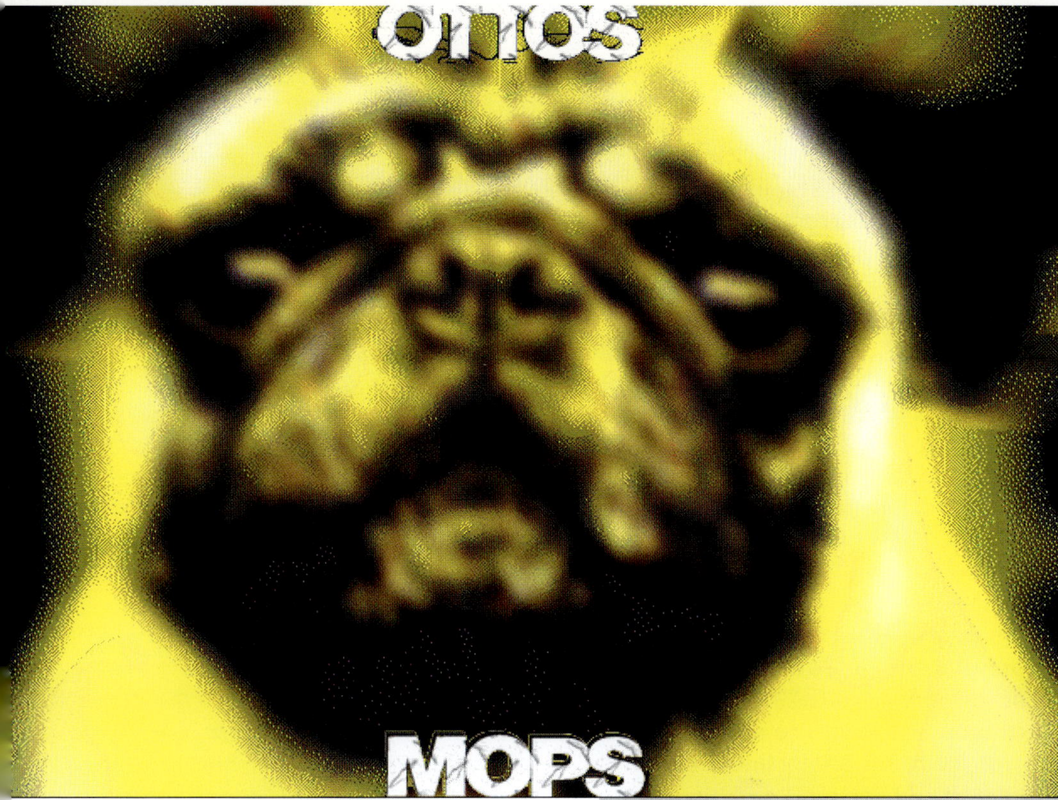

means that you have a lifetime to find the JANDL. But as in real life, the longer you play, the faster the time elapses. Finally, at the moment of your birth, the game is over – although only for the time being, of course. Your progress is indicated on the right part of the screen, and your secret referee occasionally has a comment to make on the course of your quest.

den, andererseits aber – und das ist wie im echten Leben – vergeht die Zeit umso schneller, je länger man spielt. Schließlich – im Moment der Geburt – ist das Spiel zu Ende, aber natürlich nur für dieses eine Mal. Der Fortschritt des Spielers wird im rechten Teil des Bildschirms angezeigt, und der geheime Schiedsrichter gibt von Zeit zu Zeit seinen Kommentar über den Fortschritt bei der Suche ab.

KP Ludwig John (D), born 1961, works as a freelance "interactivist" in Munich. He studied photography and art and media technology in Leipzig and Utrecht, and was co-founder of the Media Biennale Leipzig.

Bertram Quosdorf (D), born 1967, studied music in Dresden and worked in various ensembles and orchestras, and also as a theater music composer. He has been an ensemble member of the "statt-theater Fassungslos" since 1987, and much of his work focuses on texts by Ernst Jandl.

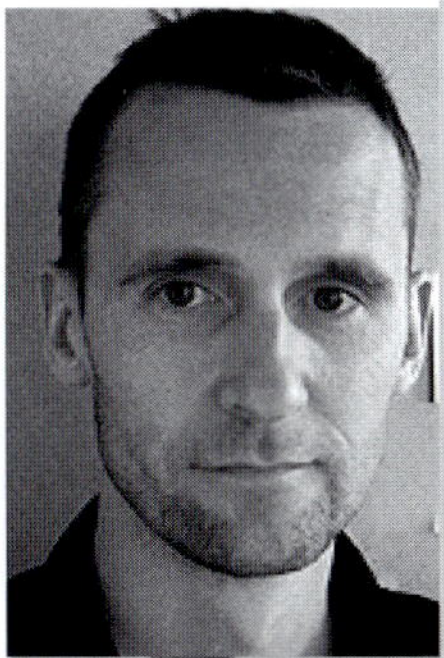

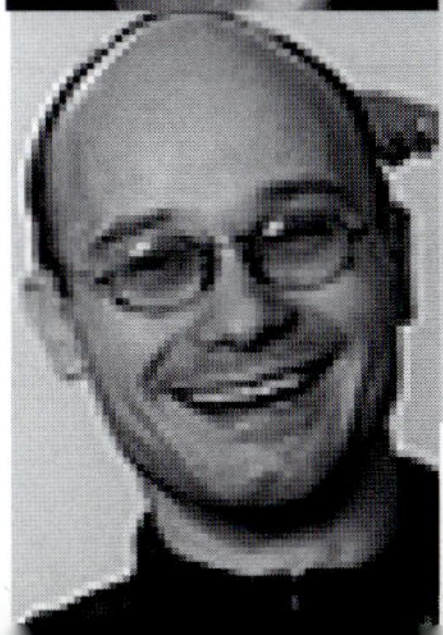

KP Ludwig John (D) geb. 1961, arbeitet als „Interaktivist" freiberuflich in München. Er studierte Fotografie und Kunst- und Medientechnologie in Leipzig und Utrecht und war Mitbegründer der MedienBiennaleLeipzig.

Bertram Quosdorf (D), geb. 1967,.studierte Musik in Dresden und arbeitete in verschiedenen Ensembles und Orchestern sowie als Komponist von Bühnenmusik. Seit 1987 ist er Ensemblemitglied im „statt-theater Fassungslos" und beschäftigt sich schwerpunktmäßig mit Texten von Ernst Jandl.

The Information Age

Jutta Kirchgeorg

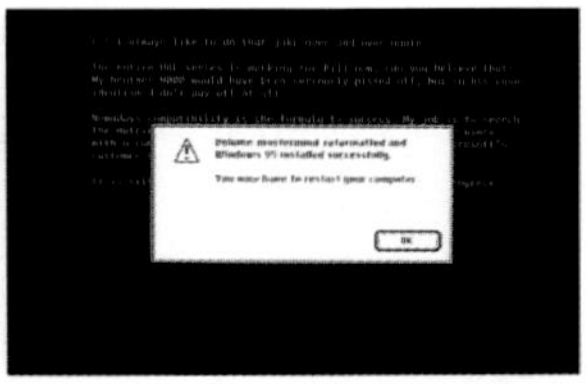

In "The Information Age," current technology is both subject matter and medium. Wrapped as an educational CD-ROM product, it stages a situation where the viewer experiences the computer's deficiency as an intelligent interface for human interaction and his gradually vanishing power of control over the machine. It is a hybrid project comprising two entities: an interactive Director movie, which functions as the primary user interface, and a website, which is being accessed from within the movie. The Director movie is an interactive presentation about major inventions and achievements in the field of electronic communications. The navigation within the piece is intentionally linear and fairly limited.

Apart from this "educational" presentation, the project also involves another level of functionality which becomes apparent only after a certain extent of user interaction: various counters monitor the user's interaction and trigger certain events which are beyond the user's control. At the beginning, the events are alerts and error messages, instructing the user to follow certain instructions in order to proceed within the piece. Parallel to the advancement towards newer technologies within the presentation, the user experiences problems of increasing frequency and fatality until control is taken completely out of his hands. Interaction turns into reaction, and finally the user is confronted with some intelligent entity which is communicating by typing chatty messages onto the screen. As the "entity" discloses all the data it has collected about the user and sends him hacking into the Pentagon webserver to shut down the Internet, the user finds himself a powerless marionette who has unknowingly handed over control to a machine.

In „The Information Age" ist die Technologie der Gegenwart sowohl Thema als auch Medium. Verpackt in eine Bildungs-CD-ROM inszeniert es eine Situation, in der der Betrachter das Versagen des Computers als intelligentes Interface für menschliche Interaktion und seinen eigenen schrittweisen Kontrollverlust über die Maschine erlebt. Es handelt sich um ein hybrides Produkt, das aus zwei Einheiten besteht: aus einem interaktiven Film, dem primären User-Interface, und aus einer innerhalb dieses Films zugänglichen Web-Site. Der Film ist eine interaktive Präsentation der wichtigsten Erfindungen und Errungenschaften im Bereich der elektronischen Kommunikation. Die Navigation in diesem Teil ist bewußt linear und eingeschränkt. Daneben umfaßt das Projekt eine zweite Funktionsebene, die erst nach einer gewissen Beschäftigung mit dem Werk in Erscheinung tritt: Verschiedene Zähler überwachen die Interaktion des Benutzers mit dem Programm und lösen Ereignisse aus, die sich der Kontrolle des Benutzers entziehen. Anfänglich handelt es sich dabei um Fehlermeldungen, die den User auffordern, bestimmte Anweisungen zu befolgen, um im Programm fortzufahren. Während der Benutzer sich zu immer neuen Technologien vorarbeitet, stößt er immer öfter auf ernste Probleme, bis er endgültig die Kontrolle verliert. Die Interaktion reduziert sich auf reine Reaktion, und zuletzt sieht sich der Benutzer mit einer intelligenten „Einheit" konfrontiert, die mit ihm kommuniziert, indem sie wortreiche Nachrichten auf den Schirm schreibt. Wenn die „Einheit" alle Daten enthüllt, die sie über den Benutzer gesammelt hat und ihn dazu zwingt, in den Webserver des Pentagons einzuloggen, um das Internet herunterzufahren, erkennt sich der Benutzer als machtlose Marionette, die ihre Steuerung unbewußt einer Maschine übertragen hat.

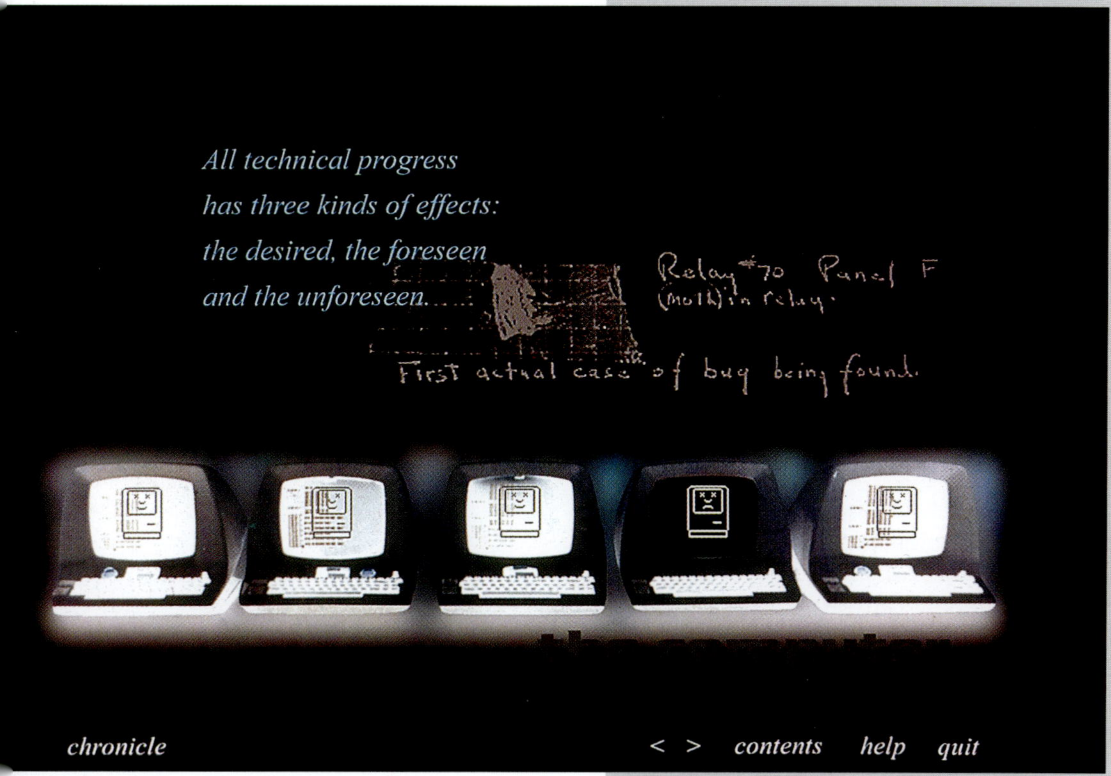

Jutta Kirchgeorg (D), born 1969, studied Communication Design at the Fachhochschule Rheinland Pfalz, majoring in Telecommunications. 1995 she won a Fulbright Scholarship to study Computer Art at the School of Visual Arts in New York City, where she focused on Interactive Multimedia and Telecommunications. This spring she completed the program with a MFA degree. The submitted work is her thesis project.

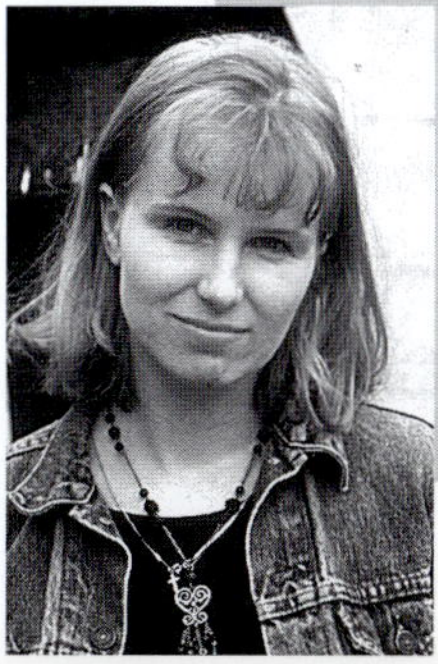

Jutta Kirchgeorg (D), geb. 1969, studierte Kommunikationsdesign mit Schwerpunkt Telekommunikation an der Fachhochschule Rheinland-Pfalz. 1995 erhielt sie ein Fulbright-Stipendium zum Studium der Computerkunst an der School of Visual Arts in New York City, wo sie sich auf Interaktive Multimedia und Telekommunikation konzentrierte. Sie schloß das Programm im Frühjahr 1997 mit dem MFA ab, das vorliegende Werk ist ihre Abschlußarbeit.

Ambitious Bitch

Marita Liulia

"Ambitious Bitch" is an update of western femininity at the turn of the millennium. I chose some of the main subjects that women's studies have brought into public discussion in recent years, and built up a multimedia presentation of 12 sections using different rhetoric and representations. I let the voices of high culture and pop, philosophy and cheap magazines talk about the same subjects simultaneously. The "buttonless" interface is specially designed for women and people who dislike computers.

It consists of 12 modules:

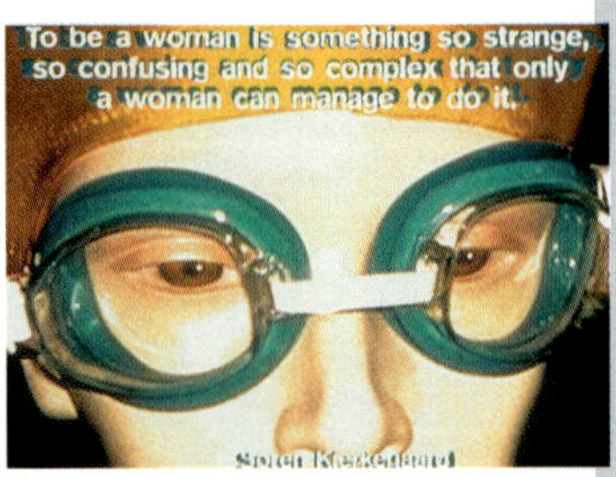

– Body-Art of Existence (questions the idea of the female body as a design object)

– Woman? Sex / Gender? (after this visually and philosophically heavily loaded section, you will not be so sure any more ...)

– Female Perversions (seven stories based on psychological case studies and some women I know).

– Erotic tales (six stories by Finnish authors describing female erotica)

– Ambitious Witch (fortune telling program)

– Waves of Modern, Postmodern & Feminisms (philosophy in a nut shell)

– Female Qualities? (questions stereotypical female characteristics)

– Traditional Wit (old and new proverbs about women)

– Ambitious Blonde (the dilemma of a perfect woman & the latest blonde jokes told by Herself)

– Fashion, Designers, Garment (authorities that create the female image)

– AB homepage (women & technology, equality) <www.edita.fi/bitch>

„Ambitious Bitch" ist eine auf den letzten Stand gebrachte Darstellung des westlichen Feminismus an der Jahrtausendwende. Ich habe einige der Hauptthemen ausgewählt, die der Feminismus in den letzten Jahren zum Thema öffentlicher Diskussion gemacht hat, und daraus eine Multi-Media-Arbeit in zwölf Sektionen mit unterschiedlicher Rhetorik und Darstellung gemacht. Hochkultur und Pop, Philosophie und billige Zeitschriften reden gleichzeitig über dieselben Themen. Das „knopflose" Interface wurde speziell für Frauen und für Leute entwickelt, die keine Computer mögen.

Das Stück besteht aus zwölf Modulen

– Body-Art of Existence (hinterfragt den weiblichen Körper als Design-Objekt)

– Woman? Sex/Gender? (Nach dieser visuell wie philosophisch stark befrachteten Sektion ist man sich da nicht mehr so sicher ...)

– Female Perversions (sieben Geschichten, die auf psychologischen Fallstudien und Frauen aus meinem Bekanntenkreis basieren)

– Erotic tales (sechs Geschichten übe weibliche Erotik finnischer Autoren)

– Ambitious Witch (ein Wahrsagerprogramm)

– Waves of Modern, Postmodern & Feminisms (Philosophie im kleinen)

– Female Qualities? (hinterfragt stereotype weibliche Charakteristiken)

– Traditional Wit (alte und neue Sprichwörter und Redensarten über Frauen)

– Ambitious Blonde (das Dilemma einer perfekten Frau und die neuesten Blondinenwitze, von ihr selbst erzählt)

– Fashion, Designers, Garment (Autoritäten, die das weibliche Image formen)

– AB homepage (Frauen und Technologie, Gleichheit) <www.edita.fi/bitch>

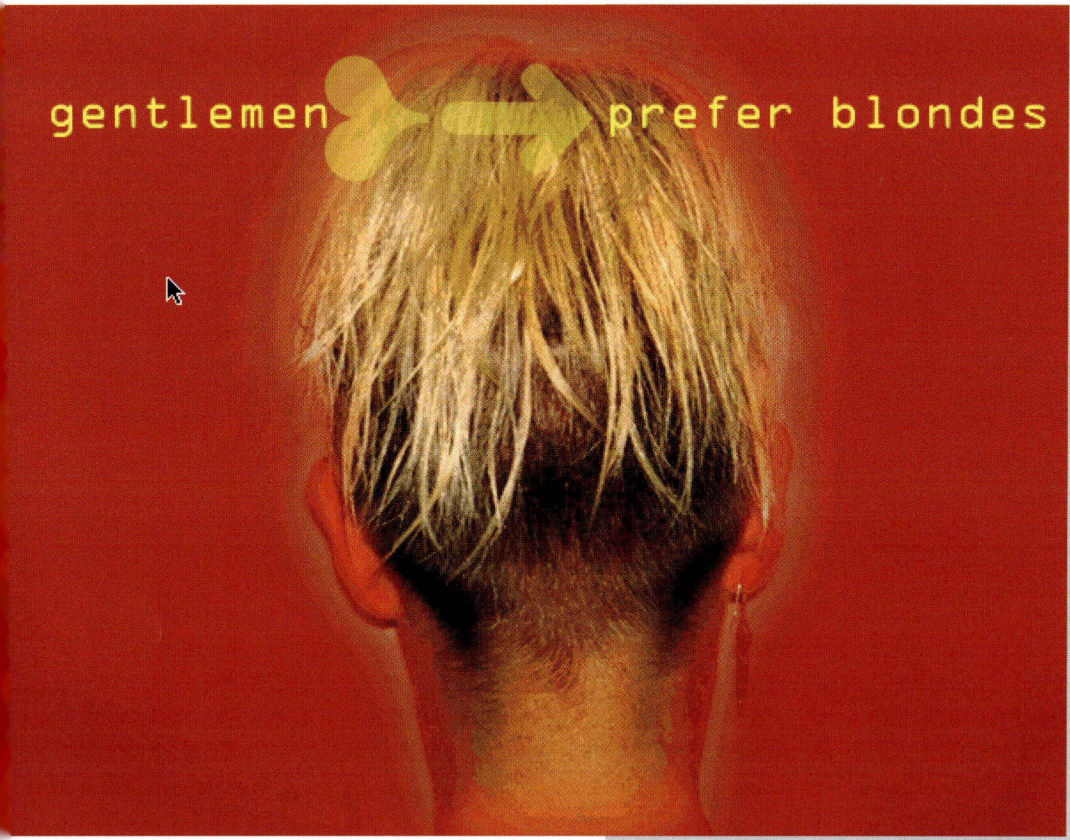

HONORARY MENTION

Marita Liulia (SF), born 1957. BA from Helsinki University and studies in drawing, painting and sculpture at University of Industrial Arts in Helsinki. Cooperation with light and sound designers Tarja Ervasti and Ilkka Volanen. Several exhibitions and performances in Scandinavia. Chairwoman of Muu r.y., a Helsinki-based organization for media art. Her CD-ROM productions were the first of their kind in Scandinavia.

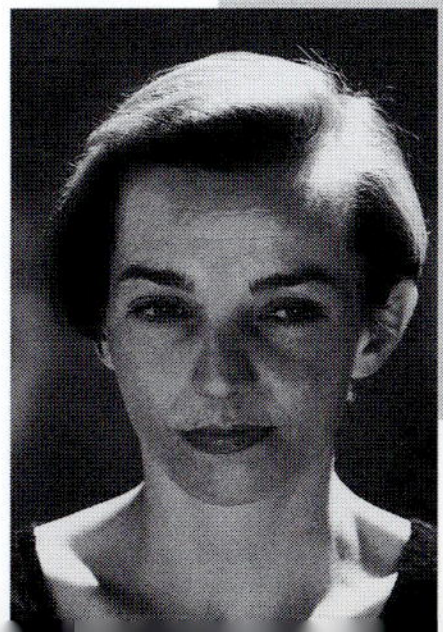

Marita Liulia (SF), geb. 1957, graduierte an der Universität Helsinki zum BA und studierte Zeichnung, Malerei und Skulptur an der University of Industrial Arts in Helsinki. Zusammenarbeit mit den Licht- und Tondesignern Tarja Ervasti und Ilkka Volanen. Zahlreiche Ausstellungen und Performances in Skandinavien. Vorsitzende von Muu r.y., einer in Helsinki ansässigen Medienkunstorganisation. Ihre CD-Rom-Produktionen waren die ersten ihrer Art in Skandinavien.

Venetian Deer

Die Veteranen

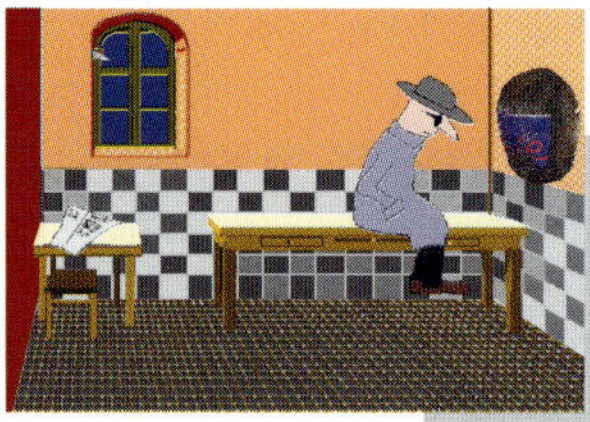

In the concept of the CD-ROM "Venetian Deer" (an anagram from "Die Veteranen"), the disc is no longer regarded as the real product, but rather the CD-ROM itself becomes the basis for multimedia (literally).

Various media were used for the production, the most diverse media could be used for presentation. To begin with, of course, there are the different computer platforms (MAC and Win). In addition, the musical works may also be played on a normal audio CD player.

But "Venetian Deer" does not work only on a personal computer. It has multiple connections to the net, which may be loaded via the Veteranen's homepage and further processed (http://www.systhema.de/veteranen). The work presented here may also be presented as an interactive live performance.

Taking interactive media work out of the narrow confines of the computer screen and turning it loose to affect "real life" is part of our global concept. For this reason, the data has been prepared and treated in such a way that it may be called for all kinds of different presentation forms.

A telephone book provides the observer with the possibility of calling up individual works directly and jumping to them from any position within the CD-ROM. As we presume that you are probably very busy, please permit us to suggest the following numbers to you as examples (type either the following number or letter codes on your keyboard, once the CD-ROM is open):

No. Code	Letter Code	Title
426	HAM	Hampel
622	MAC	Soundmachine
393	EYE	Eyeball
447	HIP	Hipkop
242	CHA	Chaos

Das Konzept der CD-ROM „Venetian Deer" (ein Anagram aus „Die Veteranen") sieht die Scheibe nicht mehr als das eigentliche Produkt, sondern die CD-ROM selbst ist die Basis für Multi-Media (im wörtlichen Sinne). Verschiedene Medien wurden zur Produktion verwendet, verschiedenste Medien kommen zur Präsentation in Frage. Als erstes natürlich die unterschiedlichen Computer-Plattformen (MAC und Win). Darüber hinaus sind die musikalischen Arbeiten auf dem normalen Audio-CD-Player abspielbar.

„Venetian Deer" arbeitet aber nicht nur am PC. Es hat mehrfache Anbindungen ans Netz, die über die Homepage der Veteranen bestückt und weiterbearbeitet werden können (http:// www.systhema.de/veteranen). Die vorgestellte Arbeit kann aber auch als interaktive Live-Performance aufgeführt werden.

Es entspricht unserem Gesamtkonzept, die interaktive mediale Arbeit aus dem engen Umfeld des Computer-Screens zu lösen und ins „richtige Leben" einwirken zu lassen. Die Daten sind daher so angelegt, daß sie für die unterschiedlichsten Präsentationsformen abrufbar sind.

Ein Telefon-Buch gibt dem Betrachter die Möglichkeit, gezielt einzelne Arbeiten aufzurufen und innerhalb der CD-ROM von jeder beliebigen Stelle aus dahinzuspringen. Da wir annehmen, daß Ihre Zeit sehr ausgefüllt ist, erlauben wir uns, Ihnen folgende Nummern als Beispiele vorzuschlagen (tippen Sie, wenn die CD-ROM geöffnet ist, auf Ihrer Tastatur folgende Ziffern- oder Buchstabencodes ein):

Zifferncode	*Buchstabencode*	*Titel*
426	*HAM*	*Hampel*
622	*MAC*	*Soundmac*
393	*EYE*	*Eyeball*
447	*HIP*	*Hipkop*
242	*CHA*	*Chaos*

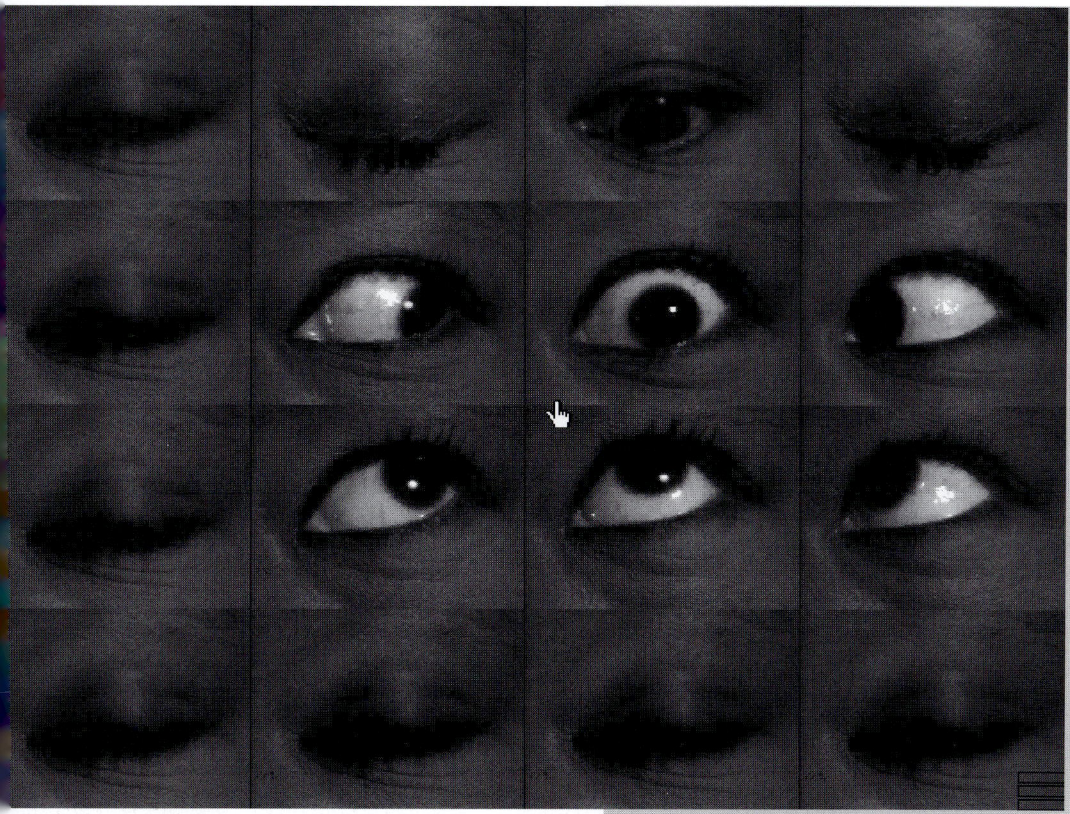

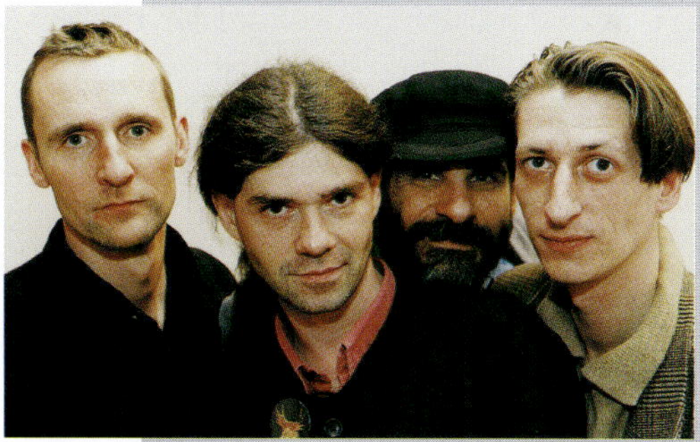

Die Veteranen (D) is a group of independent media artists: Stephan Eichhorn, teacher at the Kunsthochschule Leipzig, Tjark Ihmels, freelance media artist, KP Ludwig John, freelance media artist, and Michael Touma, teacher at the Kunsthochschule Leipzig. In addition to their work for „Die Veteranen", all the members of the group are also involved in art projects of their own.

Die Veteranen (D) ist eine Gruppe der unabhängigen Medienkünstler Stephan Eichhorn, Lehrer an der Kunsthochschule Leipzig, Tjark Ihmels, freischaffender Medienkünstler, KP Ludwig John, freischaffender Medienkünstler, und Michael Touma, Lehrer an der Kunsthochschule Leipzig. Zusätzlich zu ihrer Arbeit bei „Die Veteranen" sind die Gruppenmitglieder mit eigenständigen Kunstprojekten beschäftigt.

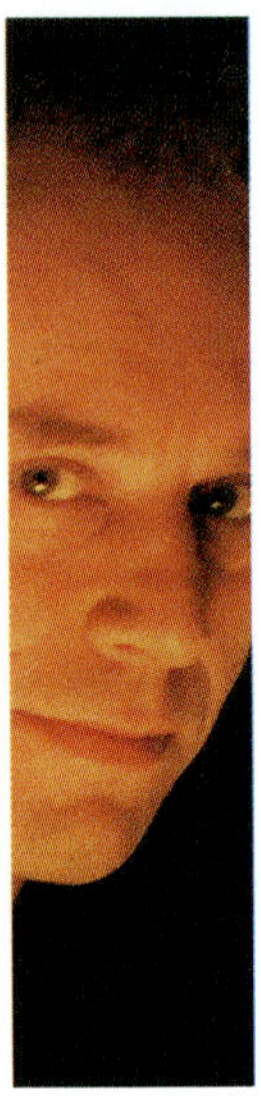

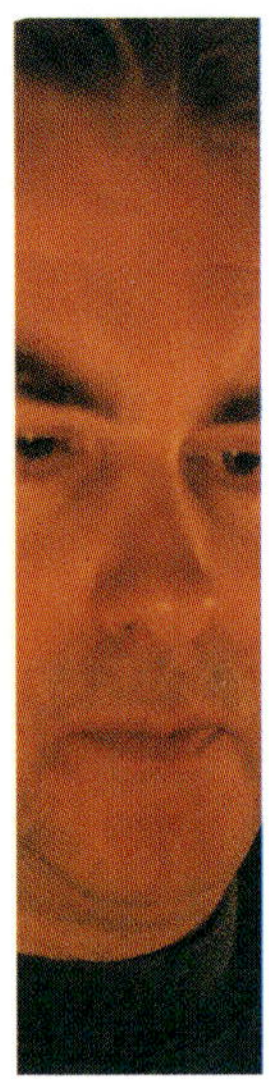

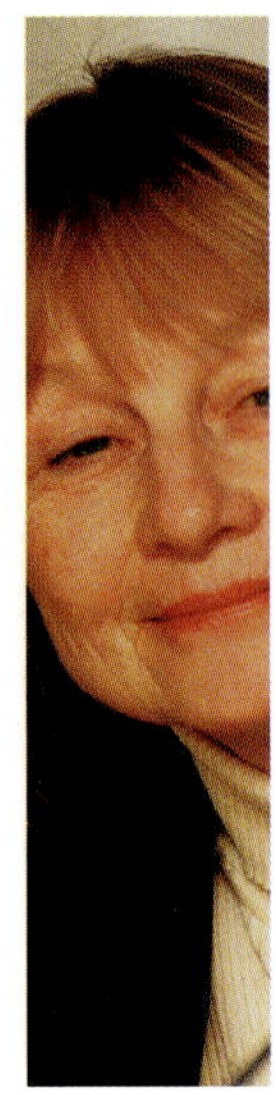

COMPUTER ANIMATION JURY

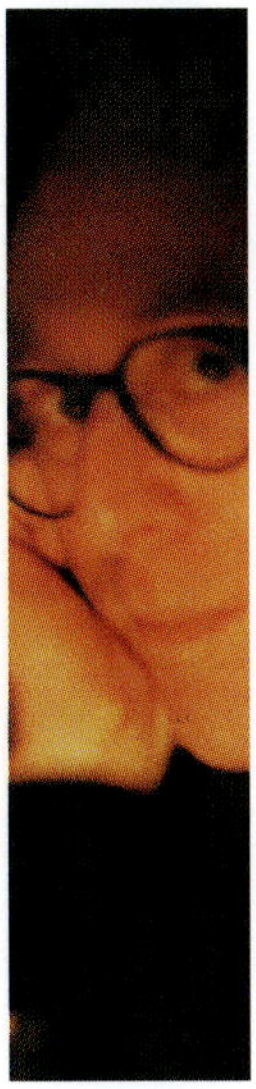

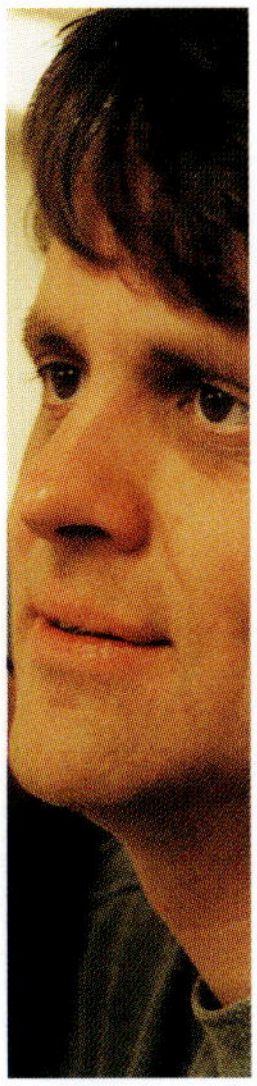

EQUIPMENT, MONEY ... OR TALENT?

GERÄTE, GELD ... ODER TALENT?

The jury consisted of Chris Wedge, Animation Director and Owner of Blue Sky Studios, Mitch Mitchell, Director of Special Effects at The Moving Picture Company, Rolf Herken, President and Director of R & D of Mental Images, Michael Wahrmann, – and myself, Mickey McGovern, Executive Producer of Millfilm, Ltd.

We looked at the submissions with the intention of awarding a first prize, two second prizes and 12 honourable mentions for advancing the science of technology in animation. We chose to look at this advance, not only with respect to technology but also in art.

It was daunting because we had 179 entries to look at in three days and we had to narrow this down to 15. The other question was the strong representation of Industrial Light and Magic. They had entered 7 of their films into the competition. ILM has been working in digital visual effects for over 20 years. They're extremely competent in this area. There were other companies represented who produced digital visual effects and there were student entries as well. But no one has the scope of money, talent, equipment, art direction and management as ILM does. It was a bit of a worry that the competition felt heavily in ILM's favour.

In spite of this we managed, after the first look through, to narrow the field to 45 possibilities. We now had 45 entries that

Die Animationsjury bestand aus Chris Wedge, dem Animation Director und Eigentümer der Blue Sky Studios, Mitch Mitchell, Leiter der Abteilung Special Effects der The Moving Picture Company, Rolf Herken, Präsident und Leiter der R&D-Abteilung von Mental Images, Michael Wahrman, und Mickey McGovern, Executive Producer von Millfilm. Ltd.

Wir haben die Einreichungen mit dem Ziel betrachtet, eine Goldene Nica, zwei Auszeichnungen und zwölf Anerkennungen zu vergeben für Werke, die eine Weiterentwicklung der Computeranimation erkennen lassen, und diesen Begriff „Weiterentwicklung" haben wir sowohl technisch wie auch künstlerisch verstanden.

Das war nicht ganz einfach, denn wir hatten 179 Einreichungen in drei Tagen zu betrachten und diese große Zahl auf 15 zu reduzieren. Und eine der Hauptfragen war die starke Präsenz von Industrial Light & Magic, die sieben ihrer Filme für den Prix Ars Electronica eingereicht hatten. ILM arbeitet seit über 20 Jahren im Bereich digitaler visueller Effekte und ist auf diesem Gebiet extrem kompetent. Natürlich gab e auch andere Unternehmen, die digitale Visual Effects eingereicht haben, und eine Menge Arbeiten von Studenten. Aber niemandem stand auch nur annähernd jene Menge an Geld, Talent, Ausrüstung, Regie kunst und Management zur Verfügung, wie dies bei ILM der Fall ist. Die Jury hatte einige Sorge, daß der Bewerb stark zu Gunsten von ILM ausfallen könnte.

Trotz dieser Bedenken gelang es uns aber bereits nach der ersten Runde, das Feld auf 45 Möglichkeiten einzuengen – 4 Einreichungen, die von Stories und Charak teren getragen wurden, die Botschaften vermittelten, künstlerische Fähigkeit ausdrückten, die interessant anzusehen ware provozierend und kreativ, die Ideen auf fa zinierende Weise transportierten. Und alle

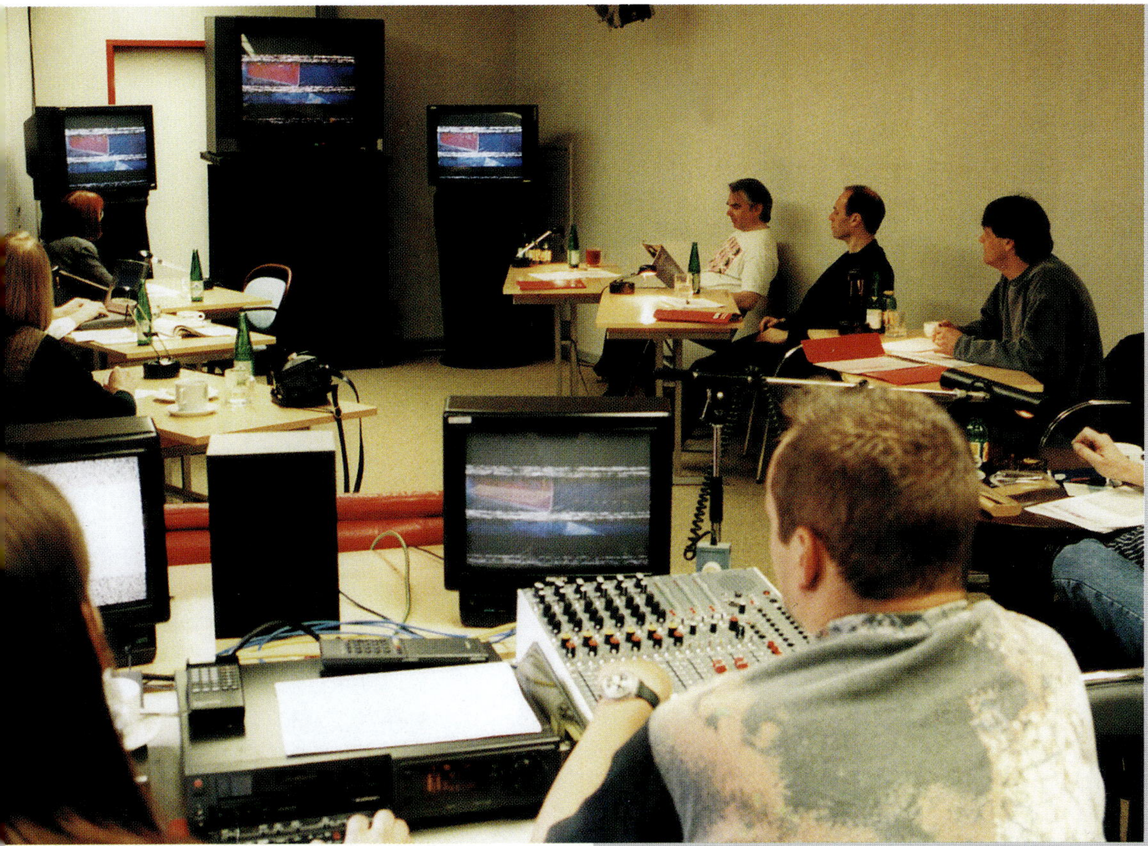

were story and character driven, had messages, showed artistic abilities, were very interesting to watch, were thought provoking and creative, and portrayed ideas in intriguing ways. They achieved this with computer technology. Art by math.

We awarded first prize to Scott Squires at ILM for "Dragonheart". The work comes from a feature film in which a computer generated dragon interacts with a live actor. The dragon moves realistically (which I can't prove because I've never seen a dragon but if I imagine a dragon it moves like this one). The dragon has very well integrated scales, textures, colours, features and snorts fire in a way that is believable. It speaks and it's mouth moves in sync to the dialogue. It has facial expressions and a twinkle in its eye like Sean Connery. It breathes. It has spirit. To make the whole

gelang dies durch den Einsatz von Computertechnologie – Kunst durch Mathematik, wenn man so will.

Die Goldene Nica wurde Scott Squires von ILM für „Dragonheart" zugesprochen. Das Werk kommt aus einem abendfüllenden Film, in dem ein computergenerierter Drache mit einem menschlichen Hauptdarsteller interagiert. Der Drache bewegt sich realistisch – was ich natürlich nicht beweisen kann, weil ich noch nie einen Drachen gesehen habe, aber wenn ich mir einen Drachen vorstelle, dann bewegt er sich genau wie dieser! Seine Schuppen, Oberflächenstrukturen, Farben und sonstigen Merkmale sind überzeugend herausgearbeitet, und er spuckt auf eine äußerst glaubhafte Weise Feuer. Wenn der Drache spricht, bewegt sich sein Mund lippensynchron zum Text, er hat eine echte Mimik, und seine Augen zwinkern wie die von Sean Connery. Er

thing more difficult for Scott Squires and his crew, the dragon interacts with water. Water doesn't scale, it drips, it splashes, it clings, it's wet and we know what it looks like and how it acts. You can't fool us there. Draco, the dragon is an incredible, beautiful and amazing work of art.

The second prize went to Chris Wedge for "Joe's Appartment". This small, independent feature film has dancing and singing cockroaches in a Busby Berkeley style. They are hilarious personalities. They are beautifully rendered so they appear photoreal. I mean they looked like real cockroaches who have been taught to sing, dance and give inspiring speeches. They completely captivate you and draw you in as they sing and dance around a "funky towel" and then dive into a toilet like Esther Williams in a production number.

We had 13 honourable mentions because we couldn't come to a unanimous decision about the other second place. We were torn between some strong performance pieces with wonderful writing, beautiful digital images, examples of creative story telling and the way characters were presented. There was a variety of intriguing styles that were all accomplished by computer graphic technology.

I hope that next year there are more submissions from every possible source. I

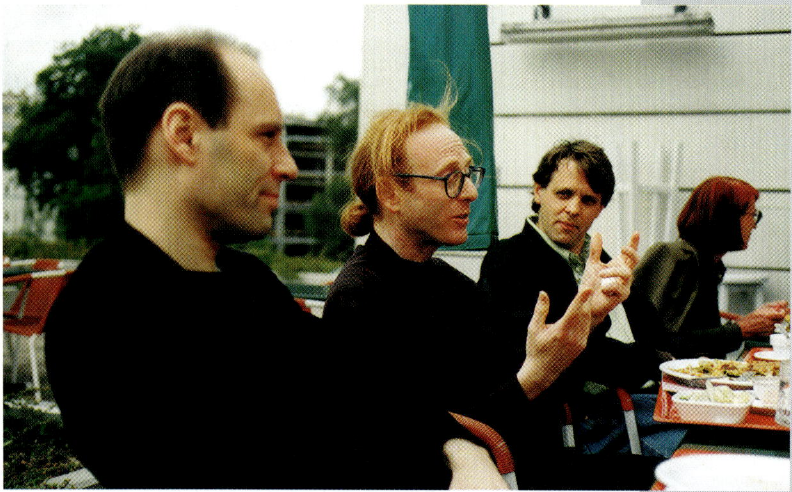

atmet. Er hat Geist, und um die Sache für Scott Squires noch schwieriger zu machen – er muß auch mit Wasser interagieren. Wasser fällt nicht einfach, es tropft, es spritzt, es bleibt hängen, es ist feucht, und wir wissen, wie es aussieht und wie es reagiert. Da kann uns keiner was vormachen. Draco, der Drache, ist ein wunderschönes, erstaunliches, unglaubliches und doch glaubwürdiges Stück Filmkunst.

Eine Auszeichnung wurde Chris Wedge für „Joe's Apartment" zuerkannt. Diese kleine unabhängige, ebenfalls abendfüllende Filmproduktion bringt tanzende und singende Küchenschaben im Stil von Busby Berkley – und es sind eindeutig fröhliche Persönlichkeiten. Das Rendering ist sehr schön, die Schaben sind absolut photorealistisch – das heißt, sie sehen aus, als hätte man Küchenschaben dressiert, ihnen das Singen und Tanzen und das Halten von flammenden Reden beigebracht. Sie fesseln einen vollständig und gewinnen jede Sympathie, wenn sie rund um ein schmuddelig-speckiges Handtuch singen und tanzen und in der Toilettenmuschel ein Wasserballett à la Esther Williams aufführen!

Wir haben insgesamt 13 Anerkennungen ausgesprochen, weil wir uns nicht auf eine zweite Auszeichnung einigen konnten. Wir waren hin- und hergerissen zwischen den zahlreichen Werken mit wunderbarer Regie und großartigen digitalen Bildern, mit Beispielen kreativen Geschichtenerzählens und beeindruckender Charakterdarstellung – kurzum, mit einer Vielzahl kreativer Stile, die alle durch Einsatz von computergraphischer Technologie entstanden sind.

Für das nächste Jahr wünsche ich mir noch mehr Einreichungen aus jeder nur erdenklichen Quelle. Ich weiß, daß die Auf-

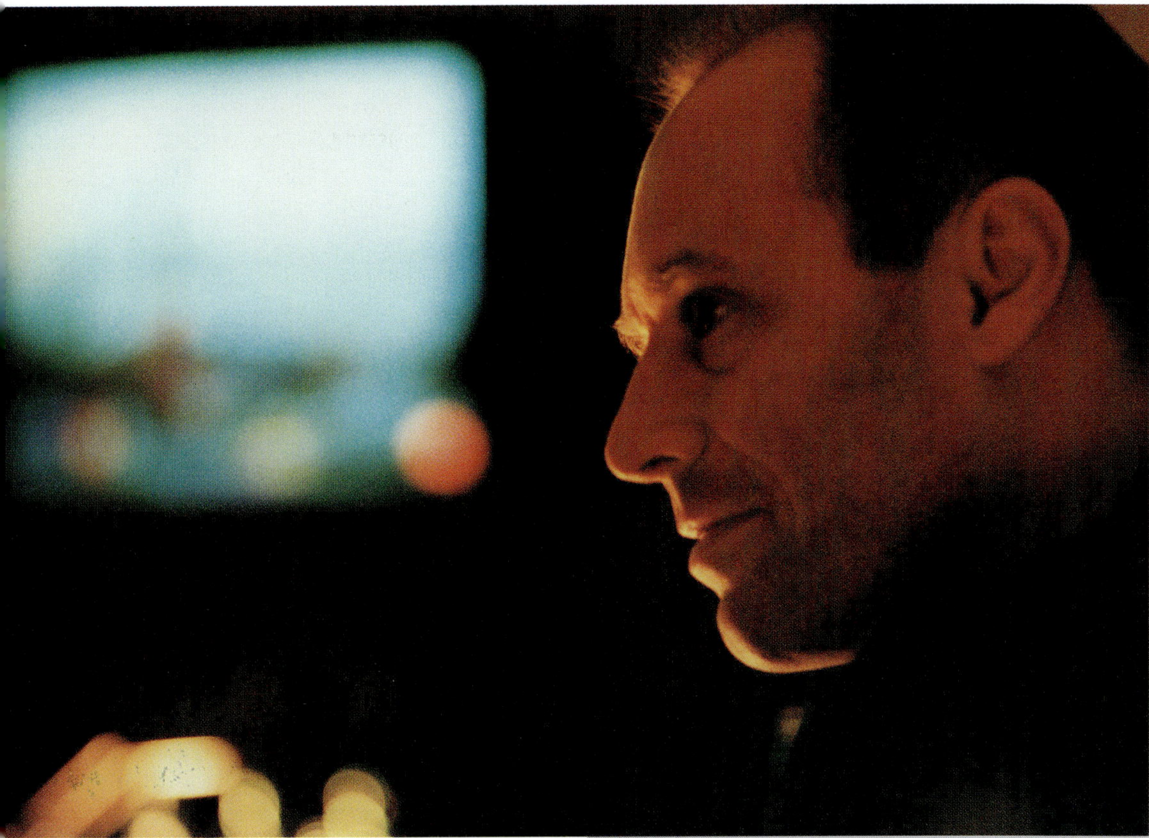

know that the judging will be more difficult because the level of quality is getting higher every year. ILM is one of the major reasons that the level of quality is getting higher because they endeavour to top themselves with every film they do.

gabe der Beurteilung wieder schwerer werden wird, weil die Qualität der Einreichungen von Jahr zu Jahr weiter steigt. Und ILM ist einer der Gründe, warum die Werke immer besser werden – selbst die Spitze gibt sich Mühe, sich bei jedem neuen Film selbst zu übertreffen.

Dragonheart

Scott Squires/ILM

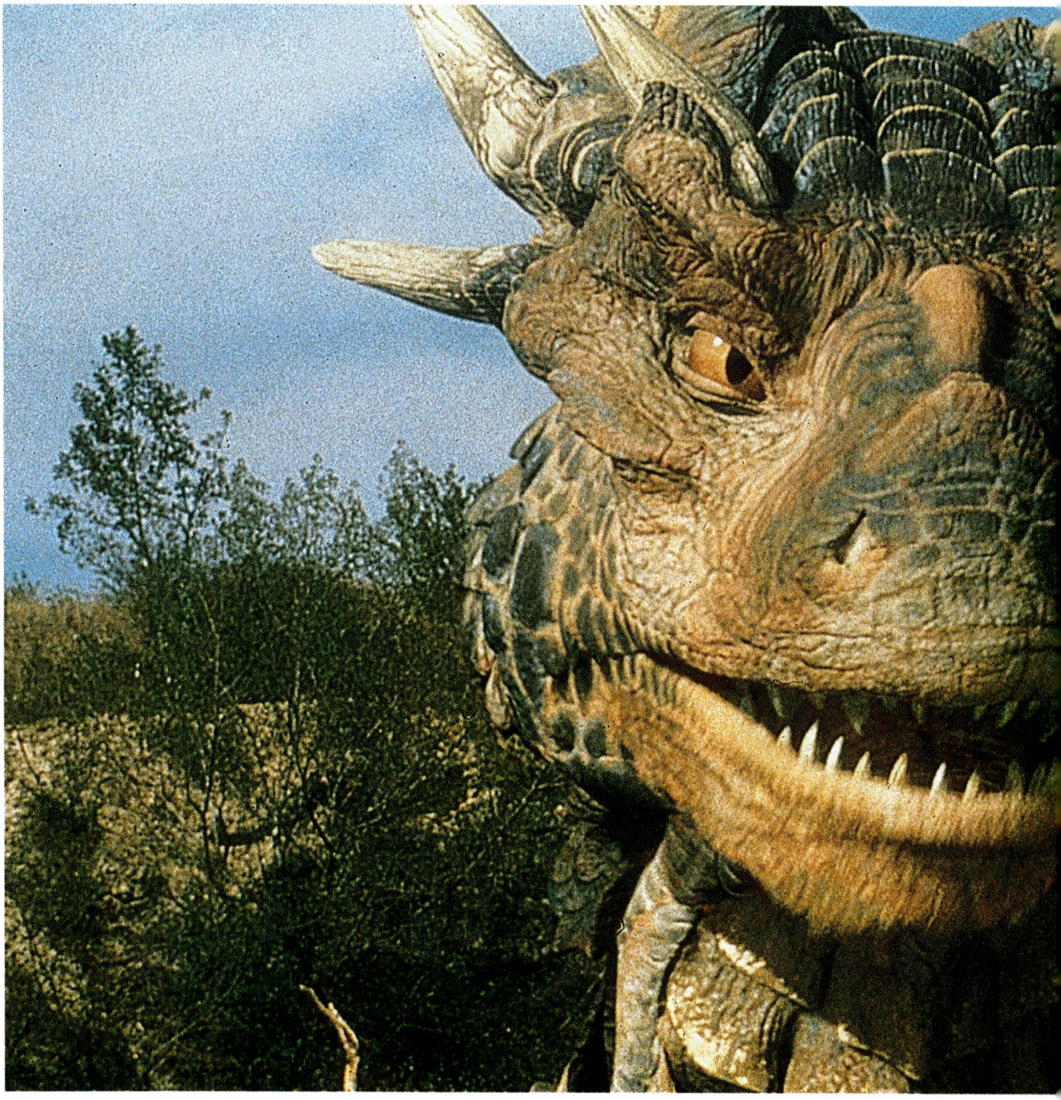

When I came on board to supervise the effects for "Dragonheart" in May of 1994, director Rob Cohen and the ILM production team were already developing storyboards and a dragon model was being designed by Phil Tippet. Our task was to create a major film actor in the form of "Draco", a mythical dragon. This meant bringing to life one of the most complex CG characters ever seen on film. We knew that the audience wouldn't easily accept this beast and

Als ich im Mai 1994 als Supervisor de Special Effects für „Dragonheart" zur Trup pe stieß, waren Regisseur Rob Cohen und das Produktionsteam von ILM schon fleiß dabei, Storyboards zu entwickeln, und auch ein Drachenmodell von Phil Tippet war bereits fertig. Unsere Aufgabe bestan darin, den Hauptdarsteller in der Form de mythischen Drachen „Draco" zu erzeuger Dies bedeutete nichts anderes, als einen der komplexesten computergenerierten

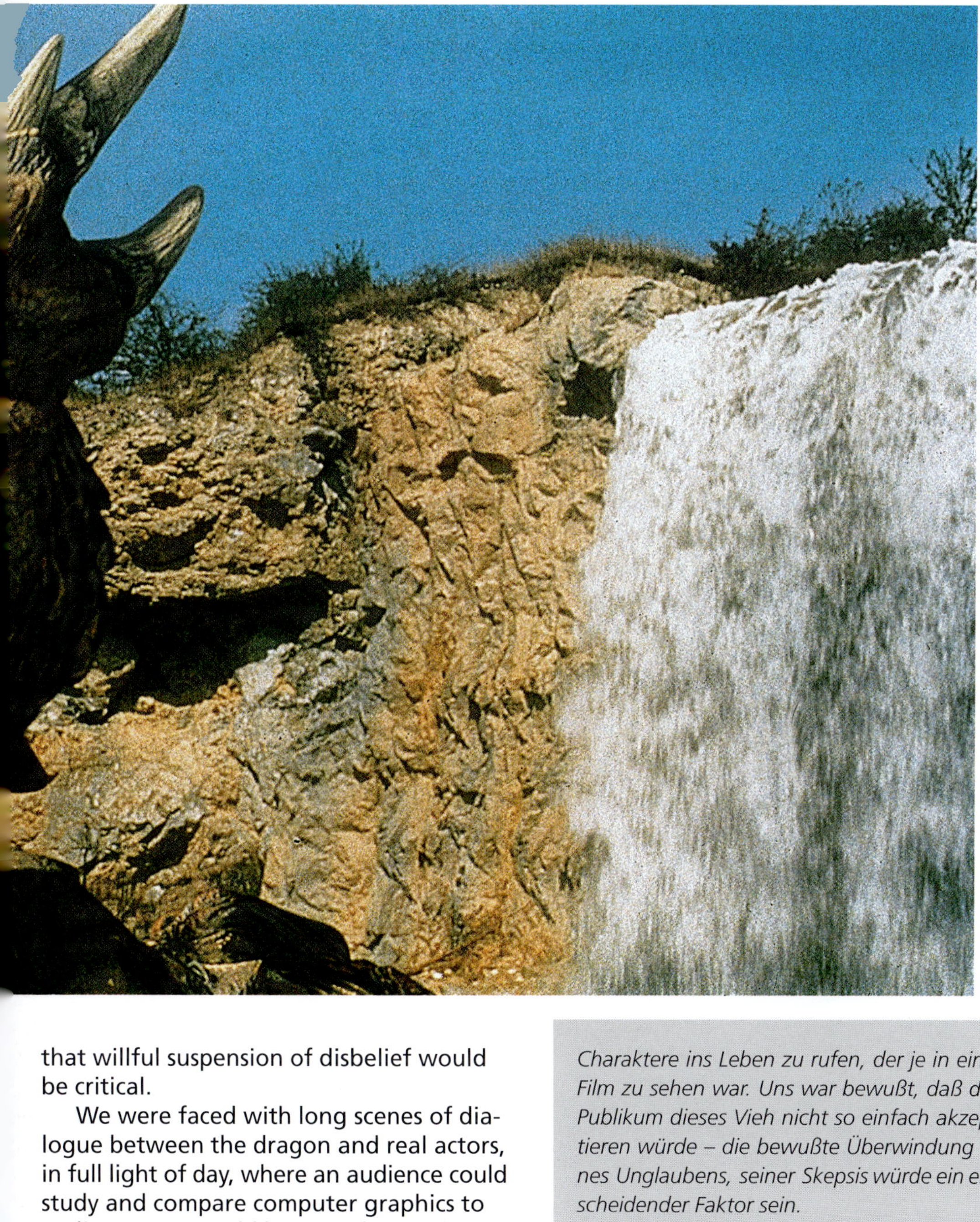

that willful suspension of disbelief would be critical.

We were faced with long scenes of dialogue between the dragon and real actors, in full light of day, where an audience could study and compare computer graphics to reality. Draco would be seen during the day, at night, in firelight, walking through grass, flying, splashing in water, swimming underwater, walking in the rain, holding an actor in his mouth and breathing fire.

Charaktere ins Leben zu rufen, der je in einem Film zu sehen war. Uns war bewußt, daß das Publikum dieses Vieh nicht so einfach akzeptieren würde – die bewußte Überwindung seines Unglaubens, seiner Skepsis würde ein entscheidender Faktor sein.

Wir standen vor der Aufgabe, lange szenische Dialoge zwischen dem Drachen und echten Schauspielern zu erzeugen, und zwar im vollen Tageslicht, so daß die Zuseher die Möglichkeit hatten, die Computeranimation

Most importantly, we needed to make him act, and try to ensure that the audience would develop an emotional tie to him.

The "Dragonheart" crew of over 70 people, produced 182 Dragon shots (covering 23 minutes of screen time), and 30 non-dragon shots. The entire post-production schedule was one year.

James Straus led our animation team. The amount and complexity of the required facial animation required us to write special in-house software known as Caricature (written by Cary Philips). This allowed the animators more interactivity when dealing with subtle facial expressions and lip sync issues. In addition, the tongue had to move correctly to provide the proper enunciation. Animators studied videotapes of Sean Connery (voice of Draco) and worked with Rob Cohen to make Draco "act".

mit der Wirklichkeit zu vergleichen. Draco sollte bei Sonnenschein zu sehen sein, bei Nacht, im Feuerschein, er sollte durchs Gras gehen, fliegen, ins aufspritzende Wasser stürzen, tauchen, im Regen gehen können, einen Schauspieler im Maul tragen und natürlich auch Feuer spucken. Und was am wichtigsten war: Er mußte auch schauspielen können, um dem Publikum die Möglichkeit zu geben, eine emotionale Beziehung zu ihm zu entwickeln.

Die über 70köpfige „Dragonheart"-Crew produzierte insgesamt 182 Drachen Einstellungen – 23 Minuten Filmzeit – und 30 Einstellungen ohne Drachen. Als Post-Production-Zeitraum war ein Jahr vorgesehen.

James Straus leitete unser Animations team. Die Menge und Komplexität der erforderlichen Gesichtsanimation zwang

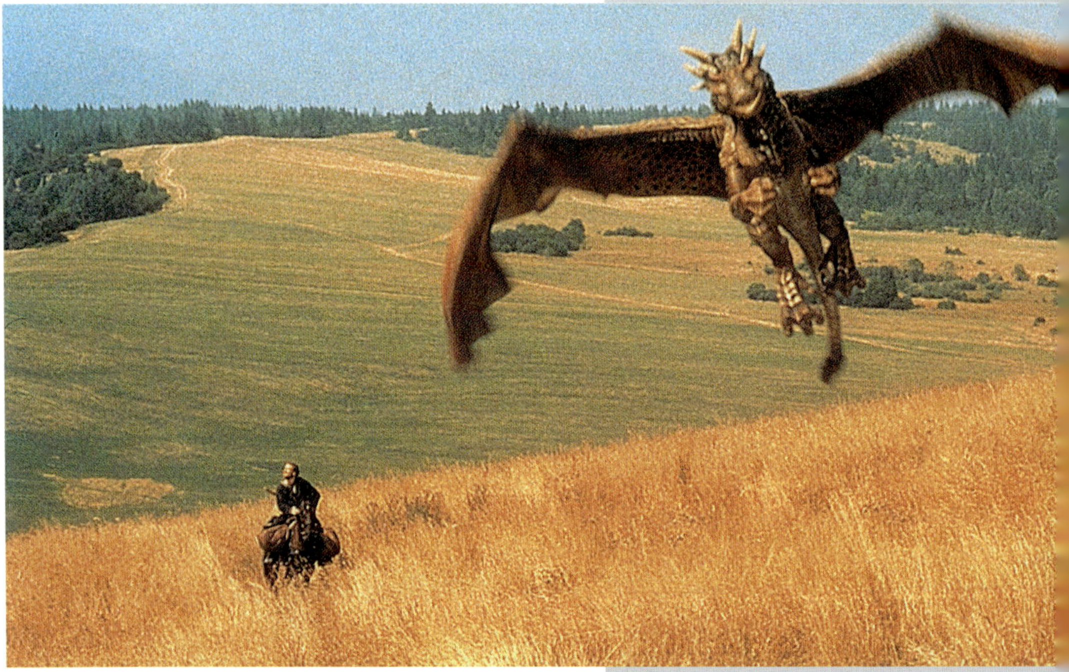

An enormous amount of detail was put into the Draco computer model as he was going to be seen up close, filling the full screen in broad daylight. He had to have detailed facial muscles in addition to body muscles. This resulted in a model several

uns dazu, eine eigene Software zu schreiben, die hausintern „Caricature" genannt wird und von Cary Philips stammt. Mit ihrer Hilfe konnten die Animatoren interaktiv arbeiten, wenn sie die subtile Mimik und die Probleme der Lippensynchro-

times more complex than those designed for "Jurassic Park". In addition to skin texture, we also had hard skin areas, horns, spikes, scales, teeth and gold scales. Each of these areas had to be painted, skinned and treated differently by the rendering software.

Draco's eyes had iris' (that animated with lighting changes), veins and wetness. His claws had to be fully articulated as in a real actor. His wings needed to be translucent when backlit (with a texture map that shows veins and other details) and iridescent in certain light. They had to be able to fold up and unfold, as well as be animated for him to fly. All the subtle animations had to continue through all shots to keep him alive. This meant creating and animating the nostrils as well as the breathing in his chest. We composited in additional details such as saliva when he opens his mouth wide. Reflections of surrounding scenes, the campfire, or Dennis Quaid ("Bowen") were also added to his eyes. Shooting for "Dragonheart" took place on location in Slovakia over a five months period. All but two weeks was shot outdoors with various weather conditions and terrain. Although most of the camera plates are moving camera shots, no motion control was used. To allow the director, and the director of photography, as much freedom as possible (and to save time), all shots were later matchmoved on the computer.

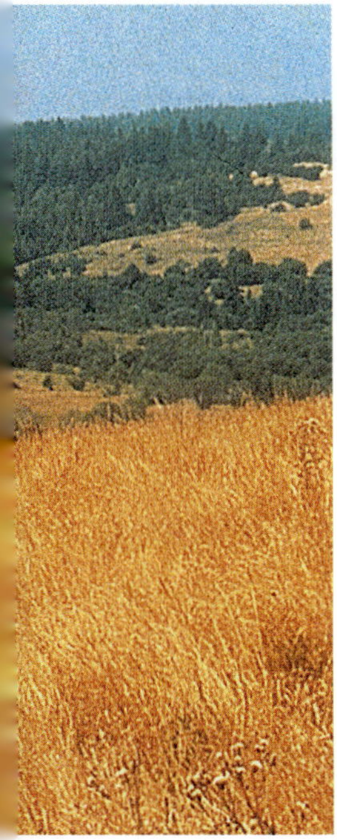

For this project we used an electronic transit system that provided accurate 3D measurements. This was critical since we

nität behandelten. Dazu hatte sich auch noch die Zunge korrekt zu bewegen, denn nur so konnte eine glaubwürdige Aussprache erreicht werden. Die Animatoren studierten Videobänder von Sean Connery (der Draco seine Stimme leiht) und arbeiteten gemeinsam mit Rob Cohen daran, den Drachen „schauspielen" zu lassen.

Eine enorme Menge an Detailarbeit mußte auch deswegen in das Computermodell Dracos gesteckt werden, weil er auch in Nahaufnahme zu sehen sein sollte, leinwandfüllend, im vollen Tageslicht. Er brauchte nicht nur detaillierte Körper-, sondern auch Gesichtsmuskeln, was zu einem vielfach komplexeren Modell im Vergleich zu jenen der Saurier in „Jurassic Park" führte. Zusätzlich zur Hautstruktur benötigte er Hornhautstellen, Hörner und Spitzen, Zähne, einfache und Goldschuppen. Und jede einzelne dieser Flächen mußte unterschiedlich bemalt, mit Haut versehen und von der Rendering-Software verarbeitet werden.

Die Augen des Drachen haben eine Iris (die auf Lichtveränderungen reagiert), Adern und sind feucht. Seine Klauen mußten ebenso exakt gearbeitet werden wie für einen echten Schauspieler. Seine Schwingen sollten nicht nur transparent sein (mit einer Texture-Map, die auch Adern und andere Details zeigt), sondern in bestimmten Lichtverhältnissen irisieren, sie mußten ein- und ausgefaltet werden können und Draco erlauben, zu fliegen. Und all diese subtilen Details der Animation mußten durch alle Einstellungen hinweg fortgeführt werden, um ihn lebendig zu erhalten. Dies bedeutete, nicht nur z. B. die Nüstern zu schaffen und zu animieren, sondern auch die Atembewegung seines Brustkorbs. Wir haben letztlich auch zusätzliche Details eingearbeitet, wie etwa Speichel, wenn er sein Maul weit aufreißt. Und in seinen Augen spiegelt sich die Szenerie um ihn, das Lagerfeuer ebenso wie Dennis Quaid als „Bowen".

Die Dreharbeiten für „Dragonheart" fanden in der Slowakei statt und dauerten fünf Monate. Bis auf zwei Wochen wurde im Freien gedreht, bei unterschiedlichen Wetter- und Geländebedingungen. Obwohl die meisten Einstellungen Kamerafahrten und Schwenks ent-

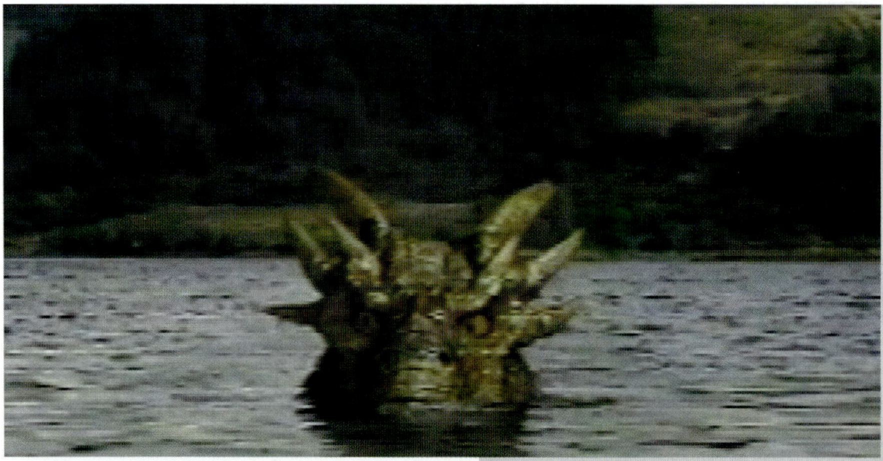

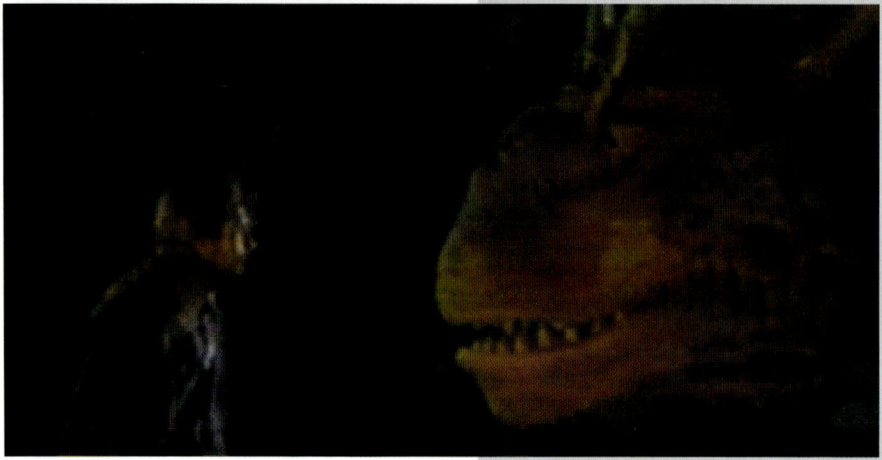

were locking a slow-moving or sitting dragon to a slow-moving camera. Any misalignment would be obvious, much more than the sky replacements or the fast camera moves we would also be doing. This information was recorded on a PowerBook and then entered into the animation system for matchmoving. This cut down the time to do post matchmoving enormously.

Each shot was different, so we set up an 18-inch posable dragon in front of the camera for the director and actors to see a facsimile of the finished shot in addition to the storyboards. Various targets were used so Dennis and the other actors would have the correct eyeline whether Draco was sitting, standing, etc. Various markers were placed in the scene and measured so these could be used for matchmoving later. All

halten, wurde auf Motion Control verzich-tet. Um Regisseur wie Kameramann maxi-male Bewegungsfreiheit zu lassen (und um Zeit zu sparen), wurden alle Einstellungen später auf dem Computer angeglichen.

Für dieses Projekt haben wir ein elek-tronisches Transit-System eingesetzt, das exakte 3D-Messungen erlaubt. Dies war insofern kritisch, als wir einen sich langsam bewegenden oder sitzenden Drache[n] mit einer langsamen Kamerabewegung koordinieren mußten. Der geringste Fehl[er] in der Ausrichtung würde da sofort ins Auge stechen, viel mehr als Fehler beim Ersetzen eines Himmels oder bei schnelle[n] Kamerafahrten, die wir auch gemacht haben. Die Information vor Ort wurde au[f] einem PowerBook aufgezeichnet und spä

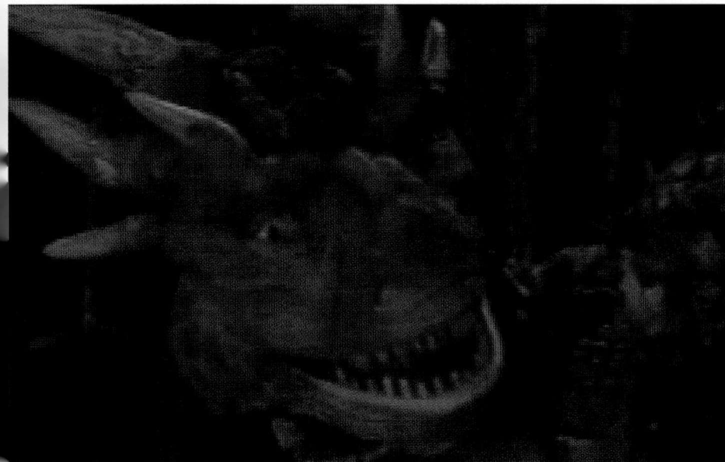

scenes with actors in front of Draco were composited using rotoscoping techniques and no bluescreen was used.

ter für das Matchmoving in das Animationssystem eingespielt. Dies hat die für das Post-Matchmoving erforderliche Zeit enorm reduziert. Jede Aufnahme war anders, weshalb wir eine 18-Zoll-Drachen-Figurine vor die Kamera montiert haben, um Regisseur und Schauspielern als Ergänzung zu den Storyboards ein Faksimile der fertigen Einstellung zu bieten. Mehrere Zielmarker wurden aufgestellt, damit Dennis und die anderen Schauspieler eine korrekte Augenhöhe einhalten konnten, je nachdem, ob Draco sitzt, steht oder sich bewegt. Markierungspunkte wurden über die Szene verteilt, um das spätere Matchmoving zu ermöglichen. Alle Szenen, in den sich Schauspieler vor dem Drachen befanden, wurden rotoskopiert, Bluescreen-Verfahren wurden nicht eingesetzt.

Scott Squires (USA) co-founded Dream Quest Images in 1979 and was a visual effects supervisor, as well as the company's president, for six years. Dream Quest projects included "Blue Thunder", "The Adventures of Buckaroo Banzai", "Deal of the Century", "One From the Heart" and "Blade Runner". Squires joined Industrial Light & Magic in 1985. In 1994, Squires received a Scientific and Engineering Award from the Academy of Motion Picture Arts and Sciences for his pioneering work in the area of film input scanning. He received an Oscar nomination for best achievement in visual effects for "The Mask" and recently received his second nomination for "Dragonheart".

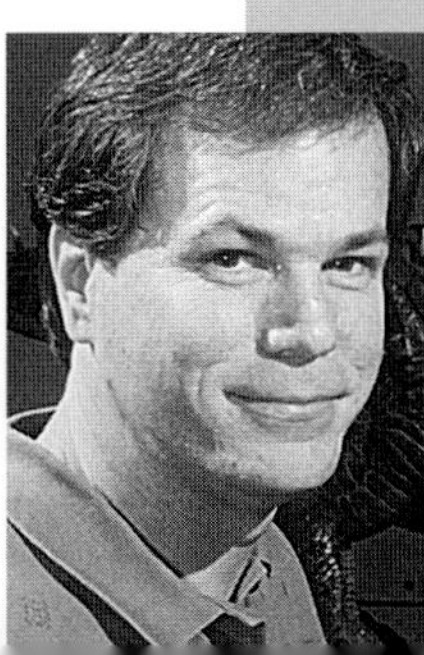

Scott Squires (USA) ist Mitbegründer von Dream Quest Images (1979), wo er sechs Jahre als Visual Effects Supervisor und Firmenchef tätig war. Von Dream Quest stammen u. a. die Effekte zu „Blade Runner". 1985 stieß Scott Squires zu ILM. 1994 wurde er von der Academy of Motion Picture Arts mit dem Scientific and Engineering Award für seine Pionierleistung im Scannen von Film-Input geehrt. „The Mask" wurde in der Kategorie „Beste Visual Effects" für den Oscar nominiert, mit „Dragonheart" erfolgte kürzlich die zweite Nominierung.

FUNKY TOWEL

CHRIS WEDGE

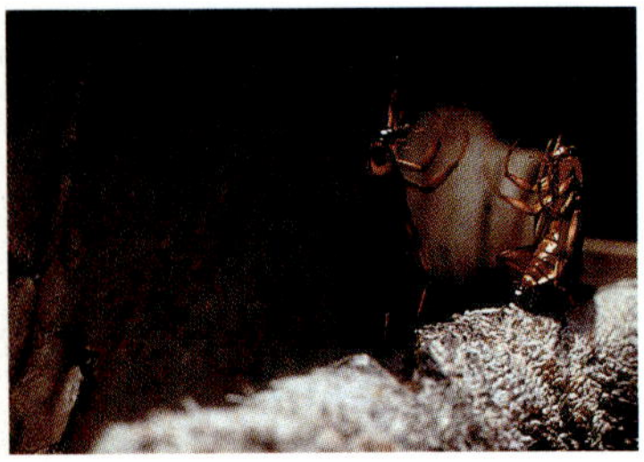

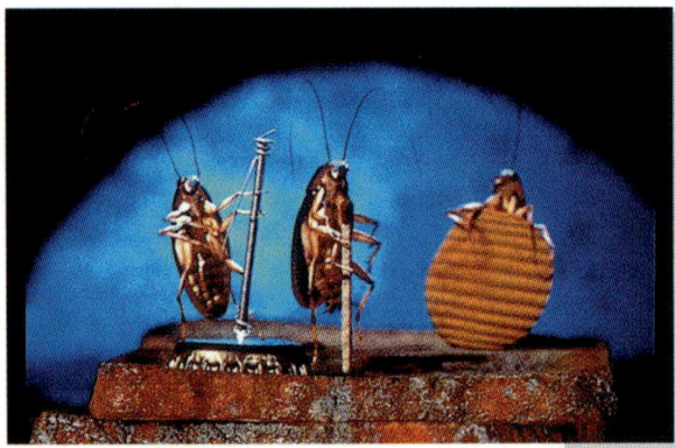

It's not your average love story. Afterall, the heros of the feature length motion picture "Joe's Apartment" are a bunch of mischevious roaches. When it comes to those with the rare talent for song and dance, it is the computer generated bunch, created by Blue Sky Studios, that clearly stand out. "Funky Towel", a two-minute homage to Busby Berkley, takes place in Joe's ripe and inspiring bathroom. The chorus of roaches sings, swims, waltzes and breakdances introducing the audience to their "funky" world.

These 3D photorealistic animated roaches interact seamlessly and credibly with live action actors and environments. It is difficult to say what was the most important element in creating the incredibly real-looking roaches. To replicate nature's own greasy creatures, CG modelers studied tanks full of live ones. Animators studied the twitchy movements of scampering roaches and merged those characteristics with the performances so they could stand up, talk, sing and dance.

Using Softimage's motion module the animators brought the stars of "Joe's Apartment" to life. To integrate the roaches into their scenes, the lighting had to be perfect. Using Blue Sky's proprietary software, CGI Studio™, the CGI world merged with the real world in seamless unity.

Blue Sky produced over 13 mintues of animation that makes you laugh at and believe in the world of talking roaches.

Das ist keine der üblichen Liebesgeschichten, denn schließlich sind die Helden des abendfüllenden Kinofilms „Joe's Apartment" ein Rudel boshafter Küchenschaben. Aber was jene betrifft, die ein offensichtliches Talent für Musik und Tanz haben – das ist ein von den Blue Sky Studios geschaffenes computergeneriertes Ensemble, das sich besonders hervortut. „Funky Towel", eine zweiminütige Hommage an Busby Berkley, findet in Joes ungepflegtem und sehr inspirierendem Badezimmer statt. Die Küchenschaben-Compagnie singt, schwimmt, tanzt Walzer und Breakdance und entführt das Publikum in ihre „schmuddelige" Welt.

Die dreidimensional und photorealistisch animierten Schaben interagieren nahtlos und glaubwürdig mit den Filmschauspielern und ihrer Umgebung. Es ist schwer zu sagen, welche die wichtigsten Elemente bei der Schaffung dieser unglaublich echt aussehenden Schaben waren. Um die schmierigen Kreaturen möglichst naturgetreu nachbilden zu können, haben die Animatoren ganze Terrarien voll mit echten Schaben studiert. Beobachtet wurden vor allem die wackelnden Laufbewegungen der umhersausenden Tiere, um diese mit den Performances zu verknüpfen und die Schaben so stehen, reden, singen und tanzen zu lassen.

Die Stars von „Joe's Apartment" wurden mit Hilfe von Softimages Bewegungsmodulen zum Leben erweckt. Die Integration in ihre Szenen erforderte perfekte Ausleuchtung. Blue Skys eigene Software CGI Studio™ ließ die Computergraphik-Welt mit der gefilmten Wirklichkeit zu einer nahtlosen Einheit verschmelzen. Insgesamt entstanden über 13 Minuten Animation für den Film, die den Besucher über die Welt der sprechenden Schaben lachen und ihn an sie glauben lassen.

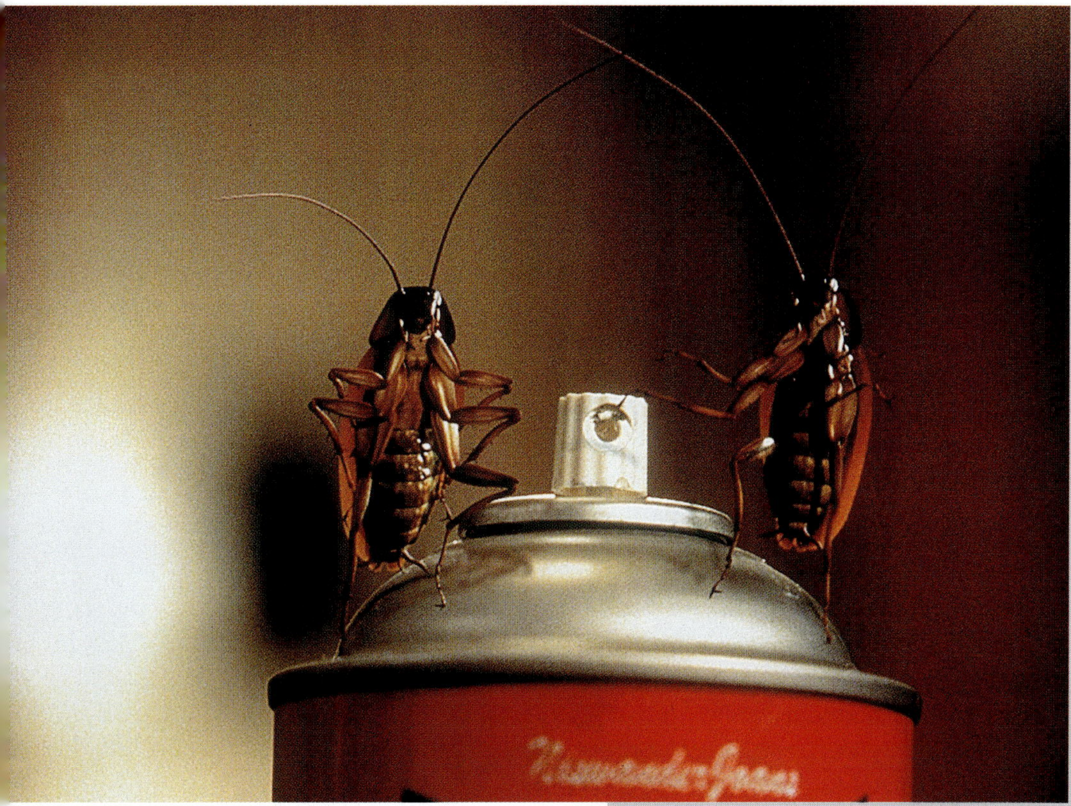

Chris Wedge (USA) has a BFA from the State University of New York and an MA in Computer Graphics and Art Education from Ohio State University. He is Vice President and Creative Director of Blue Sky Studios, Inc. and is considered one of the top computer animation directors in the world. He also was Director of Animation at MAGI/Syntha Vision, where he directed and/or animated projects for television commercials and sequences for the Walt Disney feature film "Tron".

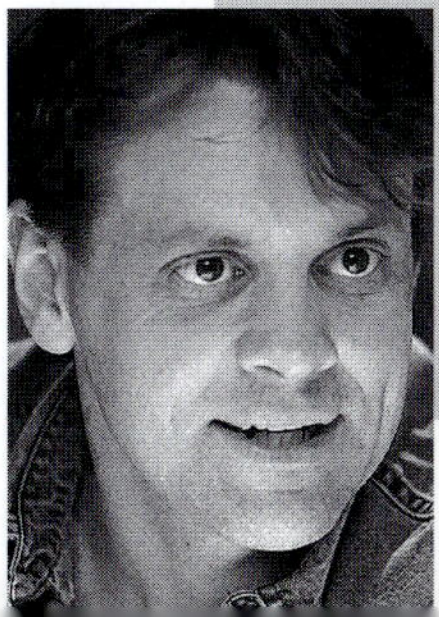

Chris Wedge (USA) graduierte an der State University of New York zum BFA und an der Ohio State University zum MA aus Computergraphik und Kunsterziehung. Er ist Vizepräsident und Kreativdirektor der Blue Sky Studios, Inc., und gilt als einer der führenden Regisseure für Computeranimation weltweit. Er arbeitete auch als Director of Animation bei MAGI/Syntha Vision, wo er für Regie und/oder Animation von TV-Spots sowie Sequenzen des Walt-Disney-Films „Tron" verantwortlich war.

Mars Attacks!

Jim Mitchell / ILM

"Mars Attacks!" was initially planned as a stop-motion project, much like director Tim Burton's earlier work "Nightmare Before Christmas". But this time the animated characters would need to exist in the real world with the likes of Jack Nicholson, Sarah Jessica Parker and Pierce Brosnan, among others. The plan was for ILM to composite the 15" armatured Martian puppets over various live-action plates. Meanwhile, technical hurdles kept popping up for the stop-motion unit, so I pursued doing a test of Martians which would be completely built and animated inside a computer. The test was successful and showed Tim that computer animation, and the array of digital tools we had to offer, would be the most effective way to blend his Martians into the live-action. Most importantly, it presented computer animation as viable a method as any other for creating complex character animations.

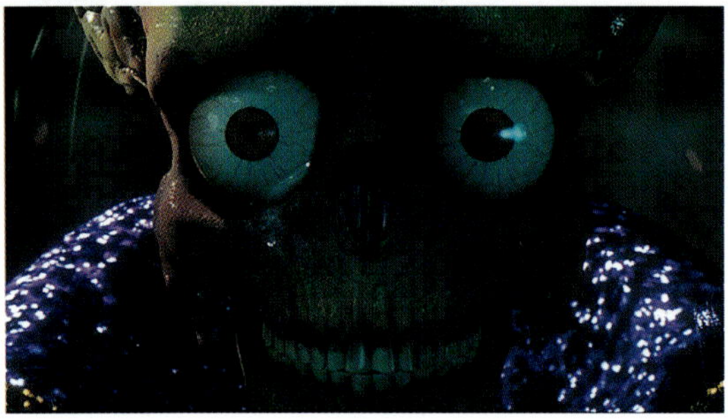

In a little over 8 months, we created 304 visual effects shots, including 246 with 3D Martian character animation. Crucial to the work at all phases was the believability of the Martian personalities, their actions and the result of those actions. A team of over 100 artists worked together to bring Tim's strange imagery to life. Certainly, a project like this requires the talents of many. Collaborating with Mike Fink and Michael Lantieri helped us to achieve the needed continuity in the character of the film.

In the 3 months of principle photography, we shot VistaVision plates along-

„Mars Attacks!" war ursprünglich als Stop-Motion-Projekt gedacht, ähnlich wie Regisseur Tim Burtons frühere Arbeit „Nightmare Before Christmas". Aber diesmal mußten die animierten Charaktere in der realen Welt neben Schauspielern wie Jack Nicholson, Sarah Jessica Parker und Pierce Brosnan bestehen. Ursprünglich war geplant, daß ILM die 15 Zoll großen Marsmenschen-Puppen über verschiedene Live-Action-Filmszenen montieren sollte. Aber im Laufe der Arbeit tauchten immer mehr technische Hürden bei der Stop-Motion-Einheit auf, deswegen arbeitete ich probeweise mit Marsianern, die komplett computergeneriert und -animiert waren. Der Test überzeugte Tim davon, daß die Computeranimation und die digitalen Werkzeuge, die uns zur Verfügung standen, wohl die effizienteste Möglichkeit waren, die Marsmännchen in den realen Film einzubauen. Und – fast noch wichtiger – er bewies, daß Computeranimation ein ebenso gangbarer Weg ist wie jeder andere, wenn man komplexe Charakter-Animationen herstellen will.

In etwas mehr als acht Monaten haben wir 304 Einstellungen mit Visual Effects, davon 246 mit 3D-Character-Animation der Marsmenschen geschaffen. In allen Phasen der Animation war die Glaubwürdigkeit der Mars-Persönlichkeiten, ihrer Handlungen und der Ergebnisse dieser Handlungen entscheidend. Ein Team von über hundert Künstlern arbeitete zusammen, um Tims seltsame Bildwelt lebendig werden zu lassen. Klarerweise benötigt ein Projekt wie dieses die Talente vieler Mitwirkender. Die Zusammenarbeit mit Mike Fink und Michael Lantieri half uns, die erforderliche Kontinuität im Charakter des Films zu bewahren.

In den drei Monaten der eigentlichen Dreharbeiten haben wir Seite an Seite mit der ersten Unit-Production-Crew VistaVision-Bilder geschossen. Entscheidend für

side the first unit production crew. Crucial to the integration of the CG Martians into the live-action plates was the comprehensive measurements of all the sets, props and camera positions. To help the actors visualize where the Martians would be, we employed "stand-ins" similar in size to the Martians. They were dressed in crude versions of their space suits, so we were able to use them as reference for lighting our CG versions. All plates used were shot clean (no "stand-ins"), requiring the fabrication of all interactive shadows and reflections. No motion control equipment was used on location.

die Integration der computergenerierten Marsianer war eine umfassende Vermessung aller Kulissen, Requisiten und Kamerapositionen. Um den Schauspielern bei der „Visualisierung" der später einzufügenden Marsianer zu helfen, haben wir „Statisten" von annähernd passender Größe eingesetzt. Sie waren in grobe Imitationen ihrer Raumanzüge gekleidet, so daß wir sie als Refernz für die Ausleuchtung unserer Computer-Versionen verwenden konnten. Die Hintergrundbilder wurden „rein" geschossen, also ohne die Modelle, weshalb wir alles, auch Schatten und Reflexionen, selbst generieren mußten. Bei den Aufnahmen vor Ort wurde kein Motion-Control-Equipment eingesetzt.

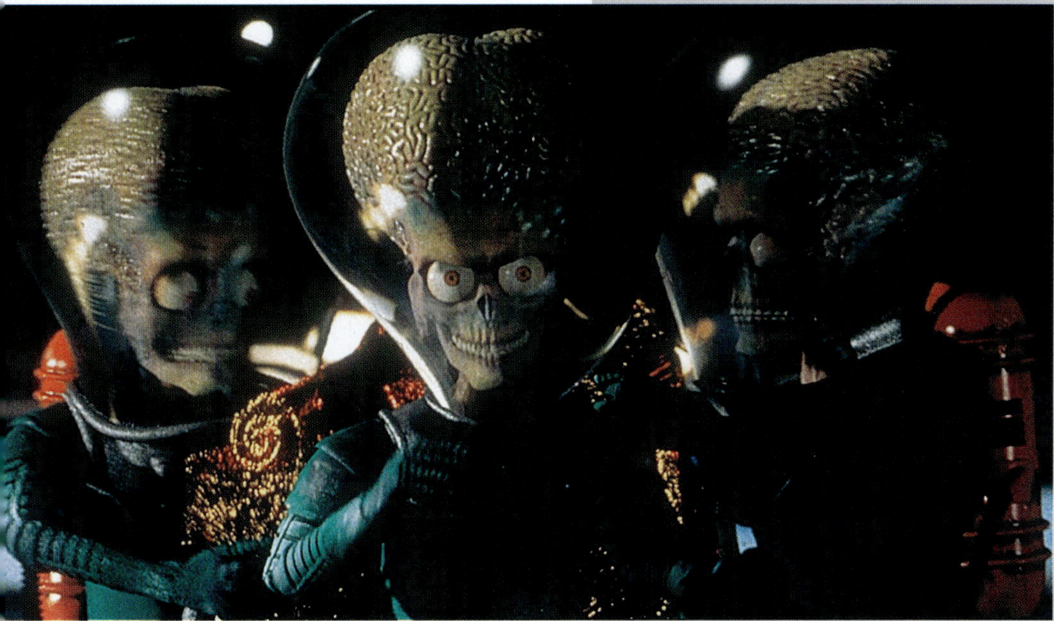

Jim Mitchell (USA) joined Industrial Light & Magic in 1990. He has played an integral role in the creation of the computer graphics imagery in "Jurassic Park", "Death Becomes Her" and "Terminator 2: Judgement Day", each of which received Academy Awards for Best Achievement in Visual Effects.

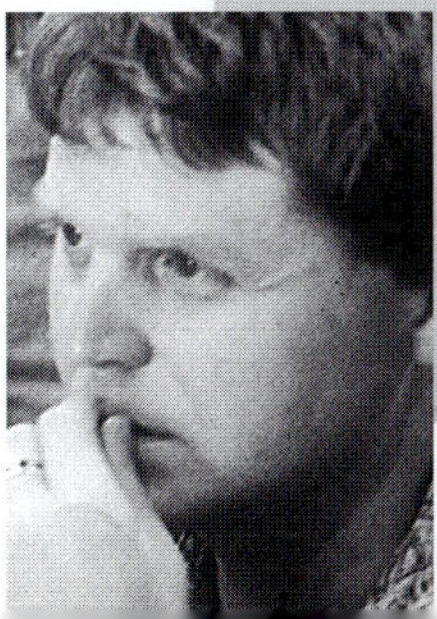

Jim Mitchell (USA) stieß 1990 zu Industrial Light & Magic und war entscheidend an der Schaffung der computergraphischen Bildwelt von „Jurassic Park", „Death Becomes Her" und „Terminator 2: Judgement Day" beteiligt, die alle jeweils den Oscar für die besten Visual Effects erhalten haben.

Twister

Stefen Fangmeier / ILM

The visual effects work in "Twister" required the creation of various tornadoes that would interact with (and tear up) the ground, buildings, trees, cars, tankers, cows, tractors, farm equipment and other debris. Initially, it was my responsibility to execute a test for director Jan De Bont to show that it was possible to create the tornadoes. We wanted to give him full freedom on size, motion and structure. We chose to use particle simulation on the computer to create the tornado. This resolved the issue of fully

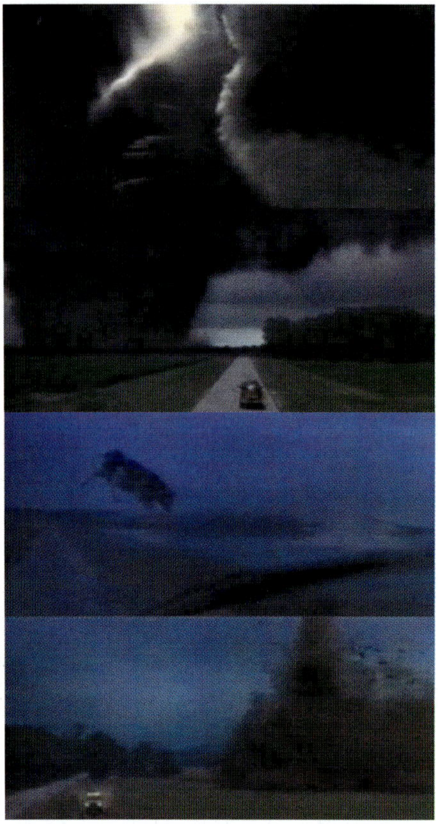

integrating the many separate passes of debris and objects into the dust, allowing for seamless compositing and camera matching of the tornado.

Since this was a whole new territory, we created a 400 frame test showing a driver's POV of an F5 tornado shredding a barn as

Die Visual Effects in „Twister" erforderten die Schaffung mehrerer Windhosen, die mit dem Boden, den Gebäuden, Bäumen, Autos, Tankwagen, Kühen, Traktoren, landwirtschaftlichen Geräten und anderen Dingen interagieren und sie aufsaugen sollten. Meine Aufgabe war ursprünglich, ein Testobjekt für Regisseur Jan De Bont zu erzeugen, um ihm zu beweisen, daß es möglich ist, solche Tornados zu generieren. Wir wollten ihm jede Freiheit hinsichtlich Größe, Bewegung und Struktur lassen und beschlossen, Partikelsimulation im Computer einzusetzen, um den Tornado zu erzeugen. Auf diese Weise wurde auch das Problem gelöst, wie die vielen unterschiedlichen angesaugten Teile und Objekte in die wirbelnde Staubbewegung integriert werden konnten, wodurch eine nahtlose Anpassung des Tornados an die Kamerabewegungen möglich wurde.

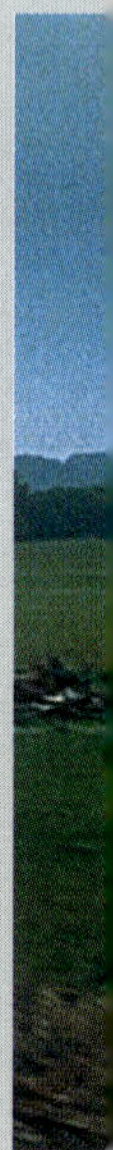

Da wir damit völliges Neuland betraten, wurde zunächst ein aus 400 Kadern bestehender Testfilm geschaffen, in dem – aus dem Blickwinkel eines Fahrers – ein Tornado der Stärke 5 eine Scheune zerlegt und einen Traktor vom Himmel fallen läßt, von dem ein Rad auf die Windschutzscheibe zufliegt. Anfangs war es fraglich, ob wir überhaupt in der Lage wären, eine glaubwürdige Szene mit der für einen Film erforderlichen Menge an Details zu erzeugen. Und selbst wenn dies möglich wäre, wie sollten wir die einzelnen Partikel so ausleuchten, daß sie aufeinander Schatten werfen, wie dies auch Wolken tun? Würde die benötigte Menge an Teilchen überhaupt in die Rechner passen? Und wäre die Verarbeitung überhaupt schnell genug, daß wir rechtzeitig liefern könnten? Der Test war erfolgreich,

a tractor falls out of the sky, sending a tire into the windshield. Whether we could create a believable effect with the right amount of detail for film was unknown. Even if this were possible, how would we light the particles to shadow themselves like clouds do? And would the number of particles needed fit in the computers? Would the process be fast enough for us to deliver on time? The test was successful and the movie was green-lighted. The last 2 seconds of the shot were used in the "Twister" trailers. Once the project began, my role was to work with the software group to develop the particle renderer and implement the larger tornadoes, devise a technique to shred structures and train and supervise a group of 30 effects animators to execute the shots.

und dem Filmprojekt wurde grünes Licht gegeben. Die letzten beiden Sekunden des Teststreifens wurden übrigens im Trailer für „Twister" eingesetzt.

Als das eigentliche Projekt lief, bestand meine Aufgabe darin, mit der Software-Gruppe einen Renderer für die Partikel zu entwickeln und die größeren Windhosen zu implementieren, sowie eine Technik zu erfinden, die größere Strukturen zerreißen konnte, und schließlich die 30 Animatoren, die die einzelnen Einstellungen produziert hat, zu trainieren und zu leiten.

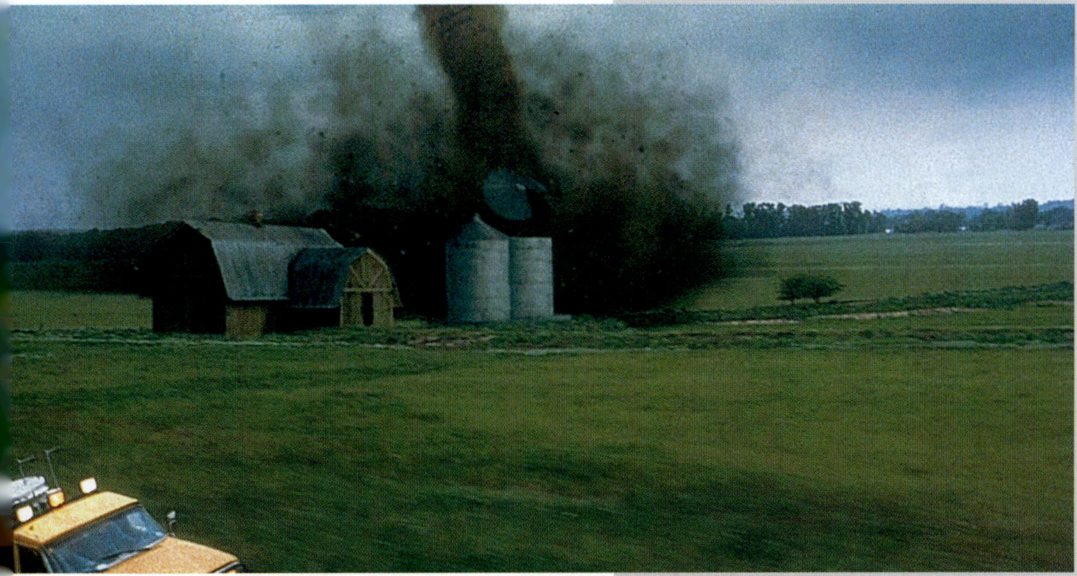

Stefen Fangmeier (USA) joined the computer graphics department at Industrial Light & Magic in 1990. His first major project was "Terminator 2: Judgement Day". Since then he has worked as a computer graphics supervisor on "Hook" and "Jurassic Park", the digital character co-supervisor on "Casper" and visual effects supervisor on "Twister". Fangmeier received his degree in computer science from California State University of Dominguez Hills in 1983.

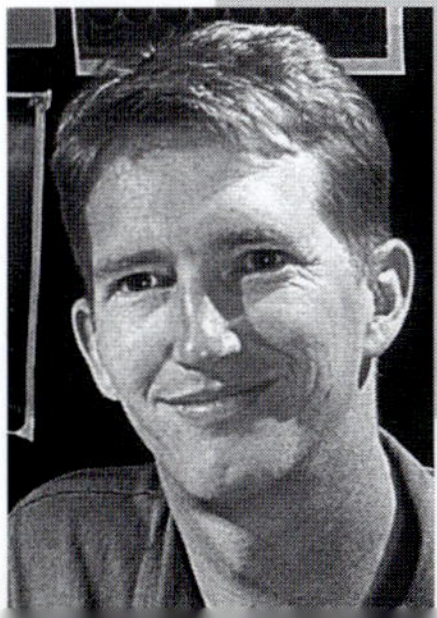

Stefen Fangmeier (USA) kam 1990 zum Computer Graphics Department von Industrial Light & Magic. Sein erstes größeres Projekt war „Terminator 2: Judgement Day". Seit damals arbeitete er als Computergraphik-Supervisor an „Hook" und „Jurassic Park", als Co-Supervisor für digitale Charaktere an „Casper" und als Supervisor für Visual Effects an „Twister". Fangmeier graduierte 1983 an der California State University in Dominguez Hills in Computerwissenschaften.

Huzzah

Larry Lamb

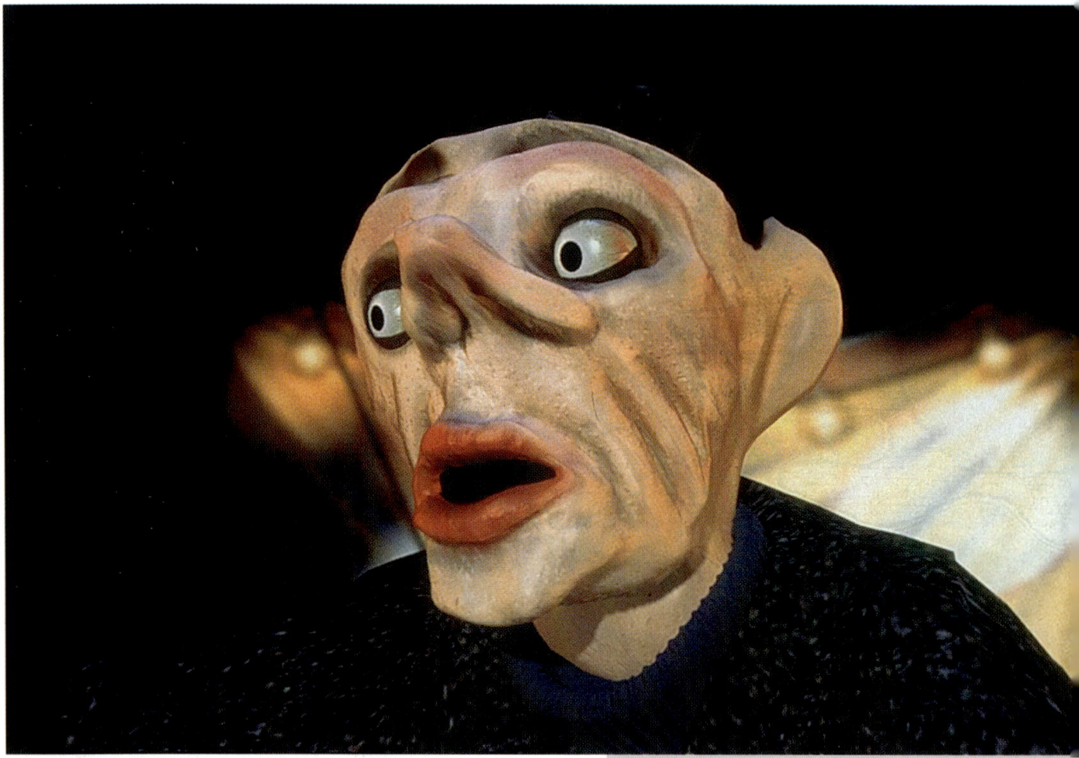

The four-minute long "Huzzah" introduces us to Darren, the mishappen carnival barker, and his muse, the sideshow freak, Bobaloo, the Beast Boy. Created at Lamb & Company from an original story by playwright/actor Kevin Kling, "Huzzah" demonstrates the dawn of "expression mapping" with full body performance motion capture.

We are constantly challenged to create highly distinctive CG characters. With so much creative talent available to us in the artistic community we wanted to reveal performance beyond voice over. By making full use of the actor's creative expression and continuity of motion, we enable Kevin to become Darren. "Huzzah" is very much an exercise in giving our character a life "out of the computer". As the industry makes greater use of CG characters, and calls on fine actors like Sean Connery to portray Draco in "Dragonheart", or Tom Hanks to give life to Woody in "Toy Story",

Der vierminütige Streifen „Huzzah" macht uns mit Darren, dem mißgestalteten Jahrmarktschreier, und seiner Muse Bobaloo, the Beast Boy, einer Attraktion des Panoptikums, bekannt. „Huzzah" entstand bei Lamb&Company nach einer Geschichte des Schauspielers und Autors Kevin Kling und demonstriert die Anfänge eines „Expression Mapping" im Zuge einer Ganzkörper-Motion-Capture-Sequenz.

Wir stehen ständig vor der Herausforderung, hochspezialisierte und sehr persönliche Computergraphik-Charaktere zu entwickeln. Angesichts des enormen zur Verfügung stehenden kreativen Talents innerhalb der Künstlergemeinde wollten wir über die klassische Animation mit darübergelegtem Sprechertext hinausgehen. Dadurch, daß wir die gesamte Bandbreite von Ausdruck und Bewegung des Schauspielers in die Animation umgesetzt haben, konnte Kevin tatsächlich zu Darren werden. So gesehen ist „Huzzah"

I think we all realize these talents have more to offer characters than their voice.

Darren was designed in paper mache and wood by artist Michael Sommers and digitized into the computer. Kevin Kling's performance of Darren was captured on the Lamb & Company motion capture stage with a combination of simultaneous motion tracking devices, including an electronic glove, a magnetic full body motion capture suit, and an optical facial motion capture system. The complete system gave animators a huge palette of Kling's motions and emotions to work from, including subtle shifts in body weight and slight facial gestures.

Key to the production of "Huzzah" is LambSoft's (Lamb & Company's sister company) comprehensive software system called ProMotion which enables animators to view, filter, edit and interpret the multiple streams of captured motion data.

der Versuch, unserer Schöpfung ihr Leben „aus dem Computer" einzuhauchen. In dem Maße, in dem die Industrie immer mehr computergraphische Darsteller verwendet und so hervorragende Schauspieler wie Sean Connery einem Draco (in „Dragonheart") oder Tom Hanks dem Woody (in „Toy Story") als Vorlage dienen, wird uns wohl allen bewußt, daß diese Talente den Filmcharakteren mehr zu bieten haben als nur ihre Stimme.

Darren wurde vom Künstler Michael Sommers aus Pappmaché und Holz hergestellt und in den Computer digitalisiert. Kevin Klings Darstellung des Darren wurde auf der Lamb& Company Motion-Capture-Bühne mit einer Kombination aus mehreren simultanen Motion-Capture-Geräten aufgenommen, darunter ein elektronischer Handschuh, ein magnetischer Ganzkörper-Capture-Anzug und ein optisches System zur Aufnahme der Gesichtsbewegungen. Dies lieferte den Animatoren eine große Palette an Klings Bewegungen und Emotionen, u. a. auch subtile Gewichtsverlagerungen des Körpers und feinste Veränderungen der Mimik.

Kernstück der Produktion von „Huzzah" ist das umfassende Softwaresystem namens ProMotion von LambSoft (einem Schwesterunternehmen von Lamb&Company), das den Animatoren das Betrachten, Filtern, Editieren und Interpretieren der multiplen eingefangenen Bewegungsdatenströme ermöglicht.

Larry Lamb (USA) is director and head of the Minneapolis based Lamb & Company Inc. The Company is specialized in animation, special effects and post production needs for TV, film, video, games and forensic presentations.

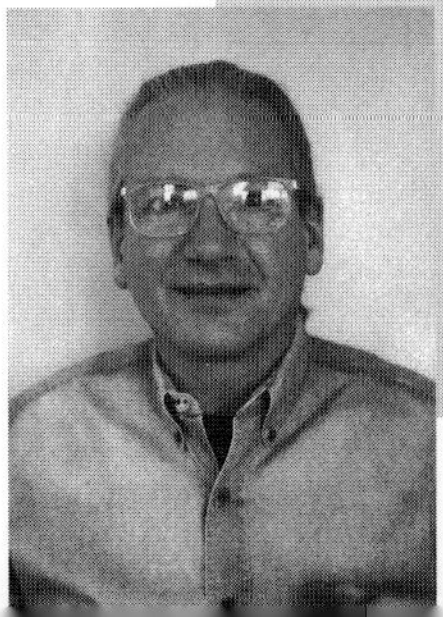

Larry Lamb (USA) ist Regisseur und Chef bei Lamb & Company Inc. in Minneapolis. Diese Firma hat sich auf Animation, Special Effects und Nachbearbeitungsverfahren für Fernsehen, Film, Computerspiele und die Visualisierung statistischer Erhebungen spezialisiert.

Busby

Anna Henckel-Donnersmarck

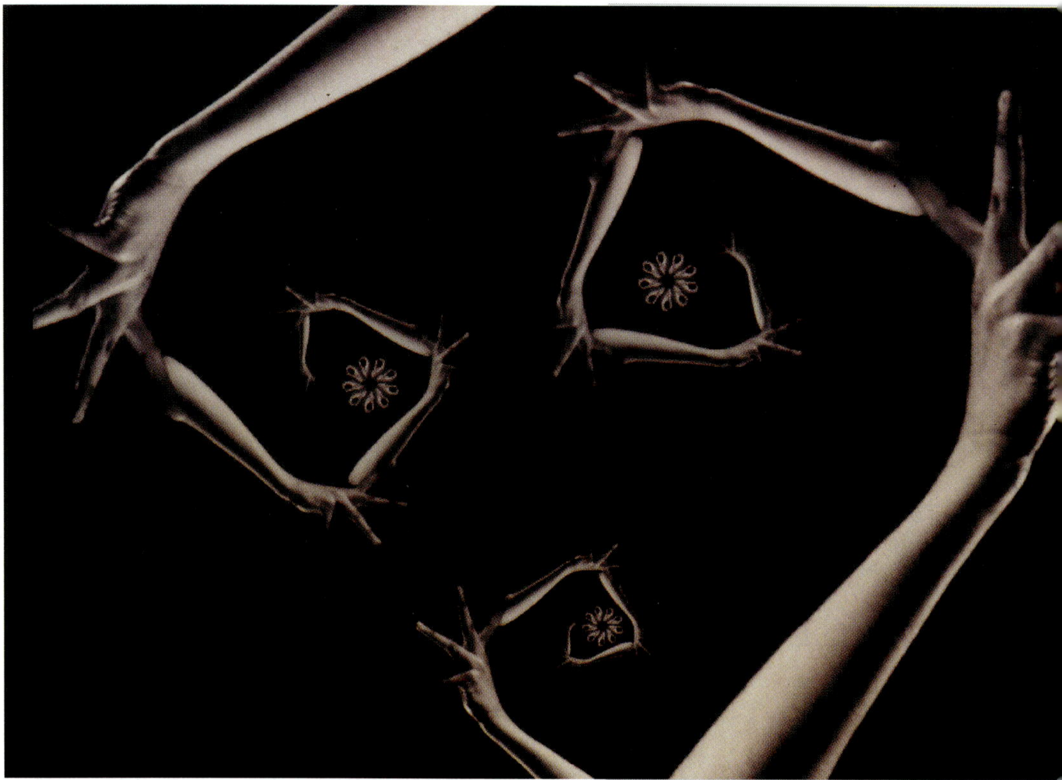

My intention was not to imitate reality using a computer (as with special effects), but rather to use the possibilities that are only available through a computer to create a new reality. The clear, mathematical world of the computer meets the soft, flowing world of human movements, and the conjunction results in a surprising diversity of ornaments and choreography. Yet at the same time, it is only by conforming to certain mathematical laws that each ornament and every choreography is able to attain a harmonious aesthetic (as it is also found in nature, if one examines the construction of blossoms, for example).

The title, but also and especially the music and the black and white aesthetic are reminiscent of the glamorous era of Hollywood musicals.

Busby Berkeley was a musical choreographer, whose break-through in Hollywood came in 1933 with his dance num-

Meine Intention war es, nicht die Realität mit Hilfe des Computers zu imitieren (wie bei Special Effects), sondern vielmehr mit Mitteln, die nur dem Computer zur Verfügung stehen, eine neue Realität zu kreieren. Die klare, mathematische Welt des Computers trifft auf die weiche, fließende Welt menschlicher Bewegungen, und durch die Verbindung entsteht eine überraschende Vielfalt an Ornamenten und Choreographien. Gleichzeitig erlangt jedes Ornament und jede Choreographie nur dann eine harmonische Ästhetik, wenn sie bestimmte mathematische Gesetze befolgt (wie es z. B. in der Natur auch zu finden ist, wenn man sich den Aufbau von Blüten etc. anschaut). Aufgrund des Titels, aber vor allem aufgrund der Musik und der Schwarz-Weiß-Ästhetik werden Erinnerungen an die glamouröse Zeit des Hollywood

bers for "42nd Street", "Footlight Parade", and "Gold Diggers". Later, he also created water ballets for Esther Williams.

Connecting isolated hand movements with 2D computer animation results in a complex, dancing ornament borne by a trace of Hollywood. Individual hand and arm movements were first shot on beta in front of blue to create a kind of basic vocabulary. In the computer (Matador), these were then released, multiplied and animated. Many of the animation ideas were already there before filming started, but many others emerged through the course of the animation process or the more in-depth exploration of the raw material. The music was already determined at the start and was used as a point of reference for the tempo and accents during the animation process. Image and music were then edited together on Avid and a dramaturgy was developed.

Musicals geweckt. Busby Berkeley war ein Musical-Choreograph, der 1933 mit seinen Tanznummern für „42nd Street", „Footlight Parade" und „Goldiggers" in Hollywood seinen Durchbruch hatte. Später entwarf er auch die Wasserballette für Esther Williams. Durch die Verbindung von isolierten Handbewegungen mit 2D-Computeranimation entsteht ein komplexes, tanzendes Ornament, getragen von einem Hauch von Hollywood.

Zunächst wurden einzelne Hand- und Armbewegungen vor Blau auf beta aufgenommen, um eine Art Grundvokabular aufzubauen. Diese wurden dann im Computer (Matador) freigestellt, multipliziert und animiert. Viele Animationsideen gab es bereits vor der Drehphase, viel ergab sich aber auch erst beim Animieren bzw. bei der genaueren Auseinandersetzung mit dem Rohmaterial. Die Musik stand von Anfang an fest und wurde während des Animationsprozesses als Referenz, was Tempo und Akzente angeht, benützt. Im Schnitt auf Avid wurden dann Bild und Musik zusammengebracht und eine Dramaturgie ausgearbeitet.

Anna Henckl-Donnersmarck (D), born 1973 in Germany, grew up in Indonesia, Japan and England; traveled extensively. Worked at odd jobs after finishing school, before studying art and design in London for a year at the Film Academy Baden-Württemberg since 1994.

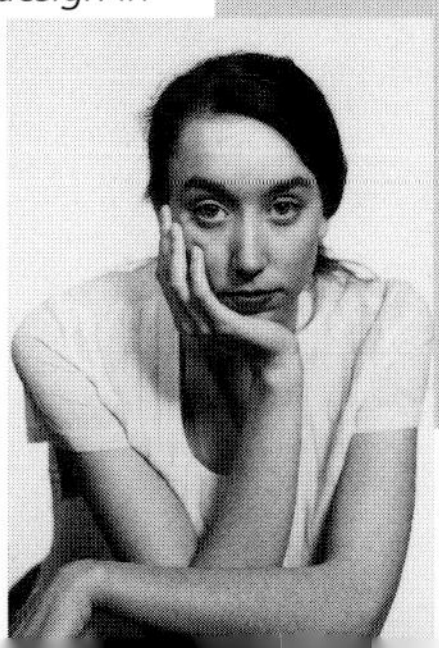

Anna Henckel-Donnersmarck (D), geb. 1973, in Deutschland, Indonesien, Japan und England aufgewachsen. Viel gereist. Nach dem Abi gejobbt, dann einjähriges Kunst- & Designstudium in London. Seit 1994 an der Filmakademie Baden-Württemberg.

An Artist

Michèle Cournoyer

"An Artist" is inspired by Article 29 of the United Nations Convention on the Rights of the Child, which states that all children have the right to develop their talents and abilities. In the current work, a young girl relegated to the kitchen escapes into a musical world of her own creation.

The director invented a story based on her personal experience. She prepared a script, storyboard and shooting script before doing the initial shooting on black and white film. A musical theme guided the gestures and attitude of the young actress. These live-action sequences were then edited and every second frame of the 16mm negative was digitized. The backgrounds and camera movements were redesigned and reproduced on the computer and the shots of the live-action characters were composited over them. Each frame was then "painted" in grey scale with an electronic felt pen. Musical notes were drawn on the images and randomly animated. The textures and coloured lighting effects were applied to the final images in batches.

„An Artist" wurde vom Artikel 29 der Konvention der Vereinten Nationen über die Rechte des Kindes inspiriert, der aussagt, daß alle Kinder das Recht auf freie Entwicklung ihrer Talente und Fähigkeiten haben. In der vorliegenden Arbeit flüchtet ein in die Küche verbanntes Mädchen in eine selbstgeschaffene musikalische Welt.

Die Regisseurin hat eine Geschichte erfunden, die auf ihrer persönlichen Erfahrung beruht. Auf der Basis eines von ihr erstellten Drehbuchs wurden die ursprünglichen Einstellungen auf Schwarzweiß-Material gedreht. Ein musikalisches Thema leitete die Gestik und Haltung der jungen Schauspielerin. Diese Live-Bilder wurden geschnitten und jeder zweite Kader des 16mm-Negativs digitalisiert. Die Hintergründe und Kamerabewegungen wurden auf dem Computer reproduziert und neu eingestellt, danach wurden die Aufnahmen der Schauspieler darübergelegt. Jeder Kader wurde mit Hilfe eines elektronischen Filzstiftes in Graustufen „bemalt". Noten wurden auf die Bilder gezeichnet und nach dem Zufallsprinzip animiert. Texturen und farbige Lichteffekte wurden gruppenweise auf die fertig gestalteten Bilder aufgetragen.

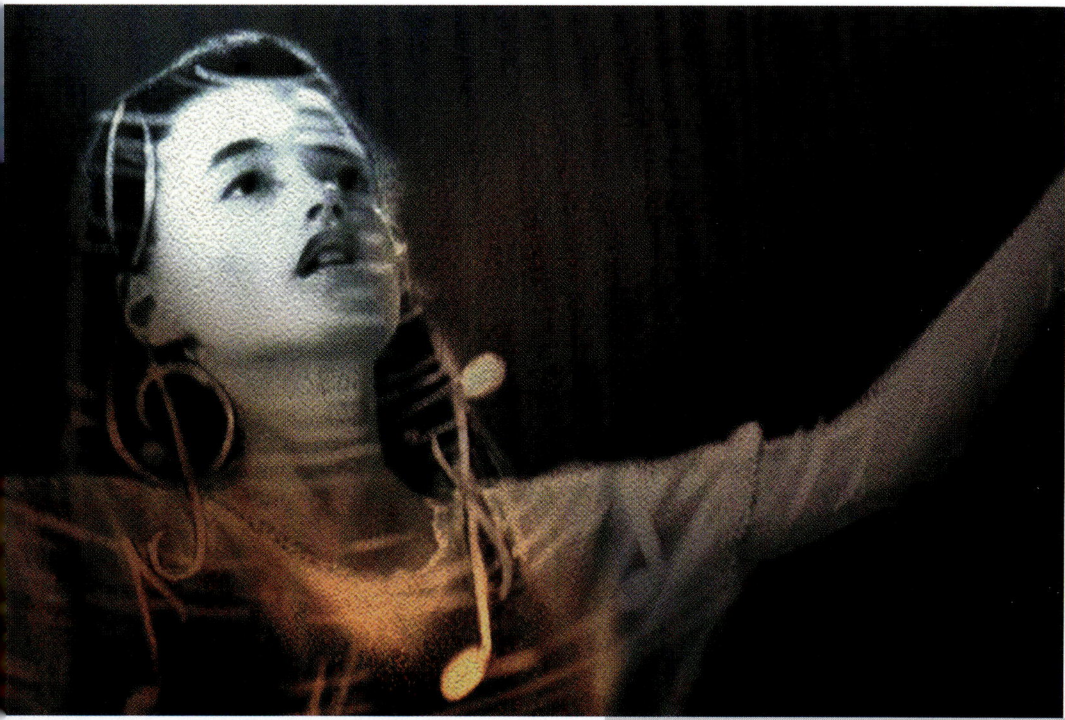

Michèle Cournoyer (CDN), born 1943, graduated from the École des beaux-arts in Quebec City and Montreal and went on to study graphic arts, photography and animation in England and Italy. Back in Quebec, she worked as a set designer, art director, costume designer and screenwriter on several important New Wave films. In 1989, she won the National Film Board of Canada French Animation Studio's 9th Cinéaste recherché(e) competition, enabling her to make her first NFB film, La Basse Cour/A Feather Tale.

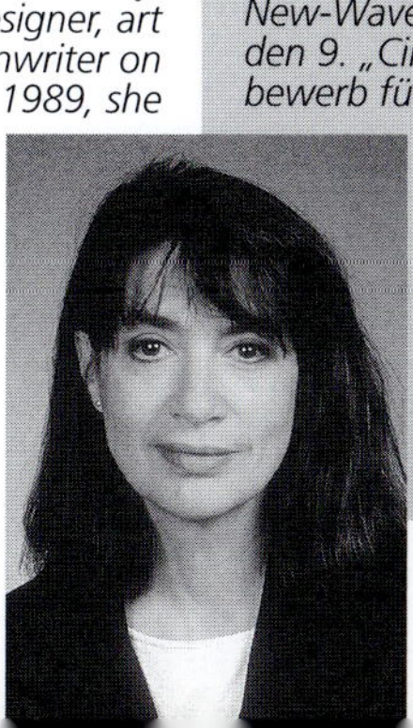

Michèle Cournoyer (CDN), geboren 1943, graduierte an der École des beaux-arts in Quebec City und Montreal und vervollkomnete ihre Studien aus Graphik, Photographie und Animation in England und Italien. Nach ihrer Rückkehr nach Quebec arbeitete sie als Regisseurin, Kostümdesignerin und Drehbuchautorin bei zahlreichen wichtigen New-Wave-Filmen. 1989 gewann sie den 9. „Cinéaste recherché(e)"-Wettbewerb für französischsprachige Animationsstudios des National Film Board of Canada, was ihr erlaubte, ihren ersten Film für das NFB, „La Basse Cour / A Feather Tale", zu drehen.

Sakuratei

Taku Kimura

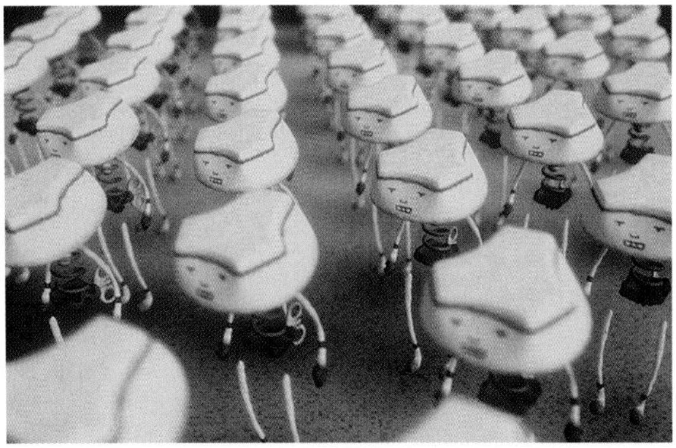

"Sakuratei" is an experimental piece which was made in co-operation with IMAGICA corporation. The work was directed by Koji Matuoka, and the music was made by Jun Miyake.

In this piece, we aimed to emphasize our unique expression. The story is set in the future, but the image itself is basically analogue, and the motif of the expression is "nostalgic and uniquely Japanese".

The following elements have contributed to the successful creation of this unique world:

* The actions in daily scenes using motion-capture (sweeping, drowsing, playing Shougi – Japanese chess, etc.).

* The natural movement using physical simulation for petals and background.

* The small and artificial space imaging a stage.

* Imagery with a small depth of field, which provides objects with specific existences.

* The shades and textures which are painted by hand (without using techniques like reflection, shade, transparency) to remove the typical touches of computer-generated images.

* Coloring infinitely close to monochrome.

* Unchangeable expressions just like Noumen (a mask used in the Japanese play No).

* The beautifully melodious music, which is performed by instruments.

„Sakuratei" entstand als experimentelles Stück in Zusammenarbeit mit der IMAGICA corporation. Regisseur war Koji Matuoka, die Musik stammt von Jun Miyake.

Wir haben versucht, für dieses Stück eine eigenständige Ausdrucksweise zu entwickeln. Die Handlung spielt in der Zukunft, aber das Bild als solches ist im wesentlichen analog, wobei wir gezielt eine „nostalgische", „typisch japanische" Anmutung erzielen wollten.

An der erfolgreichen Realisierung dieser einzigartigen Welt waren folgende Elemente beteiligt:

** Handlungen aus dem japanischen Alltagsleben (kehren, dösen, Shougispielen etc.) wurden über Motion-Capture eingefangen.*

** Die natürliche Bewegung von Blütenblättern und Hintergrund entstand durch physikalische Simulation.*

** Der kleine und begrenzte Raum stellt eine Guckkastenbühne dar.*

** Die Bildwelt mit ihrer geringen Tiefenschärfe gibt den Objekten eine besondere Existenzform.*

** Schatten und Texturen wurden von Hand gemalt (ohne Techniken wie Schattierungen, Reflexionen, Transparenz), um die typische „Computerästhetik" zu vermeiden.*

** Die Farbgebung ist so nahe wie möglich am Monochromen geblieben.*

** Gesichtsausdrücke sind wie bei einem „Noumen" (der Maske der No-Schauspieler) unveränderlich.*

** Die wundervoll melodiöse Musik wurde auf echten Instrumenten gespielt.*

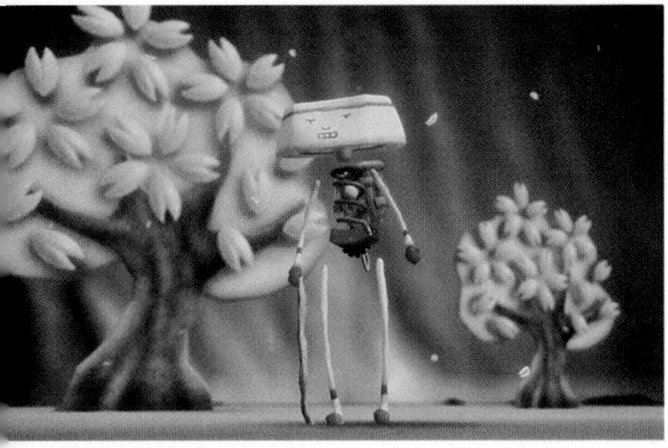

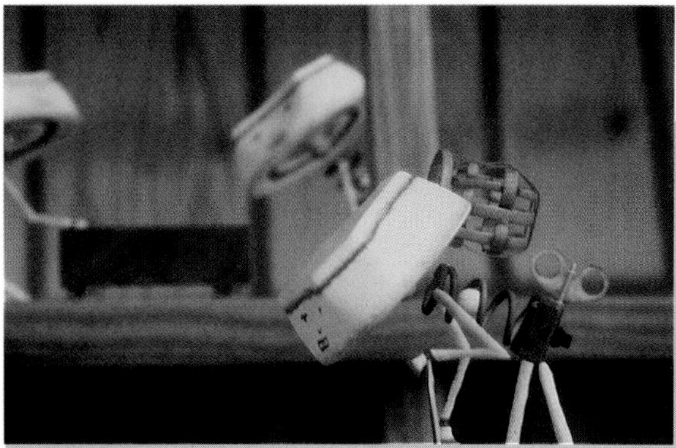

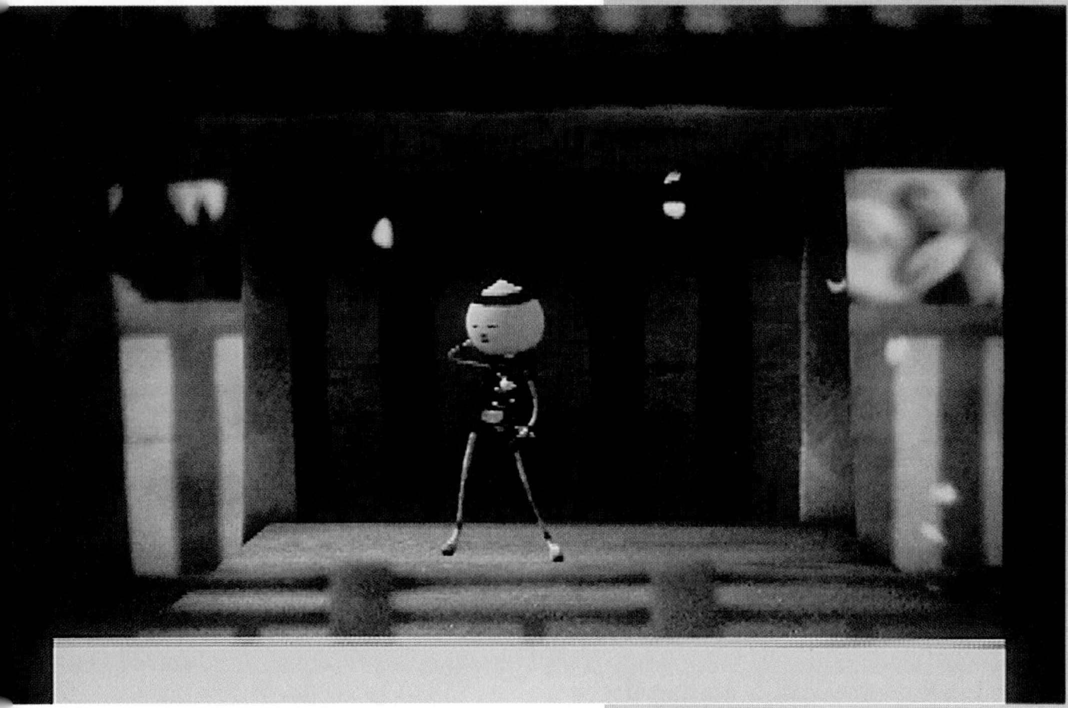

Taku Kimura (J), born 1963, studied fine art at the Art Department of Nihon University. He works as a CGI supervisor of LINKS Corporation, and he is also a part-time instructor at the Art Department of Nihon University. He was invited to IMAGINA '89 as a presenter for a lecture, and attended the exhibition "Ape Call From Tokyo" 1990 – 1992 (New York, Los Angeles, Tokyo).

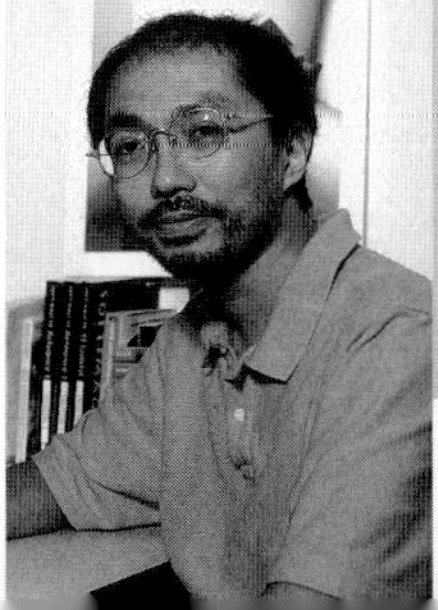

Taku Kimura (J), geb. 1963, hat Kunst am Art Department der Nihon University studiert und arbeitet als Computergraphik-Supervisor bei Links Corporation. Daneben ist er Teilzeit-Lektor an der Fakultät für Kunst an der Nihon University. Er war als Vortragender zur Imagina '89 eingeladen und hat an der Ausstellung „Ape Call from Tokyo" (New York, Los Angeles, Tokio 1990–92) teilgenommen.

The Tapir

Raquel Coelho

Rather than the typical western way of storytelling whereby the audience expects a moral statement at the end of a story, "The Tapir" has an unusual narrative. "The Tapir" presents a conflict, develops it into an action-climax situation and surprises the audience by ending in suspense. Because the narrative in "The Tapir" is a myth, it can contain numerous meanings. There is also an underlying notion that the story was always there, complete.

In my creative process I intend to search for similarities between ideas that have guided musical composition in the twentieth century and the fundamental elements of storytelling. I intend to extract from stories the essential structure of the narrative, starting with the descriptive outer layer of characters, places and situations, and ending with the inner layer where just the structure of a narrative is revealed. My sequential images can be read as a musical score, or vice versa.

There are three main reasons I have chosen contemporary music as a template for my piece: the structural aspect of the musical composition, its close connection with the studies of physics of sound, and its singular understanding of the graphic translation of sound events.

In "The Tapir" I have chosen the very same approach.

When analyzing the narrative structure of "The Tapir", the three ideas that I have chosen to guide the visual construction of the animation are: a traditional harmonic construction ending in suspension instead of repose; the simultaneous layering of events; and an atypical beginning that is filled with elements (props, characters, situations) which will all disappear as the story progresses.

In terms of narrative, the whole story is divided into nine scenes, each one corresponding to a different layer within the same space. The camera moves from the

Anders als die typisch westliche Erzähltradition, bei der das Publikum eine Moral am Ende der Geschichte erwartet, hat „The Tapir" eine unübliche Erzählform: „The Tapir" präsentiert einen Konflikt, entwickelt ihn in eine Situtation ohne Höhepunkt und endet offen und voller Spannung. Da die erzählte Geschichte eine Legende ist, kann sie zahlreiche Bedeutungen enthalten, sie impliziert auch, daß diese Legende seit jeher da gewesen ist.

Bei meinem kreativen Prozeß versuche ich, Ähnlichkeiten zwischen den Gedanken, die die musikalische Komposition im 20. Jahrhundert geleitet haben, und den grundlegenden Elementen des Geschichtenerzählens zu finden. Ich bemühe mich, aus den Geschichten den fundamentalen narrativen Gehalt herauszudestillieren, indem ich mit der beschreibenden äußeren Schicht (Personen, Ort, Situation) beginne und mit dem Kern, der den eigentlichen erzählerischen Inhalt umfaßt, ende. Meine sequentiellen Bilder können wie eine Partitur gelesen werden – und umgekehrt.

Es gibt drei Gründe, warum ich die zeitgenössische Musik als Vorbild für mein Stück genommen habe: Der strukturelle Aspekt der musikalischen Komposition, ihre enge Verbindung mit dem Studium der Klangphysik und ihr einzigartiges Verständnis der graphischen Umsetzung von Klangereignissen.

Auch im „Tapir" habe ich diesen Ansatz gewählt. Analysiert man die Erzählstruktur von „Tapir", so leiten drei Grundgedanken die visuelle Komposition der Animation: eine traditionelle harmonische Konstruktion, die in Spannung (und nicht in Auflösung und Ruhe) endet, die simultane Überlagerung von Ereignissen und ein atypischer Beginn, der mit Elementen (Requisiten, Charakteren, Situationen) überladen ist, die mit dem Fortschreiten der Geschichte verschwinden.

Vom erzählerischen Standpunkt aus ist die Geschichte in neun Szenen geteilt, jede davon nimmt eine eigene Ebene innerhalb

front layer to the last, passing by all the different events. The simultaneity of the events is depicted in the nine layers which are horizontally superimposed. From the point of view of the first layer one can see part of each one of the subsequential layers.

desselben Raums ein. Die Kamera bewegt sich von der vordersten Ebene zur hintersten und fährt so durch all die unterschiedlichen Ereignisse. Die Gleichzeitigkeit der Ereignisse wird durch die horizontale Überlagerung dieser Ebenen dargestellt. Vom Blickpunkt der vordersten aus kann man Teile aller nachfolgenden Schichten sehen.

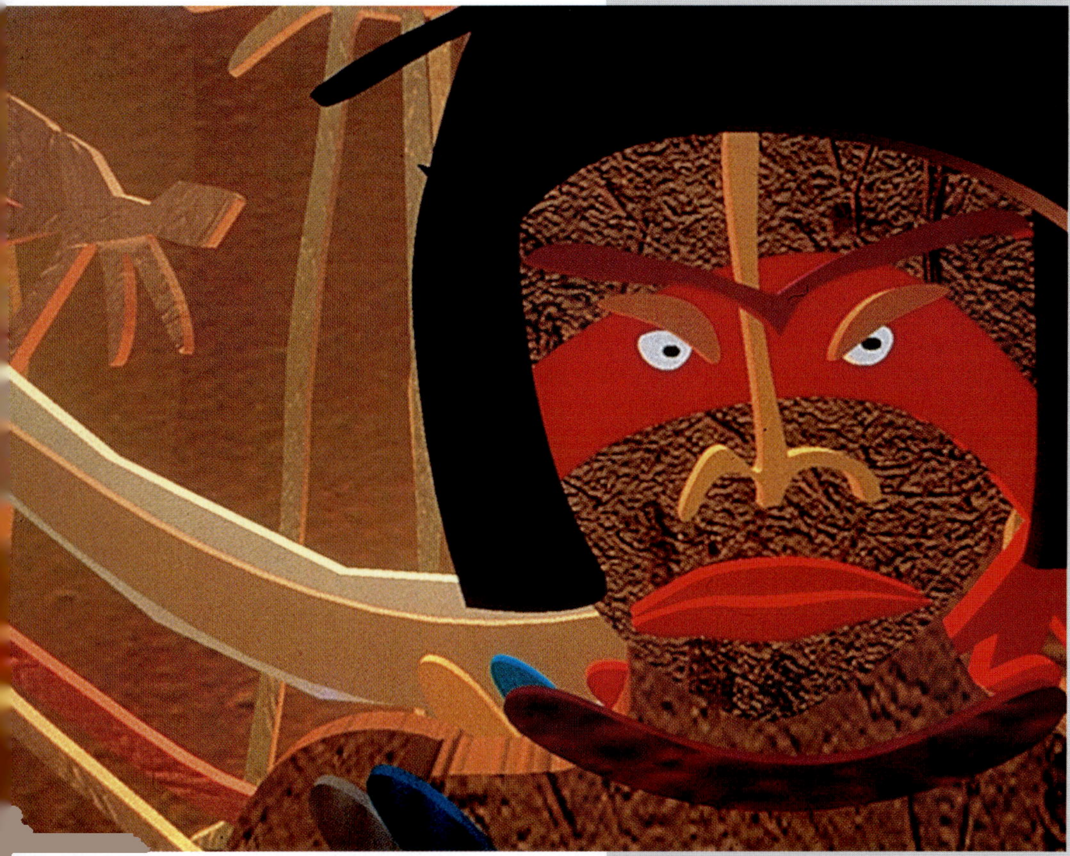

Raquel Coelho (RA), born 1965, started her career as a musician, playing viola in various orchestras and chamber groups of São Paulo. In 1986 she shifted from music to design, completing a major degree in Industrial and Graphic Design at FAAP (São Paulo) in 1990. In 1994 Raquel came to New York City interested in studying multimedia, and she started a Masters Degree course of Fine Arts/Computer Art at the School of Visual Arts. The story of "The Tapir" is part of Raquel Coelho's thesis at the School of Visual Arts.

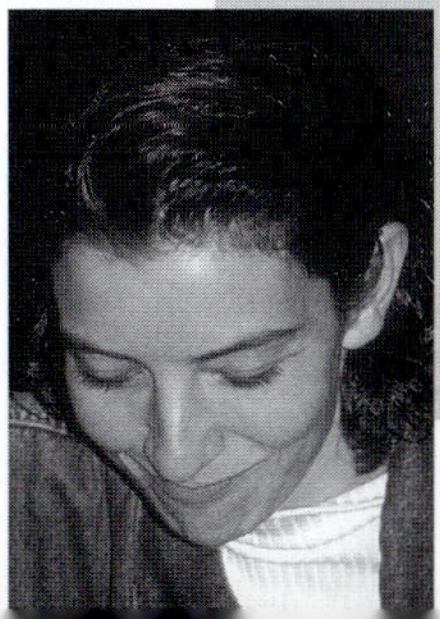

Raquel Coelho (BR), geb. 1965, begann ihre Karriere als Musikerin und spielte in verschiedenen Orchestern und Kammermusikensembles in São Paulo. 1986 wechselte sie von der Musik zum Design und schloß 1990 ihr Studium im Bereich Industrial and Graphic Design am FAAP (São Paulo) ab. 1994 kam Raquel Coelho nach New York City, um Multimedia zu studieren. Sie belegte einen Masters-Kurs in Bildender Kunst und Computerkunst an der School of Visual Arts. „The Tapir" ist Teil ihrer Abschlußarbeit zum MA.

Big Deal

Carlos Saldanha

"The Big Deal" had all of the elements that could have become a very boring commercial: a gray telephone wire against a plain white background ... and 60 seconds to fill. And that is what got Blue Sky director Carlos Saldanha excited about the project. The simplicity of the design allowed him to concentrate on developing "Jack's" personality. He wanted to create a realistic-looking character that would be fun to watch.

A considerable amount of effort went into staging and executing the wire's movements for each scene. Jack's performance was designed to highlight each feature offered by the telephone company. The trick was to animate the cord to create strong poses. For the "call waiting" scene for example, Jack taps his foot impatiently as if waiting for a call. In each scene the phone jack "head" was blocked out and animated first, creating the impetus for the animation of the phone cord. Subtle wiggles and kinks were added for realism. At one point, the wire transforms itself into the number 9, where a wave begins at the head and carries back to the end of the tail. The follow-through makes it look alive.

Some scenes were more complex than others, especially the one in which nimble Jack scrutinizes a Caller ID box. The box was modeled in Softimage, but Blue Sky's software was used to assign materials and properties to the box to make it look real. Things also got tricky when creating the three-headed wire. It took a little programming wizardry to get three wires to spring out of the main wire and look seamless. The subsequent scene, in which Jack stiffens up and imitates a periscope, was more straightforward. That scene was very simple and did not require fancy animation, but the sound of the submarine made it come to life. The sound lets you know exactly what was going on.

„The Big Deal" hat so ziemlich alles, was ein besonders langweiliger Werbespot braucht: ein graues Telephonkabel vor einer weißen Wand – und 60 Sekunden, die damit gefüllt werden wollten. Und genau das machte dem Regisseur Carlos Saldanha von Blue Sky besonderen Spaß: Die Einfachheit des Designs erlaubte ihm, sich auf die Entwicklung einer Persönlichkeit für „Jack" (jack = Stecker) zu konzentrieren. Er wollte einen realistisch aussehenden Charakter schaffen, den zu beobachten Spaß machen sollte.

Eine Menge Aufwand wurde getrieben, um die Bewegungen des Kabels für jede Szene zu gestalten. Jack sollte jede der von der Telephongesellschaft angepriesenen Eigenschaften in seinem Verhalten reflektieren. Der Trick bestand darin, die Leitung so zu animieren, daß ausdrucksstarke Posen entstanden. Für die Funktion „Anruf wartet" etwa tappt Jack ungeduldig mit dem Fuß auf den Boden, eben als würde er auf einen Anruf warten. In jeder Szene der Animation wurde Jacks Steckerkopf herausgehoben und als erster animiert und gab damit den Ausschlag für die Animation des Kabels. Zur Steigerung des Realismus wurden subtile Knicke und Kurven in die Leitung eingebaut. Einmal verwandelt sich das Kabel in die Ziffer neun, wobei eine Welle am Kopf beginnt und bis zum „Schwanzende" durchläuft. Eben diese konsequente Durchführung läßt das Kabel lebendig aussehen.

Einige Szenen sind komplexer als andere, etwa jene, in der Jack geschickt eine Anrufer-ID-Box untersucht. Die Box wurde in Softimage modelliert, aber Material und Eigenschaften der Box wurden mit Blue Skys eigener Software festgelegt, um sie realistisch aussehen zu lassen. Schwierig wurde es auch, als das dreiköpfige Kabel gestaltet werden sollte. Es bedurfte einiger programmiertechnischer Zaubertricks, um drei Kabel aus dem

Blue Sky modeled and animated with Softimage. The final render was done with Blue Sky's proprietary software, called CGI Studio™, which gave the wire its realistic look.

einen herauswachsen zu und das Ganze echt aussehen zu lassen. Die nächste Szene, in der Jack ein Periskop imitiert, war weniger aufwendig – Jack wird steif, was keine besonderen Animationstricks erfordert, und der U-Boot-Klang legt jedem klar, worum es geht.

Blue Sky modellierte und animierte mit Softimage, das endgültige Rendering erfolgte mit Blue Skys eigener Software (CGI Studio™), die dem Kabel das realistische Aussehen verlieh.

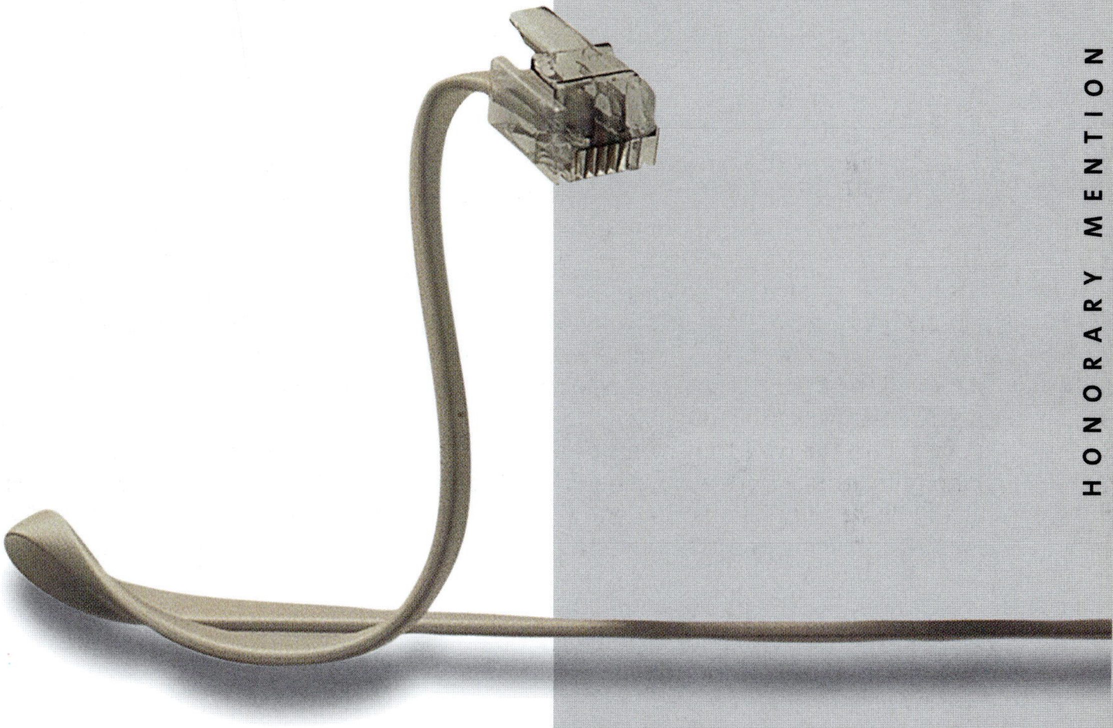

Carlos Saldanha (USA) is Animation Director at Blue Sky Studios Inc. He joined the company after receiving his Masters Degree in computer art from the School of Visual Arts New York. Carlos Saldanha has directed and animated numerous award winning commercials for Blue Sky.

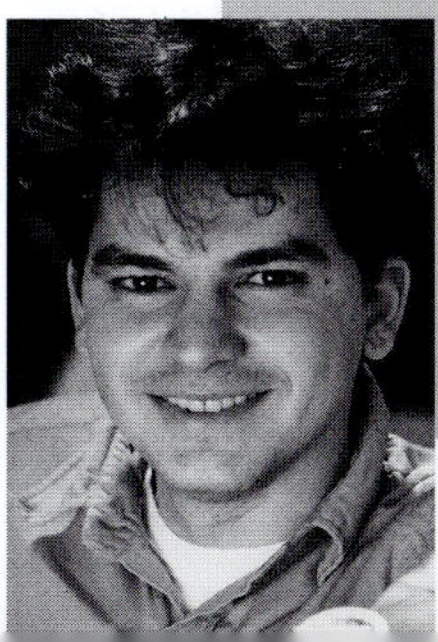

Carlos Saldanha (USA) ist Animation Director bei Blue Sky Studios Inc. Er trat in die Firma ein, nachdem er zum Master aus Computerkunst an der School of Visual Arts New York graduiert war. Carlos Saldanha hat zahlreiche preisgekrönte Animationen für Blue Sky gestaltet und geleitet.

"Levi's" Commercial

Michel Gondry / BUF

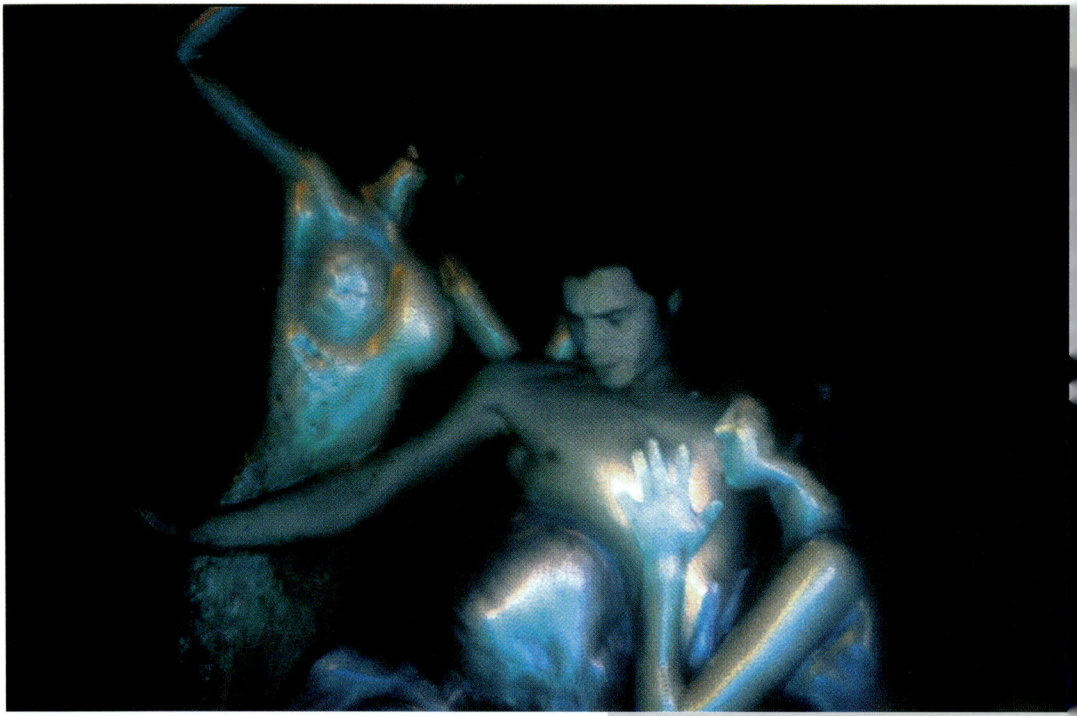

Computer graphic effects on real images

We created this method in collaboration with Michel Gondry. It requires special lighting during the shooting, which makes it possible to analyze the image and create 3D volume. Using this volume, it is possible to recreate the lighting with computer-generated virtual lights.

In order to retain a realistic aspect, we used images of real fish to make the mermaids' tails and then matched the tails to the actresses using a warping process.

We used the software Softimage.

Computergraphische Effekte auf echten Bildern

Diese Methode wurde in Zusammenarbeit mit Michel Gondry entwickelt und verlangt nach einer speziellen Beleuchtung während der Aufnahme, was eine spätere Analyse des Bildes und die Schaffung eines dreidimensionalen Volumens ermöglicht. Auf der Basis dieses errechneten Volumens kann die Beleuchtung mit computergenerierten virtuellen Lichtquellen nachgebaut werden.

Um eine realistische Gestaltung zu ermöglichen, haben wir die Schwänze der Meerjungfern aus echten Fischen konstruiert, die mittels eines Warp-Prozesses an die Schauspielerinnen angepaßt wurden.

An Software wurde Softimage eingesetzt.

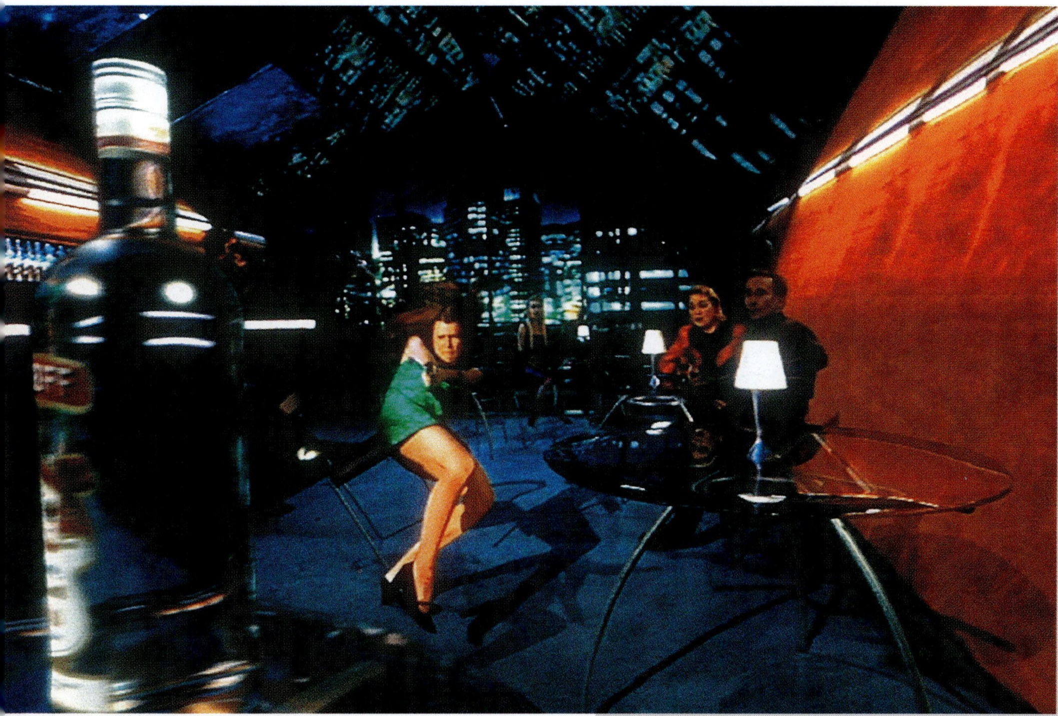

Michel Gondry (F), born 1964, musician and director of video clips, has rocketed to international recognition with a series of dazzling, jaw-dropping videos; he began animation at art school although it was not his major subject for study.

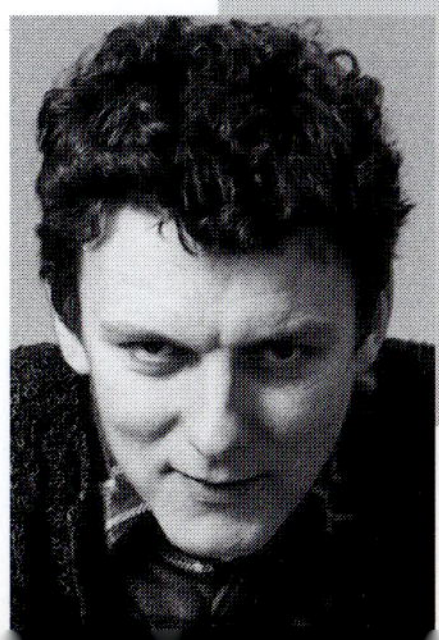

Michel Gondry (F), geb 1964, Musiker und Videoclipregisseur, hat bereits eine steile Karriere mit einer Reihe anerkannter beeindruckender Videos hinter sich; Animation hatte er während des Kunststudiums begonnen, obwohl sie nicht sein Hauptfach war.

"Coca Cola" Commercial

Michel Gondry / BUF

Computer graphic effects

We reconstructed the movement of a camera using synchronized still frames shot from different perspectives. For this we combine several layers of 2D morphing with 3D models using BUF stereo mapping.

This method is called stereo programmetrie and uses the software Softimage, Elastic reality, Tmorph, and special BUF software.

Computergraphische Effekte

Wir haben eine Kamerabewegung über Standbilder rekonstruiert, indem wir aus verschiedenen Perspektiven aufgenommene, synchronisierte Standbilder miteinander verbunden haben. Dazu wurden mehrere Ebenen von 2D-Morphing mit 3D-Modellen unter Verwendung von BUF-eigenem Stereo-Mapping kombiniert.

Diese Methode wird Stereo-Programmetrie genannt und verwendet die Software Softimage, Elastic reality und Tmorph sowie hauseigene Software von BUF.

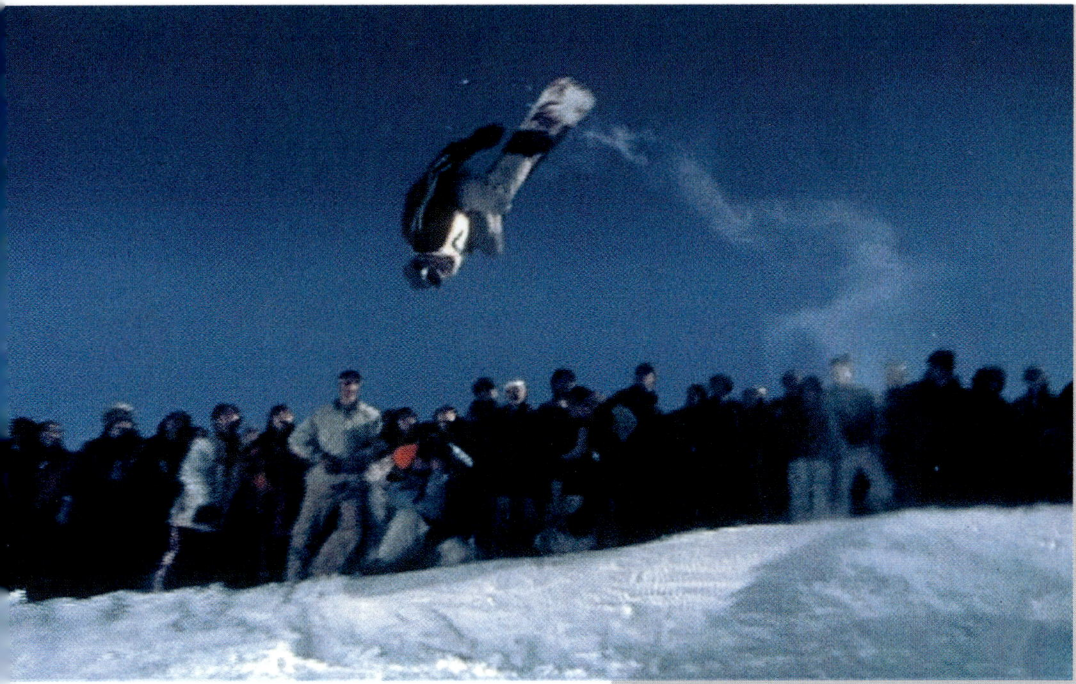

Michel Gondry (F), geb. 1964, Musiker und Videoclipregisseur, hat bereits eine steile Karriere mit einer Reihe anerkannter beeindruckender Videos hinter sich; Animation hatte er während des Kunststudiums begonnen, obwohl sie nicht sein Hauptfach war.

Michel Gondry (F), born 1964, musician and director of video clips, has rocketed to international recognition with a series of dazzling, jaw-dropping videos; he began animation at art school although it was not his major subject for study.

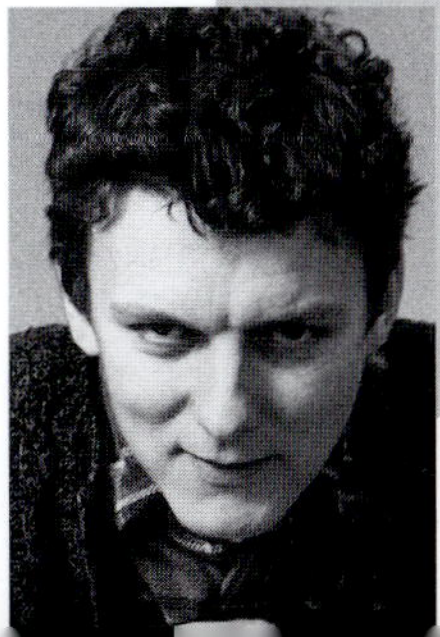

Contraption

Stephen Weston

"Contraption" is a thirty second piece directed by Stephen Weston that grabs your attention from the very first frame.

Animation House Bermuda Shorts was approached by the London Advertising Agency Kelly Weedon as they were looking for something different and striking to bring to life an idea they had for one of their clients "Fisherman's Friend". Their clever idea was to write the commercial around what some would see as the negative side of the product, a very strong throat sweet that is a little unpleasant on the first tasting.

And so the idea came about that the commercial would include a character who goes through all extremes of painful pleasure, one of these is the Fisherman's Friend lozenge, but however tortuous these moments are, he is very happy to do it again.

Stephen and Bermuda Shorts were keen to do something different with the script. Stephen wanted to create a room inside the computer that confused the viewer about its nature, using just three colours, black, white, and red, and a graphic black outline, the set took on a cartoon mood reminiscent of the oldest 2D animation tradition. Stephen then added a crazy checkerboard floor and unusual character, both of which were designed by the stop frame Director Boris Koshmehl.

To match the unusual look of the piece, Stephen looked to his 2D animation training for the storyboard and timing of the action. Rather like those old Tex Avery classics, no scene is longer than a few seconds and the viewer is carried along with the character as he is pulled and hoisted by the "Contraption" as it manhandles our character along a conveyor belt of experiences.

The final touch was a very interesting sound dub where the grunts and shouts of the character were added to the picture along with numerous sound effects, to bring the machine to life.

„Contraption" (etwa „verrückte Erfindung") ist ein 30-Sekunden-Stück unter der Regie von Stephen Weston, das vom allerersten Bild an die Aufmerksamkeit der Zuseher fesselt.

Die Animationsfirma „Bermuda Shorts" wurde von der Londoner Werbeagentur Kelly Weeden angesprochen, ob sie nicht etwas Besonderes und Griffiges produzieren könnte, um eine Idee für ihren Kunden Fisherman's Friend umzusetzen. Der Grundgedanke bestand darin, den Werbespot um etwas herum zu entwickeln, was man als die „negative Seite" des beworbenen Produkts – eines sehr starken „Rachenputzer"-Bonbons, das im ersten Moment ein wenig unangenehm schmeckt – ansehen könnte.

Und so entstand der Gedanke, daß in dieser Animation eine Figur alle Extreme peinvollen Vergnügens durchleben sollte und eines ist natürlich auch eine „Fisherman's Friend"-Pastille, aber so peinvoll die Erfahrung auch sein mag – die Gestalt will alles gleich nochmals erleben.

Stephen und Bermuda Shorts wollten das Script auch auf unübliche Weise umsetzen: Stephen schuf einen Raum im Computer, der den Betrachter hinsichtlich seiner Natur verwirrt, und verwendete nur die drei Farben Schwarz, Weiß und Rot sowie einen schwarzen graphischen Umriß. Das verlieh dem Ganzen eine Cartoon-Stimmung, die an die Veteranen der 2D-Animation erinnert. Dazu kamen noch ein wild karierter Boden und eine ungewöhnliche Figur, die beide vom Einzelbildregisseur Boris Koshmehl entworfen wurden.

Dem ungewöhnlichen Aussehen des Stückes entsprechend, griff Stephen auch für das Drehbuch und das Timing auf seine Erfahrung als 2D-Animator zurück. Wie in den guten alten Tex-Avery-Klassikern dauert keine Szene länger als ein paar Sekunden, und der Betrachter wird mitgerissen, wenn die Hauptfigur von der

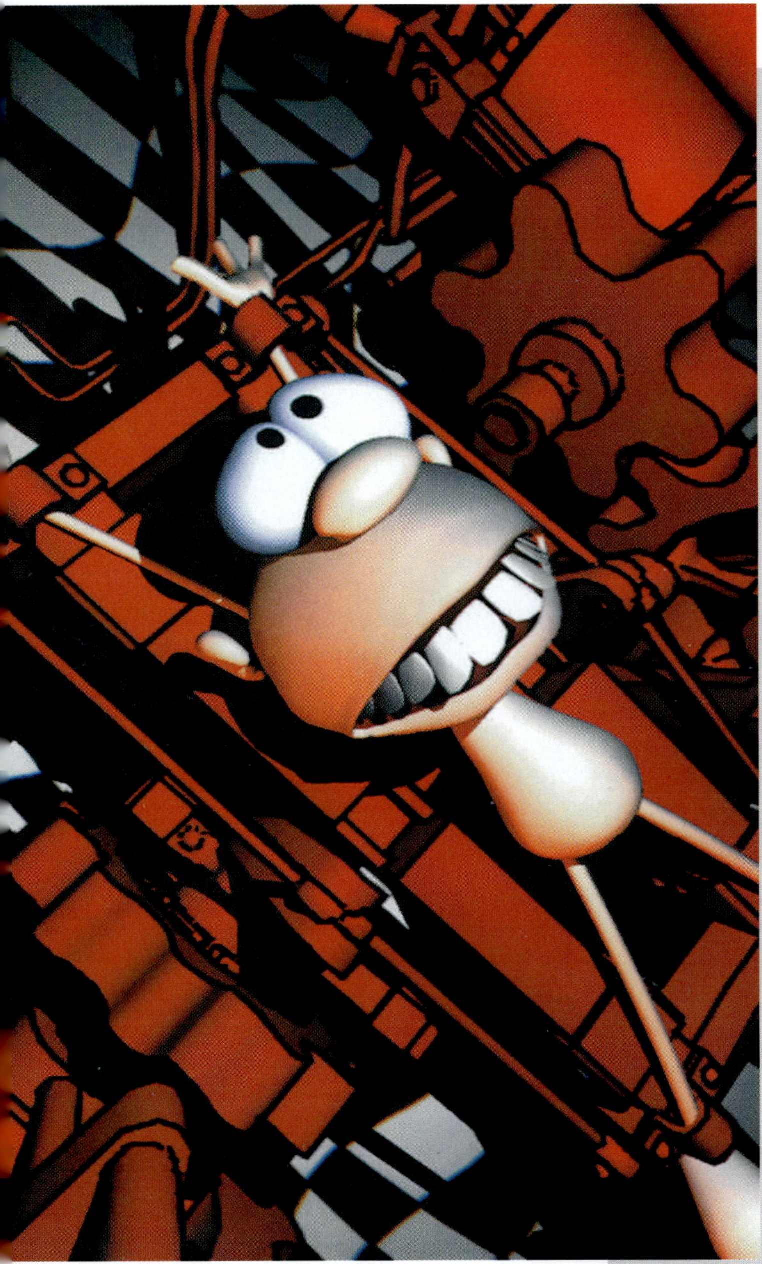

„verrückten Maschine" gezogen und gehoben und durch einen Wust von Erfahrungen gepreßt wird.
Zur Abrundung des Ganzen kam noch ein interessanter Soundtrack hinzu, in dem die Grunzer und Schreie der Figur dem Bild angepaßt und mit den zahlreichen Klangeffekten des „Klapparatismus" verwoben wurden.

Stephen Weston (CDN) graduated in classical drawn animation from Sheridan College in Canada. He contributed drawn animation sequences to several films, title sequences and commercials in Canada and Great Britain. Recently he became Director of CGI elements in a 26 part series for children.

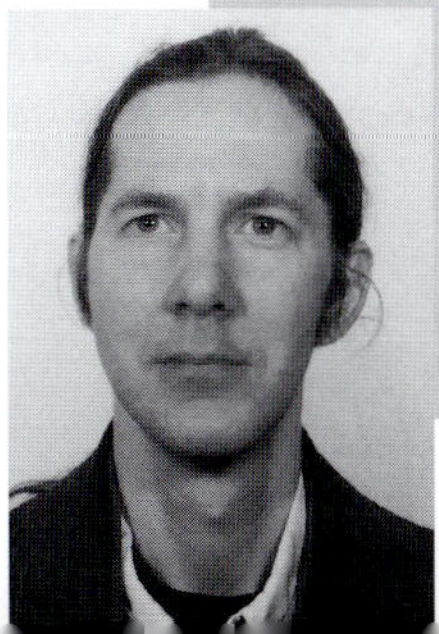

Stephen Weston (CDN) graduierte aus klassischer gezeichneter Animation am Sheridan College in Canada. Er lieferte die gezeichneten Animationssequenzen für zahlreiche Filme, Titelsequenze und Werbespots in Kanada und Großbritannien. Zuletzt war er bei einem 26teiligen Kinderfilm für die Regie der computeranimierten Teile verantwortlich.

Riante Contrée

François Vogel

In the cheerful country you disturb the tomatoes and mushrooms,

In the cheerful country beings touch and shout cries of joy

In the cheerful country I approach a graceful creature.

These two beings resemble phasmids (from gr. Phasma – apparition): wingless insects, whose long-shaped bodies resemble the twigs and leaves they live on (several types, called walking sticks, also live in the south of France) - according to "Petit Larousse".

But "Riante Contrée" tells an entirely different story ...

"It is an essay about multimedia," say some ...

"A film about the fragility of nature," say others...

"About the representations of crazy couples," you will say ...

But you are not there. Neither are the others. And neither am I.

This film was realized during the course of a post-graduate year at the computer science studio of the Ecole Supérieure des Arts Décoratifs de Paris (ENSAD). The students worked mostly on Silicon Graphics Workstations with Explore software from Alias Wavefront and common software for Macintosh, such as Photoshop, etc.

In der Lachenden Landschaft beunruhigen Sie die Tomaten und die Pilze,

In der Lachenden Landschaft berühren sich Wesen und stoßen Rufe der Freude aus,

In der Lachenden Landschaft nähere ich mich einer graziösen Kreatur.

Diese beiden Wesen ähneln den Phasmiden (von gr. Phasma – Erscheinung): Flügellose Insekten, deren länglicher Körper den Zweigen und Blättern ähnelt, auf denen sie leben (mehrere Arten, genannt Stabheuschrecken, leben auch im Süden Frankreichs) – wie es im „Petit Larousse" heißt.

Aber „Riante Contrée" erzählt von etwas ganz anderem ...

„Das ist ein Essai über Multimedia", sagen die einen ...

„Ein Film über die zerbrechliche Natur", sagen die anderen ...

„Über die Darstellung von verrückten Paaren", werden Sie sagen ...

Aber Sie sind nicht dort. Die anderen auch nicht. Und ich ebenfalls nicht.

Dieser Film wurde im Zuge eines Post-Graduate-Jahrs im Informatikstudio der Ecole Supérieure des Arts Décoratifs de Paris (ENSAD) realisiert. Die Studierenden arbeiteten zum Großteil auf Silicon Graphics Workstations mit Explore-Software von Alias Wavefront sowie mit der gängigen Software für Macintosh wie Photoshop etc.

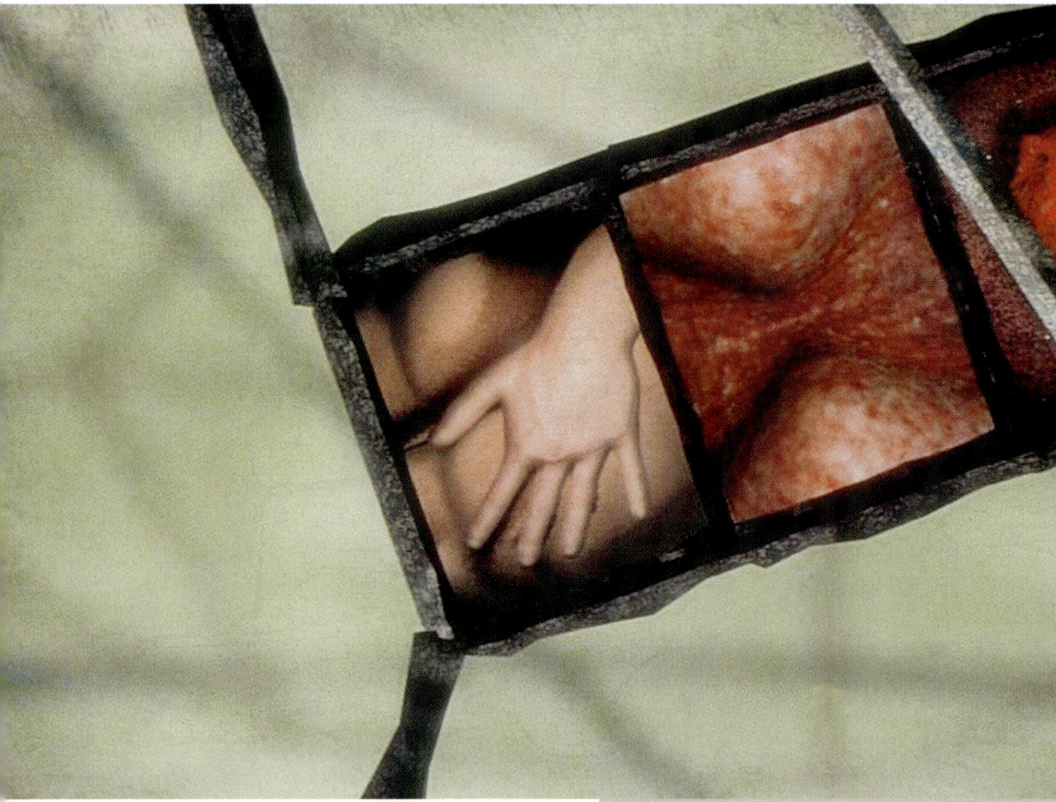

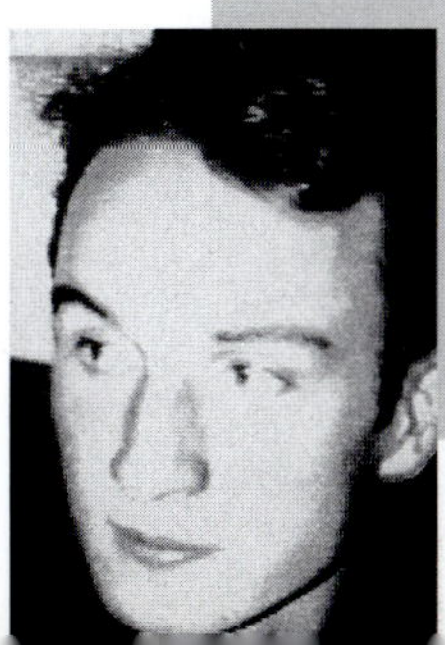

François Vogel (F) is a student at the École Nationale Supérieure des Arts Décoratifs (ENSAD) in Paris. He has worked as a video cutter and has had several short films and videoclips shown in France.

François Vogel (F) ist Student an der École Nationale Supérieure des Arts Décoratifs (ENSAD) in Paris. Er arbeitete als Video-Cutter und hat mehrere Kurzfilme und Videoclips in Frankreich veröffentlicht.

Adrénaline

Marine Poirson/Lionel Richerand

The entire city is asleep. Everything seems peaceful, yet for the inhabitants, this peace will not last long. It is a strange evening as Francine discovers to her dismay that one should not try to get money from the automatic teller when a monster from outer space is on the loose.

Frightened, she runs away as fast as she can to escape from the "Unidentified Fearsome Object". As Francine runs by screaming, her panic is passed on to the other inhabitants: a driver loses control of her car, little Chinese people's heads spin. But despite all her efforts, Francine is unable to shake off the creature from outer space. Rescue approaches in the person of a handsome stranger, who runs over the Martian with his rollerblades without even noticing, and then disappears into the night. In the meantime, Francine is so overcome by all the excitement that she faints ...

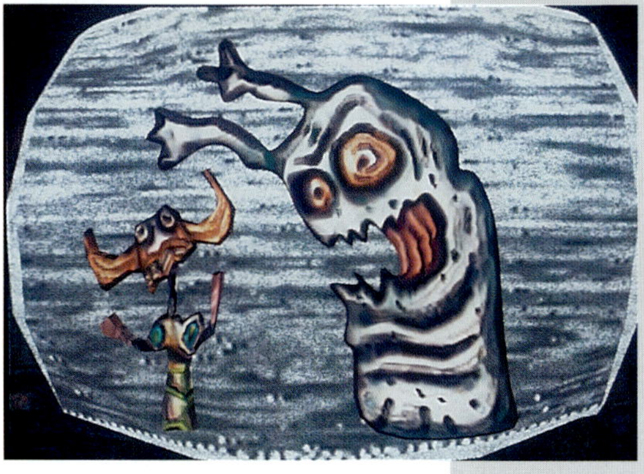

Marine Poirson and Lionel Richerand both come from the field of traditional animation. The preference for computer graphics that they share is sustained through cartoons, by Série Z and modeling clay.

The image design and surface treatment provide the space with a unique two-dimensional and yet perspectivist character, allowing them to play with visual gags that are only possible in computer animation.

This collaborative work is the result of a combination of various techniques and styles, and as such it reflects the fun and playfulness inherent to computer animation. The work was created during the course of a postgraduate year at the infor-

Die ganze Stadt schläft, alles scheint ruhig, aber diese Ruhe wird für die Bewohner nur von kurzer Dauer sein. Es ist ein seltsamer Abend, als Francine zu ihrem Schaden erfahren muß, daß man kein Geld aus dem Geldautomaten ziehen darf, wenn ein Weltraummonster umgeht.

Erschrocken flieht sie, so schnell sie kann, um diesem „Unidentifizierten Fürchterlichen Objekt" zu entkommen. Als Francine laut schreiend draußen vorbeiläuft, überträgt sich ihre Panik auf die anderen Bewohner: Eine Autofahrerin verliert die Herrschaft über ihr Fahrzeug, kleine Chinesen verdrehen die Köpfe. Und trotz aller Mühen gelingt es Francine nicht, das Wesen aus dem All abzuhängen. Ihre Rettung naht in der Person eines schönen Unbekannten, der – ohne es zu merken – das Marswesen mit seinen Rollerblades überfährt und in der Nacht verschwindet, während Francine, überwältigt von so viel Aufregung, in Ohnmacht fällt ...

Marine Poirson und Lionel Richerand kommen beide aus der traditionellen Animation. Ihre gemeinsame Vorliebe für Computergraphik nährt sich durch Cartoons, Série Z und Knetmasse.

Bildgestaltung und Flächenbehandlung geben dem Raum einen eigenartigen zweidimensionalen, aber perspektivischen Charakter, der es erlaubt, mit visuellen Gags herumzuspielen, wie dies nur die Computeranimation möglich macht.

Diese Gemeinschaftsarbeit ist Ergebnis einer Mischung unterschiedlichster Techniken und Stile und spiegelt damit die der in der Computeranimation steckende Freude und Fröhlichkeit. Das Werk entstand im Lauf eines Post-Graduate-

matics studio of the École Nationale Supérieure des Arts Décoratifs, Paris (ENSAD). Most of the film was created on Silicon Graphics workstations with Explore software from Alias Wavefront and with the aid of the usual Macintosh graphics software, such as Photoshop, etc.

Jahrs im Informatikstudio der École Nationale Supérieure des Arts Décoratifs, Paris (ENSAD). Der Großteil des Films entstand auf Silicon Graphics Workstations mit Explore-Software von Alias-Wavefront und mit Hilfe der auf dem MacIntosh üblichen Graphiksoftware wie Photoshop usw.

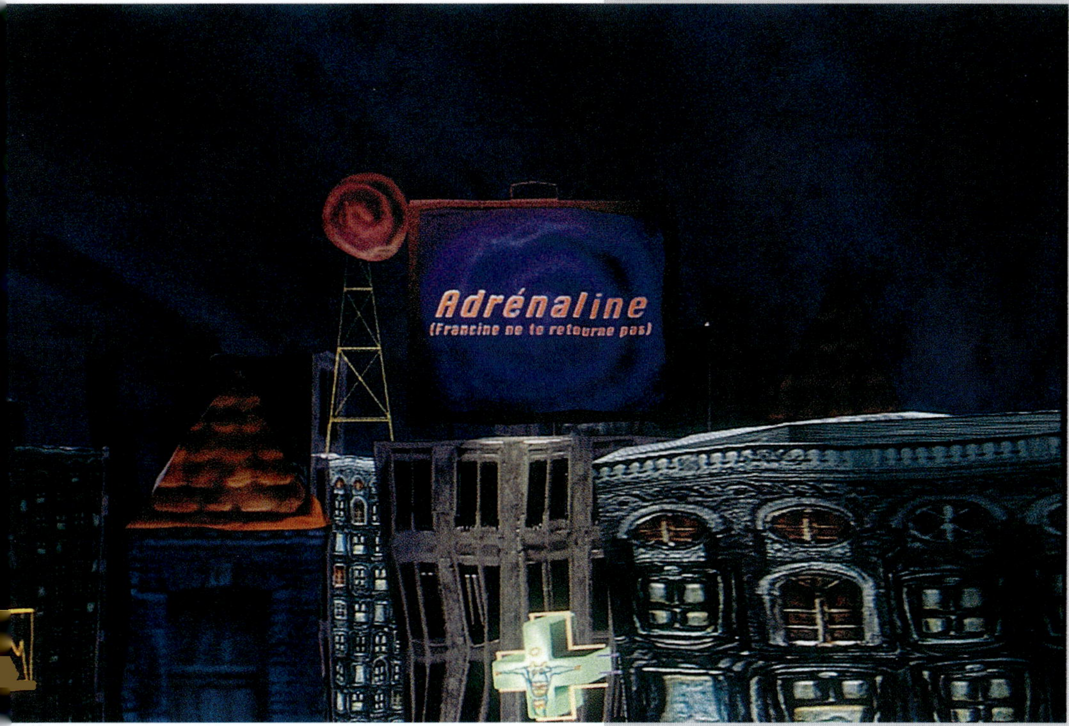

Marine Poirson and Lionel Richerand (F) are students at the École Nationale Supérieure des Arts Décoratifs (ENSAD) in Paris. Both have studied video, architecture, stage design and design. They have both had short films shown in France.

Marine Poirson und Lionel Richerand (F) sind Studenten an der École Nationale Supérieure des Arts Décoratifs (ENSAD) in Paris. Beide haben Abschlüsse in Video, Architektur, Bühnenbild und Design. Beide haben bisher Kurzfilme in Frankreich veröffentlicht.

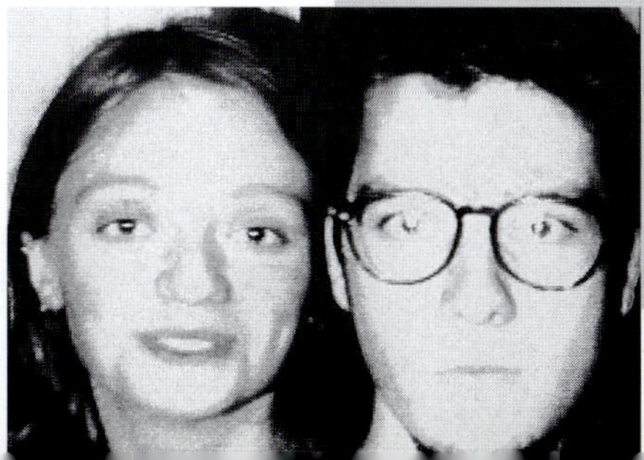

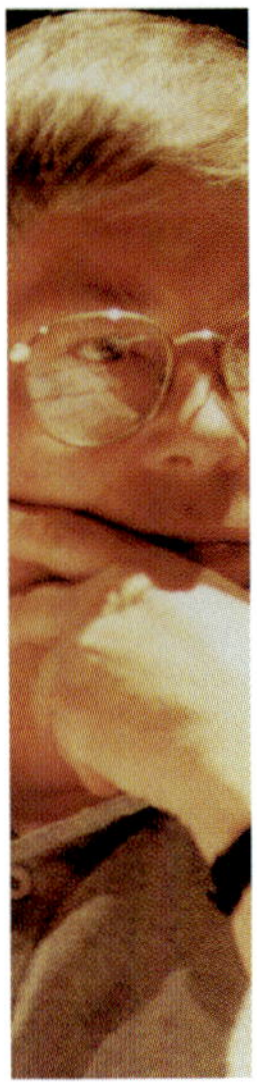
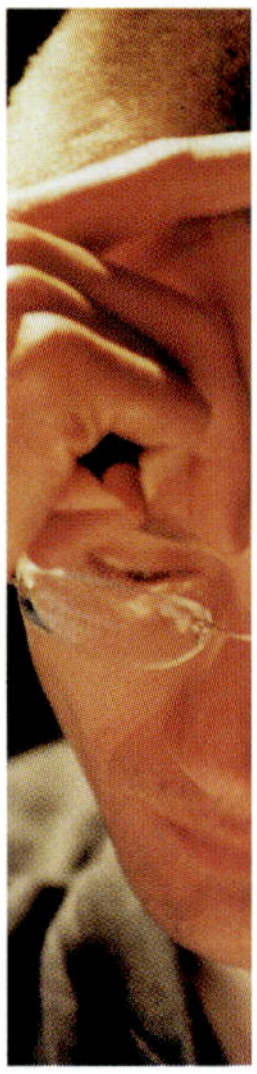
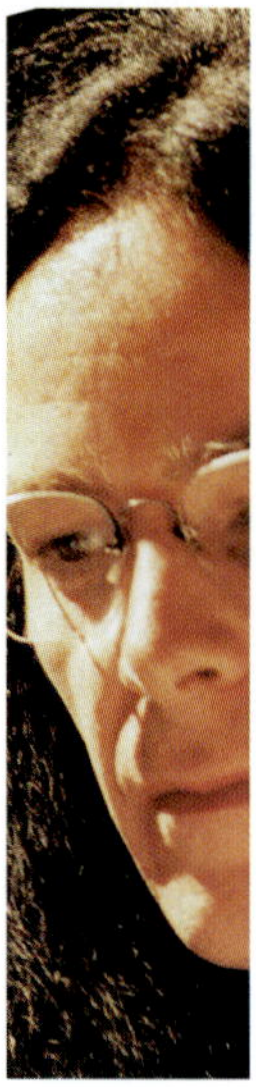

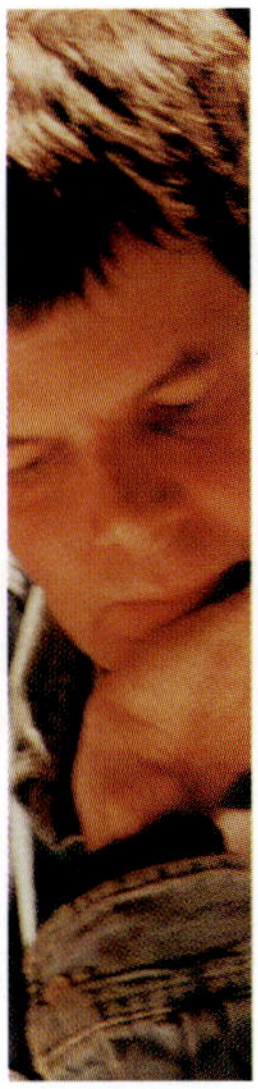

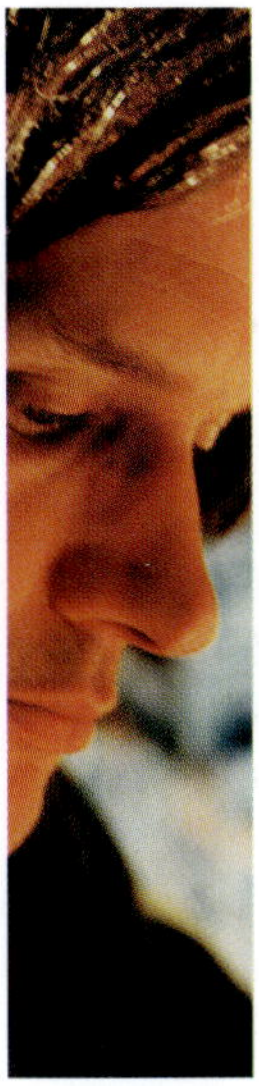

COMPUTER MUSIC JURY

Avoid those Clichés!

Vermeidet diese Klischees!

Statement of the Computer Music Jury

The Music Jury last year drew attention to some of the many categories of music and music-making which qualify to be dubbed "computer music", and to the way in which the practitioners of the genre had failed to realise that it had "matured" to such an extent that, in terms of both technique and gestural language, it had lapsed into cliché when it was clearly intending to be "serious".

True, the Jury encountered many an honourable exception, the prize-winning works among them, but listening to over 250 pieces in a short period of time made it embarrassingly clear that much computer music was becoming the oldest swinger in town.

This year's Jury comprised 80 per cent new personnel, with different backgrounds and interests, but their views were largely similar to those on which the Jury signed off last year. It too felt that originality and innovation, the qualities which the Prix Ars Electronica esteems the highest, had wilted under the pressure to master the extraordinarily pervasive set of software tools that appear to have become studio standard issue – including ProTools, GRM Tools and

Die Computermusik-Jury hat bereits im vergangenen Jahr darauf hingewiesen, daß es zahlreiche Kategorien in der Musik und im Musikschaffen gibt, die mit einiger Berechtigung „Computermusik" genannt werden können, und auch darauf, daß die Praktiker des Genres offensichtlich übersehen haben, daß diese Musik so weit „herangereift" ist, daß sie sowohl in Technik wie in Gestus in ein simples Klischee verfallen mußte, sobald sie „ernsthaft" sein wollte.

Sicher, die Jury hatte manch eine ehrbare Ausnahme angetroffen, darunter die preisgekrönten Arbeiten, aber beim Anhören von über 250 Stücken in kurzer Zeit wurde erschreckend deutlich, daß viel von der Computermusik mittlerweile zum Altbackensten weit und breit gehört.

Die Jury umfaßte heuer 80 Prozent neue Mitglieder, mit unterschiedlichem Hintergrund und Interesse, aber die An- und Einsichten entsprachen im wesentlichen jenen des vergangenen Jahres. Auch diesmal überwog der Eindruck, daß Originalität und Innovation – jene Qualitäten, die der Prix Ars Electronica am höchsten schätzt – zu kurz gekommen sind unter dem Druck, die enorm wachsende Menge an Werkzeugen einzusetzen, die scheinbar zum Studio-Standard geworden sind – ProTools, GRM Tools und Sound Hack, um nur einige zu nennen. Es wäre unsinnig, diese großartigen Werkzeuge zu verdammen, und das ist auch nicht die Absicht der Jury. Aber das übergroße Vertrauen in sie scheint selbst gute Komponisten dazu zu verleiten, den Ausdruck ihrer eigenen, persönlichen, distinktiven Stimme aufs Spiel zu setzen. Was sie zu sagen hätten, wird überdeckt von dem, was die Technologie am einfachsten und besten kann. Diese tödliche Konformität geht weit über die

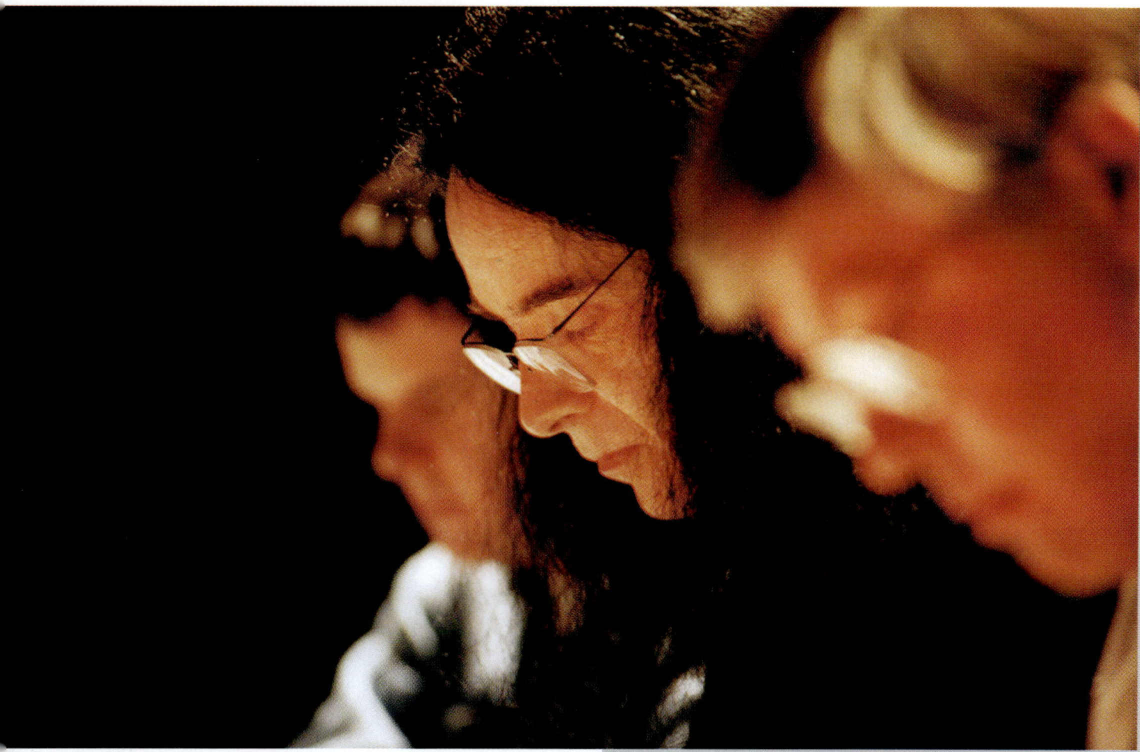

SoundHack. It would, of course, be foolish to damn these fine applications, and that is certainly not the Jury's intention. But there is something about the over-reliance on them which obliges even good composers to risk sacrificing the chance of speaking with a genuinely distinctive voice. What they have to say is somehow masked by what the technology does most easily. The deadly conformity extends beyond mere technique, which is often of admirable accomplishment, to an over-formulaic approach to gesture, pacing and mood.

These observations explain in part the Jury's (unanimous) selection of "Munich Samba" by Matt Heckert as the winner of the Golden Nica. This is a work with no electronic sound of any kind, and no loudspeaker in or out of sight. Matt Heckert is far from being alone in pursuing the development of computer-controlled acoustico-mechanical devices, but his work is among the most impressive. His work is designed not to occupy the concert hall, but spaces of a less prescriptive nature – maybe

bloße Technik – die häufig bemerkenswert ausgereift ist – hinaus und entsteht aus einem über-formelhaften Zugang zum Gestus, zum Tempo und zur Stimmung.

Diese Anmerkungen erklären schon teilweise die (einstimmige) Entscheidung der Jury für „Munich Samba" von Matt Heckert als Gewinner der Goldenen Nica. Das ist ein Werk ohne jeglichen elektronischen Klang, und weit und breit ist kein Lautsprecher in Sicht. Matt Heckert ist längst nicht der einzige, der die Entwicklung computergesteuerter elektroakustischer Geräte vorantreibt, aber sein Werk ist eines der beeindruckendsten. Seine Arbeit ist nicht dafür ausgelegt, einen Konzertsaal zu belegen, sondern viel weniger präskriptive Räumlichkeiten – vielleicht Museen oder alte Lagerhallen, vielleicht auch Einkaufsstraßen. „Munich Samba" kann automatisch betrieben werden, eher als eine Art bewegter, klangerzeugender Skulptur denn als Werk mit Exposition, Durchführung und Reprise, aber es kann auch „aufgeführt" werden, dann wird eben die Form neu strukturiert, mit Elementen relativer Statik und Aktivität, mit bewußt kalkulierten und eingesetzten Höhepunkten.

museums, maybe old warehouses, maybe shopping malls. "Munich Samba" can operate automatically, as a kind of moving, noise-making sculpture rather than a piece with a beginning, middle and end, but it can also be "performed" so that form can be shaped to produce different degrees of relative stasis and activity, with deliberately calculated and controlled high-points. "Munich Samba" does not seem to have any "meaning" in the sense of having external reference points: it is simply itself; that is, a collection of varied mechanical devices each gradually fulfilling its generic potential of movement and sound, in a huge variety of combinations and counterpoints, the basic physical principles of which are as bewitching as they are completely understandable. The other awards went to Jonty Harrison for "Unsound Objects", and Maryanne Amacher for "The Levi-Montalcini Variations".

Harrison has behind him a remarkable oeuvre of electroacoustic works, of which "Unsound Objects" is a typically virtuoso recent example. The listener is drawn into a surreal world of sonic allusion where "impossible" things are made to happen. The canvas is a dazzle of transforming colour and hectic activity – even in its more tranquil moments, the sound quality is vivid and alive. By virtue of his activities as both composer and teacher, Harrison has become one of the most influential figures in electroacoustic music.

Amacher's "The Levi-Montalcini Variations" is a completely different kind of creation, positioned somewhere between the dynamic concert-work style of "Unsound Objects" and the sculptural "Munich Samba". As much a sonic installation as a composition in the more conventional sense, it is a monumental wall of sound. It is massive, overwhelming, unrelenting and unyielding; but, once the hearing changes gear to adjust to these features, a myriad of details appear to spring from the texture, and somehow one is drawn com-

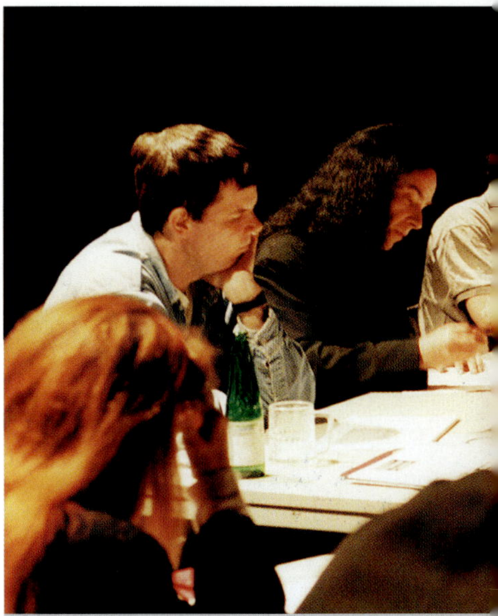

„Munich Samba" scheint nicht irgendeine „Bedeutung" im Sinne externer Bezugspunkte zu haben: Es ist, was es ist – eine Sammlung verschiedenster mechanischer Geräte, die schrittweise ihr eigenes Potential von Bewegung und Klang ausloten, in einer Vielzahl von Kombinationen und Kontrapunkten, deren zugrundliegenden pyhsikalischen Prinzipien ebenso bezaubernd wie für jedermann einsichtig sind.

Die beiden Auszeichnungen gingen a[n] Jonty Harrison für „Unsound Objects" un[d] Maryanne Amacher für „The Levi-Montalcini Variations".

Harrison kann auf ein bemerkenswertes Oeuvre elektroakustischer Arbeiten zurückblicken, unter denen „Unsound Objects" ein typisches virtuoses neueres Beispiel ist. Der Hörer wird in eine surreal[e] Welt klanglicher Anspielungen hineingezogen, in der „unmögliche" Dinge möglich gemacht werden. Das sich ergebende Bil[d] ist ein erstaunliches Durcheinander aus sich verändernden Farben und hektischer Aktivität, und selbst in seinen ruhigeren Momenten bleibt die Klangqualität lebhaf[t] und kraftvoll. Als Komponist wie als Lehre[r] ist Harrison zu einer der einflußreichsten Gestalten der elektroakustischen Musik d[er] Gegenwart geworden.

Amachers „The Levi-Montalcini Varia-

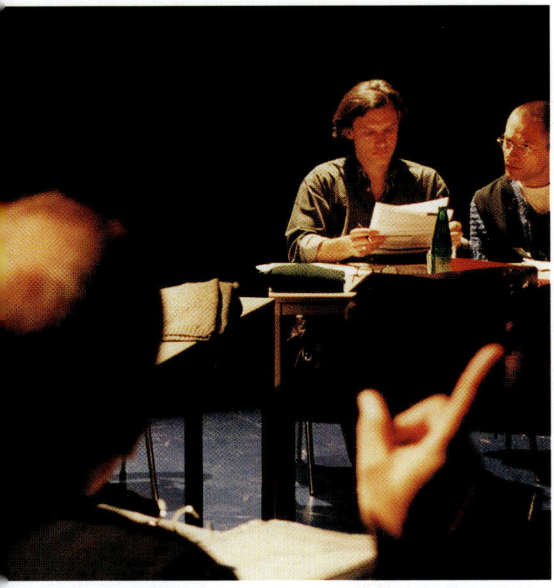

pletely into Amacher's rescaled world.

Space forbids providing here a comprehensive account of the works awarded Honourable Mentions. But let us point out a few common features which the Jury discussed.

A small group of these works (by Luigi Ceccarelli, Annie Gosfield and Mario Verandi) engage with the aesthetics (and the actual musical objects) from our ever-present musical past. Such pieces, when well executed (as these are), have an additional perspective to what is usually encountered. They seem to have been composed out of a love of past musics, but also, perhaps, out of a regret that its overpowering certainties and its ubiquity must be challenged in order to recapture a small piece of territory for the music of our time to occupy. In the right hands, computer music, because of its ability to remould and recontextualize existing, widely recognisable sounds, is especially well suited to sustain the challenge.

Two more characteristics caught the jury's ear: concision and humour. Anyone familiar with the world of computer music will instantly recognise that these are rare jewels, much to be treasured when encountered. So "Bravo!" to Ambrose Field, Annie Gosfield and Mark Wingate for

tions" ist ein völlig anderer Typus von Werk und liegt irgendwo zwischen den konzertanten „Unsound Objects" und dem skulpturhaften „Munich Samba". Als klangliche Installation wie auch als Komposition im konventionellen Sinn erzeugt es eine monumentale Wand aus Klängen. Es ist massiv, überwältigend, unnachgiebig und hartnäckig, aber wenn sich das Gehör auf diese Eigenschaften einmal eingestellt hat, wird es eine Myriade von Details in dieser Textur entdecken, und letztlich wird man komplett in Amachers „Welt im anderen Maßstab" hineingezogen.

Aus Platzgründen kann hier keine detaillierte Beschreibung der mit einer Anerkennung bedachten Werke erfolgen, aber einige Gemeinsamkeiten und Besonderheiten seien dennoch herausgegriffen:

Ein kleiner Teil dieser Arbeiten (die von Luigi Ceccarelli, Annie Gosfield und Mario Verandi) beschäftigt sich mit der Ästhetik (und den tatsächlichen musikalischen Objekten) unserer allgegenwärtigen musikalischen Vergangenheit. Wenn solche Stücke gut ausgeführt sind (und diese sind es), bieten sie eine zusätzliche Perspektive zu jener, der man normalerweise begegnet. Sie scheinen aus Liebe zur „klassischen" Musik komponiert zu sein, aber anscheinend auch aus Trauer darüber, daß deren alles überrollende Gewißheit und Allgegenwart in Frage gestellt werden muß, damit sich die Musik unserer Zeit wenigstens ein kleines Stück des Territoriums zurückerobern kann. In den richtigen Händen kann die Computermusik, eben wegen ihrer Fähigkeit zur Umformung und Umstrukturierung existierender und leicht erkennbarer Klänge, dieses Anliegen gut unterstützen.

Zwei weitere Charakteristika haben bei der Jury offene Ohren gefunden: Kürze und Humor. Und wer die Welt der Computermusik kennt, weiß, daß es sich dabei um seltene und kostbare Funde handelt. Deshalb ein „Bravo!" an Ambrose Field, Annie Gosfield und Mark Wingate, die dieses möglicherweise so fruchtbare Feld beackert haben.

Abschließend muß man noch feststellen, daß bei den Werken der „puren" Computer-

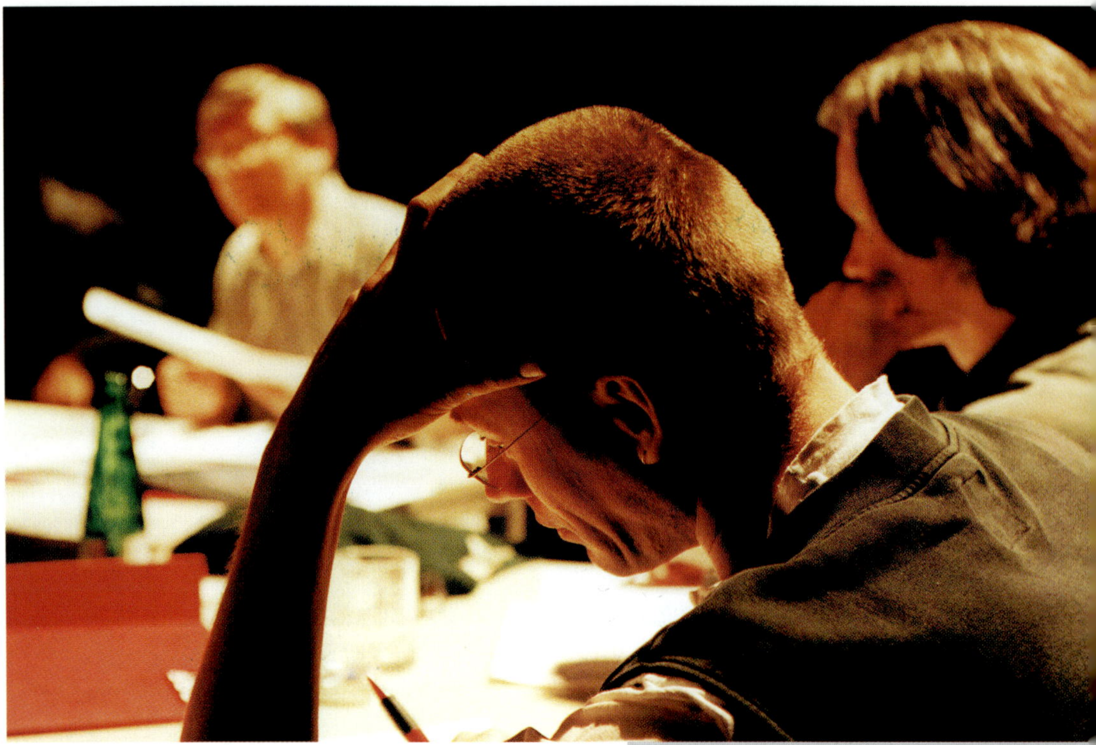

digging successfully in these potentially rich mines.

Finally, one can see that among the entries of "pure" computer music on tape (a category which covered about 85% of the entries this year), some works continue to establish a high quality reference in the genre. The pieces by Patrick Ascione, Christian Calon, Ake Parmerud and John Young clearly demonstrate that there are still some new paths to explore in this field and that, when artists of imagination and skill try to overcome the obviousness prescribed by the tools, their works continue to enlarge the sonic soundscape and artistic perception in general.

Last year's Jury identified four categories from which it hoped an increased number of future submissions would come: works designed for alternative modes of presentation (i.e. non-concert works), works with wider aesthetic range (avoid those clichés!), works by people without institutional affiliations (perhaps operating in a real, unsubsidized marketplace), and works by women. The new Jury paid no

musik auf Band (und das waren auch heuer rund 85 Prozent der Einreichungen) einige Arbeiten weiterhin einen hohen Standard in diesem Genre setzen. Die Werke von Patrick Ascione, Christian Calon, Ake Parmerud und John Young beweisen, daß es auch auf diesem Gebiet noch immer neue Wege zu gehen gibt und daß – wenn Künstler mit Imagination und Geschick die von ihren Werkzeugen vorgefertigte Offensichtlichkeit zu überwinden versuchen – ihre Arbeiten die Klanglandschaft und die künstlerische Warhnehmung generell durchaus noch erweitern können.

Die Jury des vergangenen Jahrs hat vier Kategorien herausgearbeitet, in denen sie in Zukunft noch gerne mehr Einreichungen sehen wollte: Mehr Werke für alternative Präsentationsmodi (d. h. nichtkonzertante Arbeiten), Werke mit größeren ästhetischer Bandbreite (vermeidet diese Klischees!), Arbeiten von Leuten ohne institutionelle (An-)Bindung (vielleicht aus nicht-unterstützten Märkten), und Werke von Frauen. Die Jury des heurigen Jahres

conscious attention to these hopes, but independently chose to make awards as though it had. The truth is, however, that the proportion of submissions from these categories was no greater this year than last. The message for next year must be the same: the Jury will seek to reward genuine innovation – of means, context and aesthetics – alongside manifestations of quality and accomplishment.

Among this year's submissions were a welcome handful which seek to explore the musical potential of the World Wide Web. While the Jury felt that there was no outstanding exemplar on this occasion which would have justified an award, it certainly felt that the Web is now beginning to offer an opportunity to develop entirely new modes of musical creation. One clear difficulty is limitations on bandwidth and the absence of guaranteed quality of network service which impede the use of high-quality audio in the networked environment. So there are two challenges to those considering submitting network "pieces" to the Prix Ars Electronica in the future. Firstly, is it possible to produce an award-winning work within current technical limitations? Secondly, is it possible to push the boundaries and exploit for artistic ends the experimental high-performance networking technologies which are now becoming more widely available?

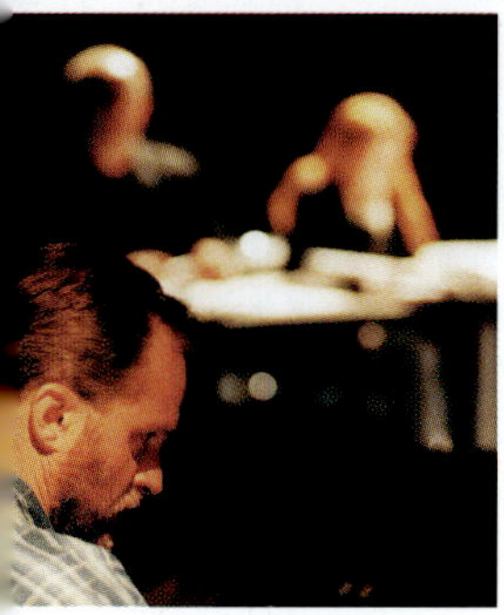

hat zwar kein besonderes Augenmerk auf diese Kategorien gerichtet, aber die Anerkennungen wurden so vergeben, als hätte sie hier spezielle Schwerpunkte gesetzt. In Wirklichkeit war der Anteil dieser Einreichungen nicht größer als im vorigen Jahr, und die Botschaft für 1998 muß die gleiche bleiben: Die Jury wird sich bemühen, echte Innovation auszuzeichnen – im Bereich der Mittel, des Kontexts und der Ästhetik –, und natürlich weiter nach hervorragenden Leistungen in Qualität und Durchführung suchen.

Unter den diesjährigen Einreichungen waren auch einige, die versuchten, das musikalische Potential des World Wide Web zu erforschen. Wenn auch die Jury kein so herausragendes Beispiel gefunden hat, das einen Preis oder eine Anerkennung verdient hätte, so ist sie doch zur Überzeugung gelangt, daß das Web beginnt, völlig neue Wegen musikalischen Schaffens zu ermöglichen. Sicherlich gibt es Schwierigkeiten aufgrund der Bandbreite und der unsicheren Qualität der Netzwerkdienste, denn sie verhindern die Verwendung von hochqualitativem Audio-Equipment im Netz. Aber es gibt zwei Herausforderungen für alle, die überlegen, bei zukünftigen Ausgaben des Prix Ars Electronica „Netz-Werke" in der Kategorie Computermusik einzureichen. Erstens: Können überhaupt preiswürdige Arbeiten innerhalb der derzeitigen technischen Beschränkungen entstehen? Und zweitens: Können die jetzt verfügbar werdenden experimentellen Hochleistungs-Netzwerktechniken zu künstlerischen Zwecken eingesetzt und damit die Grenzen weiter hinausgeschoben werden?

Munich Samba

Matt Heckert

After working for 8 years building machines for performances and producing soundtracks to accompany them, in 1986 I began to experiment with the idea of using the machines themselves to produce some of the sound, then all of the sound. In 1988, I began work on the "Mechanical Sound Orchestra".

The criteria for this project were:

– To build machines that were "playable," in that each would have a variety of timbres or rhythms they could produce;

– They would be remote controlled via a computer interface;

– Performances would consist only of sounds produced by the machines at the time: no tapes, samples or any other auxiliary sound would be used. This was very important, as I wanted to create a situation where all motion produces sound, and any sound could be traced to its source at that time. I wanted live performances to have the immediacy of an improvisational situation.

I found that when a mechanical device performs a repeated task by remote control, an observer tends to believe the device is somehow expressing some kind of autonomous emotion – frustration, desire, etc. – thus creating the mystique of an intelligent or sentient machine. This is not a sound issue, of course, but one which deals with the audience/performer relationship and the audience's perception that some sort of non-verbal communication is taking place.

With the machine's movements controlled by a Macintosh computer running MIDI sequencing software, a finer degree of control is possible than if it were controlled by manual switches; in addition, timing structures can be used that otherwise would be physically impossible. For instance, in one beat I can send a servo motor from zero to full speed, and back to zero 4 times, using a 120 bpm clock. This

Nachdem ich schon acht Jahre damit zugebracht hatte, Maschinen für Performances und die zugehörigen Soundtracks zu entwickeln, begann ich 1986 mit der Idee zu experimentieren, die Maschinen selbst dazu zu verwenden, einen Teil des Klanges – und später den gesamten – zu erzeugen. 1988 begann ich dann, am „Mechanical Sound Orchestra" zu arbeiten.

Die Kriterien für dieses Projekt waren:

– Maschinen zu erzeugen, die „spielbar" waren, d. h. die eine Vielzahl von Klangfarben oder Rhythmen erzeugen konnten;

– die Maschinen sollten über ein Computer-Interface fernsteuerbar sein;

– und die Klänge bei den Performances sollten ausschließlich aus den Maschinen kommen – keine Samples, keine Bänder, keine anderen zusätzlichen Klangquelle sollte eingesetzt werden. Dies war sehr wichtig, da ich eine Situation schaffen wollte, in der nur die Bewegung den Klang erzeugt und in der jeder Klang zu seinem Ursprung zurückverfolgt werden kann. Ich wollte Live-Performances mit der Unmittelbarkeit einer musikalischen Improvisation.

Wenn ein mechanisches Gerät eine ferngesteuerte repetitive Handlung ausführt, neigt meiner Erfahrung nach der Betrachter dazu anzunehmen, daß das Gerät irgendwie bemüht ist, eine eigenständige Emotion auszudrücken – Frustration, Wunsch, was auch immer –, und so die Mystik einer intelligenten oder fühlenden Maschine entsteht. Dies ist natürlich keine Frage des Klangs, hat aber dennoch eine Menge mit der Beziehung zwischen Hörer und Aufführendem und mit der Wahrnehmung des Publikums, daß hier irgendeine nicht-verbale Form der Kommunikation stattfindet, zu tun.

Wenn die Maschine über einen Macintosh Computer und eine MIDI- Sequencer-Software gesteuert wird, so ist eine

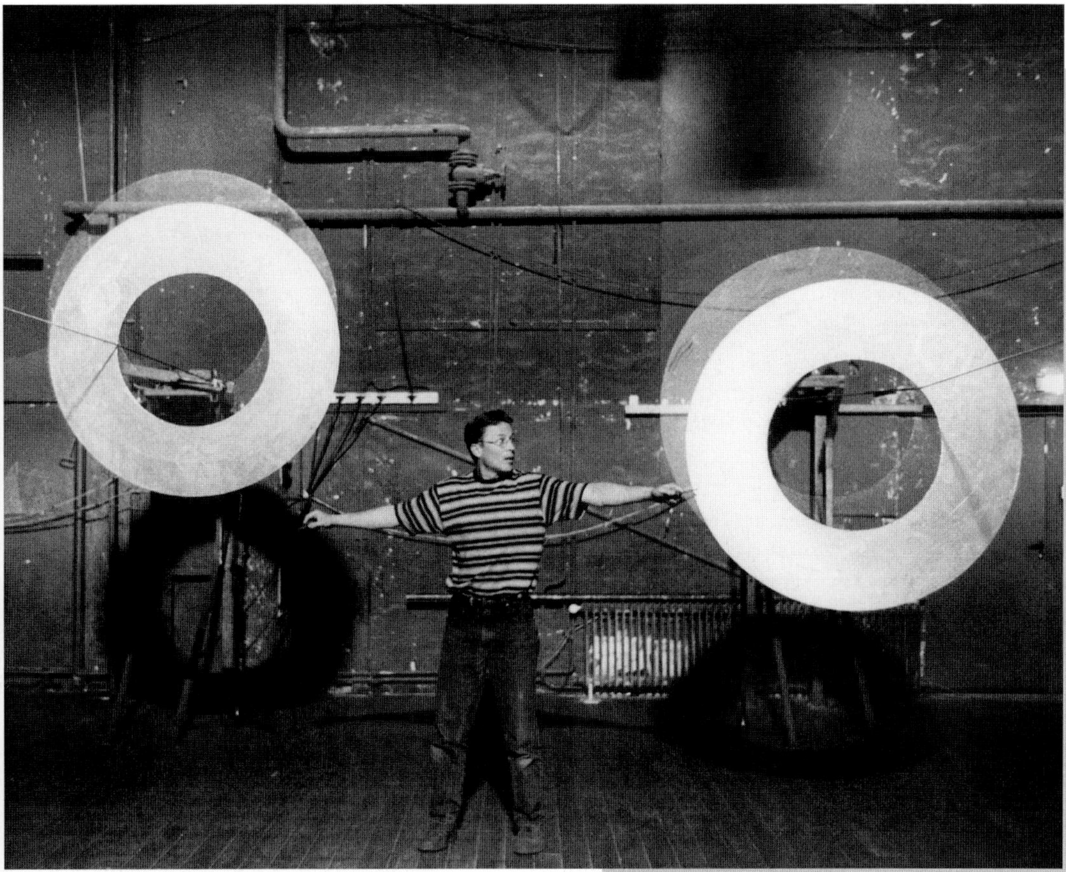

can also be repeated with accuracy, as it is the computer doing the switching.

Continuous controllers are used to vary motor speeds and note events are used for switching transistors, and it is this information that is recorded by the software. Once recorded and stored as a sequence or subsequence, any series of control commands can be used at any time. When called from the computer keyboard, control information is sent out instantly and the system responds much like any MIDI controlled electronic instrument, but the big differences are the inertial delay encountered when stirring a machine out of rest and the fact that the sound is coming live from machines.

The inertial delays vary from machine to machine, with some smaller ones being only milliseconds, to larger machines that won't get going for 1 to 8 seconds. This

wesentlich feinere Ansteuerung möglich als über manuelle Schalter; dazu kann noch ein Timing verwendet werden, das jenseits der physischen Möglichkeiten liegt. Wenn ich einen Taktgeber mit 120 Schlägen pro Minute verwende, kann ich in einem einzigen Beat einen Servomotor viermal aus dem Stand zu voller Geschwindigkeit und wieder zurück zum Stillstand bringen. Und dies kann ganz exakt wiederholt werden, da ja der Computer die Schaltung vornimmt.

Stufenlose Regler werden zur Einstellung der Motorgeschwindigkeiten verwendet, „Notenereignisse" schalten die Transistoren, und diese Information wird dann von der Software aufgezeichnet. Ist sie einmal als Sequenz oder Sub-Sequenz gespeichert, kann jederzeit jede beliebige Serie von Steuerbefehlen abgerufen werden. Wird sie über die Tastatur aufgerufen, wird diese Steuerinformation sofort abgesandt und das System regiert darauf (fast) wie jedes

gives a sort of push-pull feel to playing live that is more like working with a human-powered band. The other sounds present are what I refer to as auxiliary or secondary sounds. These are the various motor hums, solenoid clicks, metal creaks, air exhaust and so on, that are not the primary effects I was after in designing a particular machine, but the ones that come along with its operation – much like the inhale of a horn player too close to a microphone or the finger-squeak on a guitar or violin neck, my machines breathe and stretch in their own ways.

Having this type of control interface allows me to respond to the differences in the acoustic dynamics of a specific location as well as to improvise new compositions during performances.

What I hope to provide is a greater sense of immediacy for an audience, with some feeling of being inside a system or mechanism. This feeling is enhanced by the way the machines are presented to the public – not on a stage out of reach, but spread throughout the performance space so the audience can be in close proximity to the instruments.

andere MIDI-gesteuerte Instrument. Der große Unterschied liegt in der Anfangsträgheit, die merkbar wird, wenn eine Maschine aus ihrer Ruhe gerissen wird, und eben darin, daß der Klang live von den Maschinen gemacht wird.

Die Trägheitsmomente variieren von Gerät zu Gerät und liegen zwischen einigen wenigen Millisekunden bei kleineren und zwischen einer und acht Sekunden, bis die größeren Maschinen anspringen. Dies gibt einem bei einer Live-Aufführung ein Gefühl des Hin- und Hergerissenwerdens, das fast mit der Arbeit mit einem menschenbetriebenen Orchester vergleichbar ist. Die anderen Klänge bieten das, was ich als sekundäre oder „Sub-Klänge" bezeichnen möchte. Dazu gehören die Summtöne der Motoren, das Klicken von Relais, das Knacken von Metall, Druckluftzischen und so weiter – dies waren nicht Effekte, die ich bei der Entwicklung der diversen Apparate anstrebte, aber sie kamen beim Betrieb einfach dazu. Wie der Hornist, der zu nahe am Mikrophon Luft holt, oder der Geiger oder Gitarrist, der mit dem Finger auf der Saite rutscht – meine Maschinen atmen und strecken sich auf ihre Weise.

Diese Art von Steuer-Interface erlaubt mir, auf die unterschiedliche akustische Dynamik verschiedener Aufführungsorte ebenso einzugehen wie neue Kompositionen während der Aufführung zu improvisieren.

Ich hoffe, dem Publikum ein stärkeres Gefühl von Unmittelbarkeit zu vermitteln, das Gefühl, im Inneren eines Systems oder Mechanimus zu sein. Dieses Gefühl wird durch die Art der Präsentation der Maschinen noch unterstrichen – nicht außer Reichweite auf einer Bühne, sondern über den gesamten Aufführungsraum verteilt, so daß das Publikum ganz nahe bei den maschinellen Musikern sein kann.

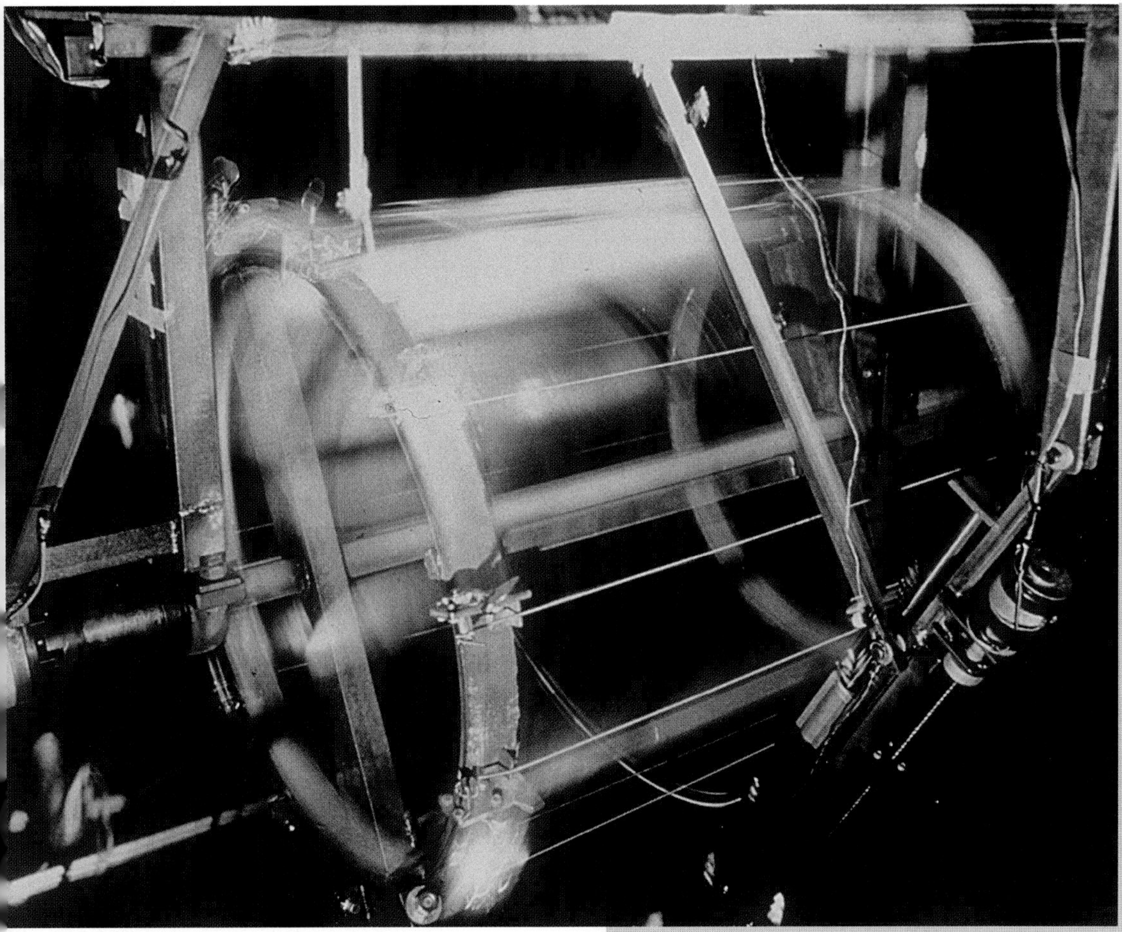

Matt Heckert (USA) has been working as a performance and sound artist since 1978. In the 80's he worked as one of the directors of Survial Research Laboratories building robots and designing soundtracks for the performances and for film. In 1989 he began working on the Mechanical Sound Orchestra, a group of computer controlled mechanical sound instruments. He has presented this in the USA and in Europe since that time, working solo and in collaboration with others.

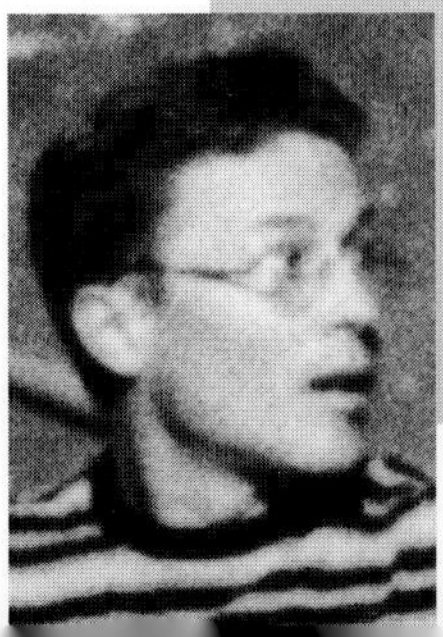

Matt Hackert (USA) arbeitet seit 1978 als Performance- und Klangkünstler. In den 80er Jahren war er einer der Leiter der Survival Research Laboratories, wo Roboter gebaut und Soundtracks für Perfomances und Filme entwickelt wurden. 1989 begann er mit dem Mechanical Sound Orchestra zu arbeiten, einer Gruppe computergesteuerter mechanischer Klanginstrumente. Er hat diese Arbeit in den USA und in Europa sowohl als Solomusiker wie in Zusammenarbeit mit anderen präsentiert.

The Levi-Montalcini-Variations

Maryanne Amacher

"The Levi-Montalcini Variations" is a new work which I am developing for emerging 3-D audio platforms anticipating the extended storage capacities of the near future. Taking VR (Virtual Realitiy), 3-D sonic imaging and graphics, telepresence, and cyberspace as a point of departure, the work examines possibilities of individualizing multi-dimensional sonic environments for listeners and for spaces. With today´s programmable immersive technologies, interactive scenarios can be created which focus on multiple perceptual viewpoints as we respond to auditory events. Immersive "aural architectures" are constructed, addressing new concepts of experience design – an approach to auditory communication with sonic perceptual information being processed by the listener, in addition to sonic information produced acoustically.

A virtual auditory environment is designed in which people navigate through sonic architectures composed of rare and unusual atmospheres. The experience of being there is simulated perceptually, by the "Transformative Modifier Scores" an interactive interface in which sonic events are transformed to sound "as if" perceived in these particular environments. An evolving narrative will be created in serialized Episodes for interacting with Lead Characters (The Sound Stars) and Supporting Sonic Characters using immersive digital technologies. The work may be produced as a series of installations customized for specific locations (with hard disc re-programming); and for the multi-platforms of emerging media and telecommunication technologies: VR, DVD, sonic theaters, interactive audio cable, home 3-D surround theaters, interactive TV/radio simulcast.

My goal is to compose environment-oriented sculpture that becomes a virtual part of the architecture of our living places. These would be multi-dimensional sonic environments accompanied by imaginative

Die „Levi-Montalcini-Variations" sind ein neues Werk, das ich für die in Entstehung begriffenen 3D-Audio-Plattformen unter Vorwegnahme der erweiterten Speichermöglichkeiten der nahen Zukunft entwickle. Ausgehend von Virtueller Realität (VR), 3D-Klangbildern und -Graphik, Telepräsenz und Cyberspace untersucht das Werk die Möglichkeiten, eine multidimensionale Klangumgebung für Zuhörer und Räume zu individualisieren. Mit den heutigen programmierbaren immersiven Technologien können interaktive Szenarien geschaffen werden, die sich – je nach der Reaktion des Hörers auf auditive Ereignisse – auf multiple Wahrnehmungsstandpunkte konzentrieren. Eine alles umhüllende „aurale Architektur" wird konstruiert, die neue Konzepte des Erfahrungsdesigns anspricht, ein Ansatz zur auditorischen Komposition als „Geographie der Wahrnehmung". Ein intelligentes Interface wird mit der vom Hörer verarbeiteten Klangwahrnehmungsinformation ebenso kommunizieren, sie verstärken und auf sie reagieren wie auf die akustisch produzierte Klanginformation.

Eine virtuelle Klangumgebung wird entworfen, in der Menschen durch eine Klangarchitektur reisen, die aus seltenen und ungewöhnlichen Atmosphären besteht. Die Erfahrung des Dort-Seins wird in der Wahrnehmung simuliert, und zwar durch die „Transformative Modifier Scores", d. h. durch ein interaktives Interface, in dem klangauslösende Ereignisse so in Klang umgesetzt werden, als würden sie in diesen speziellen Umgebungen perzipiert. Eine sich weiterentwickelnde Geschichte entsteht als Serie von Episoden, um mit den Hauptdarstellern („Sound Stars") und Nebendarstellern („Supporting Sonic Characters") mit Hilfe immersiver digitaler Technologien zu interagieren. Das Werk kann als eine Serie von Installationen produziert werden, das

scenarios for entering the interior of the music, accompanied by a score in the form of interactive software for people to construct and transform the aural architecture of these worlds. Certainly the kind of "transformative" variations which I am composing cannot be vividly experienced with the mouse or other such simplistic interfaces. Bruce Toggnazzini, founder of Apple's Human Interface Group (1985) discusses in a recent interview how the interface he helped develop for Macintosh has become an obstacle. He suggests abhandoning our keyboard-centric view of computers. Design instead for a computer that leads you away from thinking about computers as tools you use and toward thinking of them as a space you enter. Rooms, walls, and corridors that sing.

an unterschiedliche Aufstellungsorte angepaßt werden kann (jedesmal mit Umprogrammierung der Festplatte), aber auch für die in Entwicklung befindlichen Multi-Plattformen der neuen Medien- und Kommunikationstechnologien (VR, DVD, Klangtheater, Interaktive Audio-Kabel, private 3D-Surround-Theater, interaktive TV/Radio-Simulcast-Ausstrahlung etc.)

Mein Ziel ist es, umgebungsorientierte Skulpturen zu komponieren, die eine tragende Rolle in der Architektur unserer Lebensräume spielen. Diese werden aus multidimensionalen Klangumgebungen bestehen, begleitet von phantasievollen Szenarien, die zum Eintritt ins Innere der Musik einladen, und von einer Partitur in der Form interaktiver Software, mit deren Hilfe die klangliche Architektur dieser Welten konstruiert und verändert werden kann. Sicherlich kann diese Art von „transformativen" Variationen, die ich komponiere, nicht mit der Computermaus oder ähnlich primitiven Interfaces erlebt werden. Bruce Toggnazzini, Begründer von Apple's Human Interface Group (1985) erläuterte jüngst in einem Interview, daß und wie jenes Interface, das er für den Macintosh mitentwickelt hat, zu einem Hindernis geworden ist. Seiner Meinung nach sollten sollten wir unsere „Tastatur-zentrische" Sicht des Computers aufgeben. Gefragt sei ein Design, das es uns ermöglicht, den Computer nicht als ein Werkzeug, dessen man sich bedient, anzusehen, sondern als einen Raum, den man betritt: Räume, Wände und Korridore, die singen.

Maryanne Amacher (USA) is a composer and sound installation artist. Musical studies in Salzburg, Austria, and Dartington Devon, England and composition studies with Karlheinz Stockhausen. Studies in music and computer science at the University of Pennsylvania. Fellow at the Center for Advanced Visual Studies, MIT (1972–76) – Collaboration with the visual artists Scott Fisher, Luis Frangella, and the architect Juan Navarro Baldeweg. Created works with John Cage and Merce Cunningham.

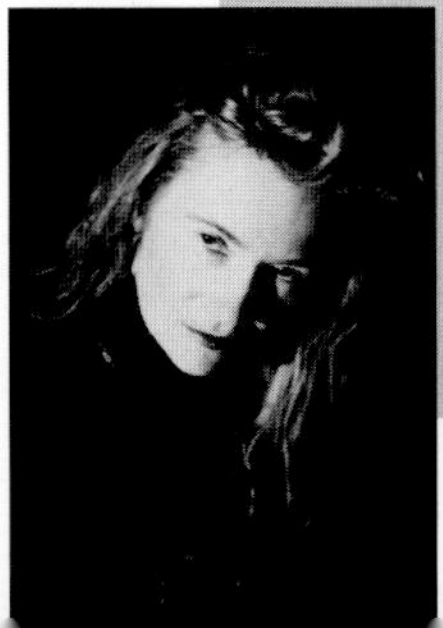

Maryanne Amacher (USA) ist Komponistin und Klanginstallations-Künstlerin. Sie studierte Musik in Salzburg und Dartington (Devon, GB) und Komposition bei Karlheinz Stockhausen sowie Musik und Computerwissenschaften an der University of Pennsylvania. 1972–76 war sie Fellow am Center for Advanced Visual Studies des MIT. Zusammenarbeit mit den visuellen Künstlern Scott Fisher, Luis Frangella und mit dem Architekten Juan Navarro Baldeweg sowie mit John Cage und Merce Cunningham.

Unsound Objects

Jonty Harrison

One of the main criteria in Pierre Schaeffer´s definition of the "sound objects" was that, through the process of "reduced listening", one should hear sound material purely as sound, divorced from any associations with its physical origins – in other words, what is significant about a recorded violin sound (for example) is that particular sound, its unique identity, and not its "violin-ness". Despite this ideal, a rich repertoire of music has been created since the 1950s which plays precisely on the ambiguities evoked when recognition and contextualisation of sound matieral rub shoulders with more abstracted (and abstract) musical structures. But as these structures themselves grow organically out of the peculiarities of the individual sound objects within them, the ambiguity is compounded: interconnections and multiple levels of meaning proliferate. The known becomes strange and the unknown familiar in a continuum of reality, unreality and surreality, where boundaries shift and continually renewed definitions are the only constant ...

In "Unsound Objects", the sounds used are all from the "real world" – recordings of actual sonic events which have undergone varying degrees of computer modification. The work operates on three levels. The first addresses what Denis Smalley has called "spectro-morphology" – a notion very close to the Shaefferian ideals mentioned above and which is the closest to what wie understand in the west as "music". The second is related to "soundscape" – recognisable, everyday sound environments which we have all experi-enced. The third level of the discourse is precisely the points of interconnection (collision, friction, impact, interpenetration?) of these two apparently contradictory worlds; it is concerned with the recontextualisation of recognisable sounds and the "meanings" released by this process – in particular, the shock of recognition caused by the familiar

Eines der Hauptkriterien in Pierre Schaeffers Definition des „Klangobjekts" ist, daß der Hörer durch einen Prozeß des „reduzierten Hörens" das Klangmaterial ausschließlich als Klang aufnehmen soll, ohne irgendeine Assoziation mit dessen physischen Ursprung – mit anderen Worten: Was an einem aufgezeichneten Violinton von Bedeutung ist, ist der besondere Klang, seine Einzigartigkeit und nicht seine „Violinhaftigkeit". Trotz dieser Idealvorstellung wurde seit den 50er Jahren ein reiches Repertoire an Werken geschaffen, das genau auf jene Unschärfe abzielt, die entsteht, wenn die Erkennung und Kontextualisierung von Klangmaterial sich an stärker abstrahierten (und abstrakteren) Klangstrukturen reibt. Aber da ja auch diese Strukturen selbst wiederum organisch aus den in ihnen enthaltenen individuellen Klangobjekten erwachsen, wird Vieldeutigkeit erzeugt: Querverbindungen und multiple Bedeutungsebenen entstehen. Das Bekannte wird fremd und das Unbekannte vertraut, in einem Kontinuum aus Realität, Irrealität und Surrealität, in dem die Grenzen verschwimmen und nur die ständig erneuerten Definitionen konstant sind.

In „Unsound Objects" stammen alle verwendeten Klänge aus der „realen Welt" – Aufnahmen von tatsächlichen Klangereignissen, die unterschiedlichen Graden der Bearbeitung mittels Computer unterworfen wurden. Das Werk operiert auf drei Ebenen. Die erste spricht das an, was Denis Smalley die „Spektro-Morphologie" genannt hat, eine der oben angeführten Schaefferschen Definition sehr verwandte Ansicht, die noch am ehesten das beschreibt, was in der westlichen Kultur als „Musik" verstanden wird. Die zweite Ebene bezieht sich auf den Begriff „Soundscape" – Klanglandschaft –, also auf erkennbare Klangumgebungen, die wir alle schon erlebt haben. Die dritte

turning up unexpectedly or, conversely, by a strange intruder in an otherwise well known sonic space. If the material in "Unsound Objects" is – literally – elemental, then this only serves to heighten the impact of sounds as universally recognisable as water (river, sea, rain), wood, fire, birds, wind, children, thunder and the human presence implied by footsteps.

This last example is a good illustration of the enormous potential of the sonic arts for blurring the distinctions between reality, unreality and surreality – only in sound can we believe in the simultaneous presence of footsteps on the different surfaces of snow, gravel and dry bracken (as occurs at ca. 5'40 of "Unsound Objects"). Because we can recognise each situation specifically, we can believe what we know to be physically impossible, even at the actual moment of hearing it.

Diskursebene umfaßt genau jene Verbindungs- (oder Kollisions-, Reibungs-, Aufschlags-, Durchdringungs-?)-Punkte dieser scheinbar gegensätzlichen Welten, sie beschäftigt sich mit der Re-Kontextualisierung erkennbarer Klänge und mit den „Bedeutungen", die von diesen ausgelöst werden, besonders mit dem Schock, der entsteht, wenn unerwartet etwas Vertrautes auftaucht, oder umgekehrt, beim Eindringen des Ungewohnten in einen ansonsten gut bekannten Klangraum. Wenn das Klangmaterial in „Unsound Objects" im wahrsten Sinne des Wortes „elementar" ist, so dient dies zur Verstärkung des Eindrucks, den Klänge hervorrufen, die allgemein erkennbar sind – Wasser (Fluß, Meer, Regen), Wald, Feuer, Vögel, Wind, Kinder, Donner und die durch das Geräusch von Schritten implizierte menschliche Präsenz.

Besonders letzteres ist ein gutes Beispiel für das gerade in der Klangkunst enorme Potential, die Grenzen zwischen Realität, Irrealität und Surrealität zu verwischen – nur dem Klang „glauben" wir die gleichzeitige Präsenz von Schritten auf so unterschiedlichen Oberflächen wie Schnee, Kies und vertrocknetem Farn (wie es bei etwa 5'40 von „Unsound Objects" geschieht). Eben weil wir jede Situation als solche gesondert erkennen, können wir auch das glauben, wovon wir selbst im Moment des Hörens schon wissen, daß es physisch unmöglich ist.

Jonty Harrison (GB), born 1952, studied at the University of York (DPhil in Composition) and worked at the National Theatre and City University. In 1980 he joined the Music Department of The University of Birmingham, where he is now Senior Lecturer, as well as Director of the Electroacoustic Music Studios and BEAST (Birmingham ElectroAcoustic Sound Theatre). He has been on the Board of Sonic Arts Network, the UK's national organisation for electroacoustic music for many years.

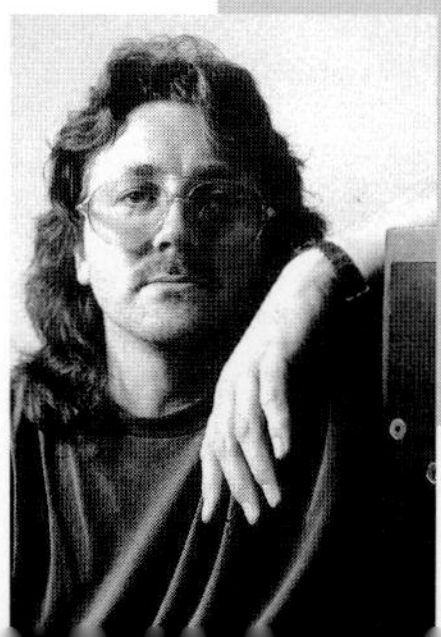

Jonty Harrison (GB), geb. 1952, studierte an der University of York und erwarb einen Dr. phil. aus Komposition. Er arbeitet am National Theater und der City University. 1980 trat er ins Music Department der University of Birmingham ein, wo er derzeit als Senior Lecturer und Leiter des Electroacoustic Music Studios und des BEAST (Birmingham ElectroAcoustic Sound Theater) tätig ist. Er ist seit vielen Jahren Mitglied des Board of Sonic Arts Network, der staatlichen britischen Organisation für elektroakustische Musik.

Primitive

Patrick Ascione

In a beautiful "style exercise" far removed from anecdotes or any type of realism, "Primitive" by Patrick Ascione presents an interesting reflection on the essence of computer-acoustic art itself. Some time ago, Pierre Boulez wanted to name one of his works after a painting by Klee: "At the Border of the Fertile Land." Listening to this piece, you actually have the impression of standing at the entrance to a land of promises – it is so rich in virtual events that you hardly know where to begin making use of them. In "Primitive", you first experience the development of completely heterogeneous sound objects, from nearly saturated, ponderous to sharp, pointed; from the most stable to the most flexible; from the most ethereal to the most rounded and voluminous. Intangible, fleeting figures seem to be constantly circling. The form seems to be hidden, and yet the direction of the whole is perceptible as being constantly present. The stress on the importance of a masterful projection is quite evident – especially shortly before the end, where you hear the space opening up before a brilliant, dynamic and devastating finale.

Jacques Bonnaure

In einer schönen „Stilübung" präsentiert „Primitive" von Patrick Ascione fernab alles Anekdotischen und jedes Realismus eine interessante Reflexion über das Wesen der computerakustischen Kunst selbst. Vor langer Zeit wollte Pierre Boulez einmal eines seiner Werke nach einem Bild von Klee benennen: „An der Grenze des fruchtbaren Landes". Wenn man hier den Reichtum des Stückes hört, hat man tatsächlich den Eindruck, sich am Eingang eines Gebietes voller Verheißungen zu befinden - so reich an virtuellen Ereignissen, daß man gar nicht weiß, wo man beginnen soll, sie auszunützen. In „Primitive" erlebt man zuerst die Herausarbeitung von völlig heterogenen Klangobjekten mit, von fast satten, schwerfälligen bis hin zu spitzen, pointierten, von den allerstabilsten bis hin zu den allerbeweglichsten, von den ätherischsten bis zu den abgerundet-voluminösen. Es ist ein ständiges Kreisen von nicht greifbaren, flüchtigen Figuren. Die Form scheint versteckt zu bleiben, und dennoch stellt sich die Führung des Ganzen als ständig präsent dar. Man muß feststellen, daß auf eine meisterliche Projektion Wert gelegt wurde - besonders kurz vor dem Ende, wo man den Raum sich öffnen hört vor einem brillanten, dynamischen und grausamen Finale.

Jacques Bonnaure

Patrick Ascione (F), born in 1953, has directed the CAP (Centre d'Art Acousmatique et de Polyphonie Spatiale) in Vierzon since 1991. Cofounder of the ABSO movement (Mouvement pour le développement de l'art acousmatique) in 1991. Work for the Groupe Expérimental de Bourges; working visits to GMEA, GES, GRM, GMVL, IRCAM, CIRM.

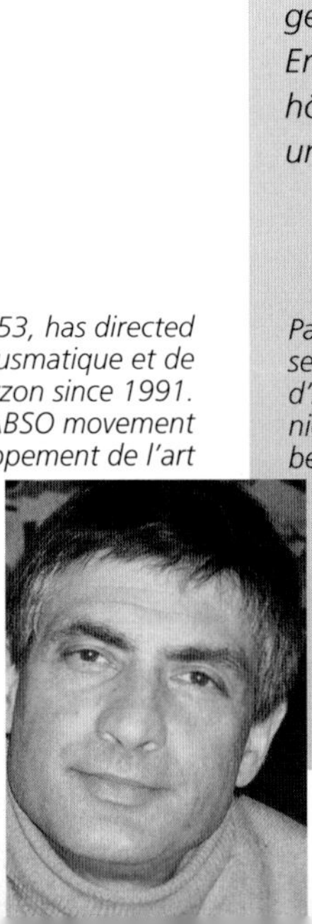

Patrick Ascione (F), geb. 1953, ist seit 1991 Direktor des CAP (Centre d'Art Acousmatique et de Polyphonie Spatiale) in Vierzon. 1991 Mitbegründer der ABSO-Bewegung (Mouvement pour le développement de l'art acousmatique), Arbeit für die Groupe Expérimental de Bourges, Arbeitsaufenthalte bei GMEA, GES, GRM, GMVL, IRCAM, CIRM.

Les Corps Éblouis

Christian Calon

It is when in a moment of sharp consciousness, Matter through the proliferation of its forms strikes us with the measure of the ephemeral. A painful and at the same time magnificent experience.

One being dissolved in the Number.

Magnificence of the abounding and of each of its fragile manifestations. Sparkling of Presence. And Time.

What could be the sense of this action, this abundance of sound, other than an answer to the amazement at the luxurious prodigality and the sumptuousness of Matter.

And this little sentence, these few words, that carried me along or guided me, I don't recall:

Then as a drapery,
Spring came
On our dazzled bodies. L. F.

It is dedicated to Chantel.

The present recast version was premiered at the Marseilles National Theatre.

Es findet in einem Moment des gesteigerten Bewußtseins statt, daß die Materie, durch die Vermehrung ihrer Gestalten, sich uns plötzlich mit dem Maß ihrer Flüchtigkeit aufdrängt. Eine schmerzhafte, doch gleichzeitig großartige Erfahrung.

Die Eins löst sich in der Zahl auf.

Die Herrlichkeit der Fülle, ihrer zerbrechlichen Manifestationen. Glanz der Präsenz. Und die Zeit.

Welchen Sinn hätte diese Handlung, diese Fülle des Klangs, wenn sie nicht eine Antwort auf das Staunen ange-sichts der luxuriösen Üppigkeit und Pracht der Materie wäre!

Und dann dieser kleine Satz, diese paar Worte, die mich getragen oder geleitet haben:

Dann kam wie ein Vorhang
der Frühling
auf unsere geblendeten Körper. L. F.

Für Chantel.

Die überarbeitete Version wurde am Théatre National von Marseille uraufgeführt.

Christian Calon (CDN) is working on the realization of volumes or virtual sound spaces to be entered by the audience. Founding member of the CEC (Canadian Electroacoustic Community); appointed to the artistic direction of the studios of the Groupe de Musique Expérimentale de Marseille, in 1991.

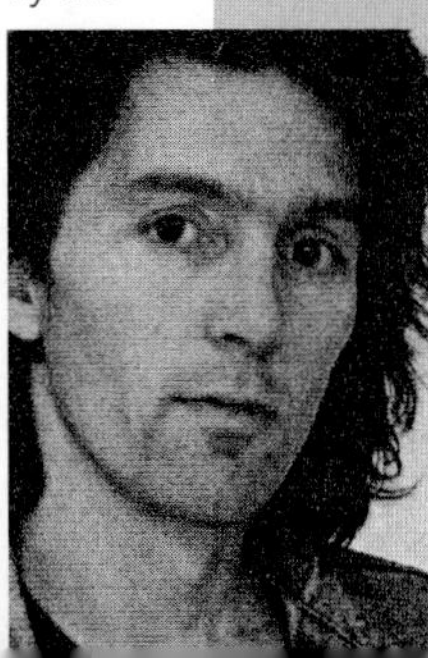

Christian Calon (CDN) arbeitet an der Verwirklichung von „Volumes", d. h. virtuellen Klangräumen, die das Publikum betreten kann. Gründungsmitglied von CEC (Canadian Electroacoustic Community), seit 1991 künstlerischer Leiter der Studios der Groupe de Musique Expérimentale de Marseille.

Exsultet

Luigi Ceccarelli

"Exsultet" uses the text and melody of "Exsultet", the ancient gregorian version of the Christian liturgy. A soloist begins the composition conventionally by singing the original version. Space, clearly artificial, is the only variant.

The voices of the other singers slowly follow one by one. However, all the vowel sounds (that is the tuned sounds) have been electronically removed and only consonants have been retained.

In antithesis to this material, a second type of treatment of the material (in which tuned vowel sound stand out) has been adopted.

Besides "Exsultet", another two chants: the "Alleluia-Pascha Nostrum" and the "Alleluia-De Profundis" have been used for this elaboration. These pieces were sung in the Fossanova Abbey and recorded in a way that greatly enhanced reverberation of the surroundings. In order to make the variations and the overlapping sounds generated by the acoustics of the Abbey more perceptible, all the sounds have been temporally time-stretched using the technique of "stretching" from eight to ten times. The time-stretching of the reverberation generates a very rich and varied sound texture. Halfway through the piece, the entire text of "Exsultet" is recited in a whisper by one singer overlapping the sound. Even this type of recitation – in which the text remains imperceptible – acquires value as a texture of quite a diverse genre.

„Exsultet" verwendet Text und Melodie von „Exsultet", der alten gregorianischen Version der christlichen Liturgie. Ein Solist beginnt die Komposition konventionell, indem er die Originalversion singt. Der Klangraum – offensichtlich künstlich – ist das einzig Variierende.

Die Stimmen der anderen Sänger fallen nacheinander ein. Allerdings wurden hier alle Vokal-Töne (also die gestimmten, klingenden Elemente) elektronisch entfernt, nur die Konsonanten blieben übrig.

Als Antithese dazu wurde eine zweite Bearbeitung desselben Materials vorgenommen, bei der nur die gestimmten (vokalischen) Töne verblieben sind.

Neben „Exsultet" wurden zwei weitere Choräle für diese Bearbeitung verwendet, „Alleluia – Pascha Nostrum" und „Alleluia – De Profundis". Diese Choräle wurden in der Abtei zu Fossanova gesungen und so aufgenommen, daß der vorhandene Umgebungshall noch wesentlich verstärkt wurde. Um die Variationen und die durch die Akustik der Abtei erzeugten, sich überlagernden Klänge besser wahrnehmbar zu machen, wurden alle Klänge mit „Stretching"-Techniken temporär auf das Acht- bis Zehnfache ihrer Länge gedehnt. Diese Zeitdehnung des Halls erzeugt eine variationsreiche Klangtextur. Etwa zur Hälfte des Stücks wird der vollständige Text des „Exsultet" von einem Sänger geflüstert rezitiert, so daß er sich über diese Klangstruktur legt. Selbst diese Art von Rezitation, bei der der Text unverständlich bleibt, gewinnt als Textur einen eigenen Wert.

Luigi Ceccarelli (I), born 1953. Completed his musical studies at the „Conservatorio G. Rossini" of Pesaro. Since 1970 he has worked as a composer using the most advanced technologies. A fundamental part of his work is tied to the coreography of Lucia Latour and the Altroteatro Dance Company. In addition he is founder of "EdiSon" studio. He is also professor of Electroacoustic Music at the Conservatory of Perugia and sound engineer at many festivals.

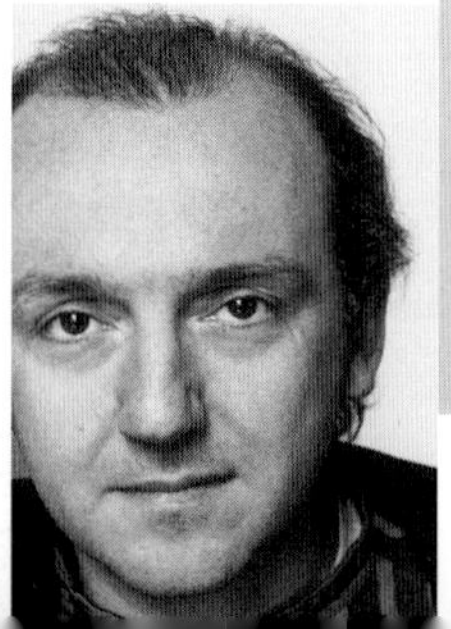

Luigi Ceccarelli (I), geb. 1953, schloß sein Musikstudium am Conservatorio Gioacchino Rossini in Pesaro ab. Seit 1970 arbeitet er als Komponist unter Verwendung der modernsten Technologien. Ein entscheidender Teil seiner Arbeiten ist eng mit der Choreographie von Lucia Latour und der Altroteatro Tanzcompagnie verbunden, daneben ist er Gründer des „EdiSon" Studios. Er ist Professor für Elektroakustische Musik am Konservatorium Perugia und arbeitet als Toningenieur bei zahlreichen Festivals.

Pyrotechnic!

Ambrose Field

Imagine a crowd at a fireworks display. As the sky lights up, they gasp and groan with satisfaction. At first, the electroacoustic transformations appear to be concerned with manipulating the acoustic impacts caused by the exploding fireworks. These materials are cleanly extracted from their original context and given their own space. While the listener concentrates on the dynamic energy of these textures, the role of the crowd changes subtly. By the end of the work the spectators react to the colours and impacts of totally abstract electroacoustic sounds!

This work is short!

Computer music is often far too long. "Pyrotechnic!" has only one message. When that has been delivered and developed, the work ends. Fini.

WARNING: Fasten your seatbelts.

High Impact electroacoustic Sounds to enjoy! Computer music must offer something original, and something special. It must speak out – offer social comment. It´s all very well making appealing sounds, but where is the message these days? "Pyrotechnic!" aims to capture the excitement of a fireworks display and offer a subtle musical twist.

Design and Function

The work is designed to open or close concerts (in a variety of ways). It ist not for the middles. You can also play it at home or in your car. Problem is, what do you play afterwards?

Man stelle sich eine Menschenmenge bei einem Feuerwerk vor: Wenn der Himmel aufleuchtet, starren die Leute und stöhnen vor Begeisterung. Zuerst scheinen sich die elektroakustischen Veränderungen auf die Manipulation der akustischen Schläge der explodierenden Feuerwerkskörper zu beziehen. Dieses Material wird aus dem Zusammenhang gerissen und erhält seinen eigenen Raum. Während sich der Zuhörer auf die dynamische Energie dieser Texturen konzentriert, verändert sich unmerklich die Rolle der Menschenmenge. Am Ende des Werkes reagieren die Zuseher auf die Farben und Geräusche völlig abstrakter elektroakustischer Klänge!

Das Werk ist kurz

Computermusik ist häufig viel zu lang. „Pyrotechnic!" hat nur eine Botschaft, und wenn die entwickelt und überbracht ist, ist das Werk aus. Ende.

Warnung: Bitte anschnallen!

Höchst eindrucksvolle elektroakustische Klänge zu genießen! Computermusik muß etwas Originelles bieten, etwas Spezielles. Sie muß soziale Anliegen haben und diese zum Ausdruck bringen. Ansprechende Töne zu erzeugen, ist gut und schön, aber wo bleibt heutzutage die Botschaft? „Pyrotechnic!" zielt darauf ab, das Aufregende an einem Feuerwerk einzufangen und eine subtile musikalische Verdrehung davon wiederzugeben.

Design und Funktion

Das Werk wurde entwickelt, um Konzerte zu eröffnen oder zu schließen (auf vielfältige Weise!). Es ist nichts für die Mitte. Man kann es auch zu Hause oder im Auto abspielen. Das Problem ist nur, was spielt man danach?

Ambrose Field´s compositional output includes electroacoustic music for tape, music education works and community music projects. In 1993 he was funded by the British Academy to embark on a PhD in electroacoustic composition with Denis Smalley at City University, London. He is currently a Lecturer in Music at the University of York.

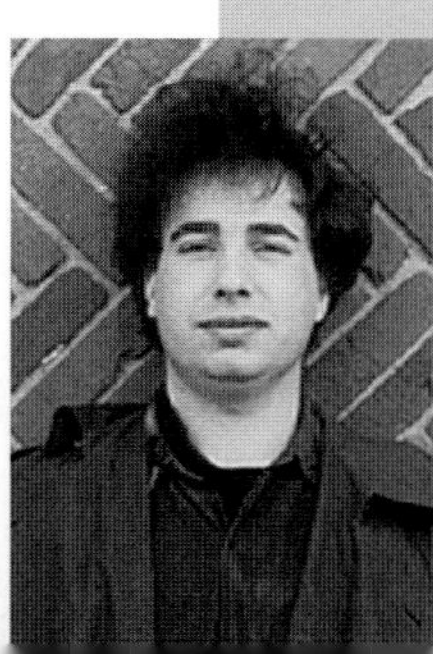

Ambrose Fields (GB) kompositorische Produktion umfaßt elektroakustische Musik für Band, musikerzieherische Arbeiten und Gemeinschaftsmusik-Projekte. 1993 erhielt er von der British Academy ein Doktorandenstipendium für elektroakustische Komposition bei Dennis Smalley an der City University, London. Derzeit arbeitet er als Lektor für Musik an der University of York.

The Manufacture of Tangled Ivory

Annie Gosfield

"The Manufacture of Tangled Ivory" is a work that explores the history of altered piano: from early prepared piano sounds, through the techniques of *musique concrète*, to state-of-the-art computer manipulation. The piece utilises altered piano sounds, which have been sampled, prepared, detuned, digitally edited and manipulated. The work explores combinations of traditional and non-traditional techniques, such as contrasting detuned and microtonally tuned piano sounds with a traditionally notated score in three movements. In this case, each note in the score refers to a sample´s position on a midi keyboard, and not to its actual sounding pitch. Multiple detuned scales are used, either one at a time, or by juxtaposing a different microtonal scale for each of several layers. Scales are constructed more by ear and from random detuning and pitch displacement than from ordered microtonal systems.

The work also incorporates samples of prepared piano and the computer equivalents of the techniques used in musique concrète. Thus, the sounds used span the history of altered piano techniques, ranging from the early years of prepared piano, through musique concrète techniques, to modern computer manipulation of the sounds.

„The Manufacture of Tangled Ivory" ist ein Werk, das die Geschichte des veränderten Klaviers untersucht: von den frühen Klängen der präparierten Klaviere über die Techniken der Musique Concrète bis hin zur zeitgenössischen Computermanipulation. Das Stück benützt abgewandelte Klavierklänge, die gesampelt, präpariert, verstimmt, digital geschnitten und manipuliert wurden, und erforscht die Kombinationsmöglichkeiten traditioneller und nicht-traditioneller Techniken, wie etwa die Gegenüberstellung verstimmter und mikrogestimmter Klavierklänge in einer traditionell notierten dreisätzigen Partitur. In diesem Fall entspricht jede Note der Partitur der Position eines Samples auf einem MIDI-Keyboard und nicht der tatsächlichen Klanghöhe. Mehrere verstimmte Skalen werden entweder einzeln oder als Gegenüberstellung von unterschiedlichen mikrotonalen Tonleitern auf einzelnen Ebenen verwendet. Die Tonskalen werden eher nach dem Gehör und durch zufälliges Verstimmen bzw. durch zufällige Tonhöhenverschiebung erzeugt denn als geordnetes mikrotonales System.

Das Werk enthält auch Samples von veränderten Klavieren und Computer-Äquivalente zu den traditionellen Techniken der Musique Concrète. So umspannen die verwendeten Klänge die Geschichte der Techniken des veränderten Klaviers von den Anfängen bis hin zur Computermanipulation des Klaviertons.

Annie Gosfield (USA) is a composer and performer currently based in New York City. Much of her work incorporates composing and improvising with detuned, prepared, and electronically altered piano sounds. Her work has been premiered at the New Music Marathon, Prague, The Kitchen and Tribeca Hall, New York. Annie has performed with David Moss, Roger Kleier, Marc Ribot, LaDonna Smith, Davey Williams, and Elliott Sharp.

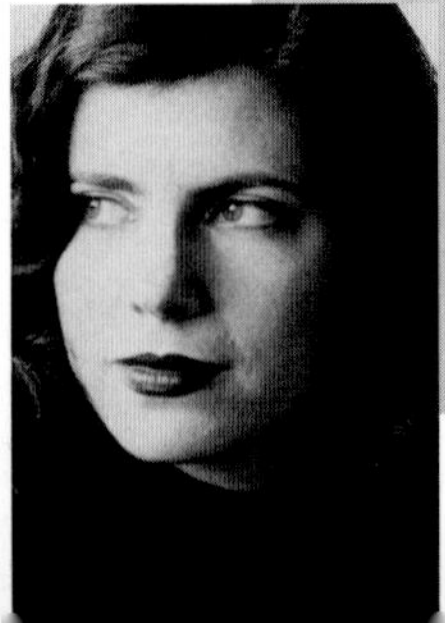

Annie Gosfield (USA), Komponistin und Musikerin. Viele ihrer Arbeiten enthalten Kompositionen und Improvisationen mit verstimmten, präparierten und elektronisch veränderten Klavierklängen. Ihre Arbeiten wurden aufgeführt beim New Music Marathon, Prag, The Kitchen und Tribeca Hall, New York. Annie ist mit David Moss, Roger Kleier, Marc Ribot, LaDonna Smith, Davey Williams und Elliott Sharp aufgetreten.

GoldGlow

Joseph Hyde

"GoldGlow" concerns the contrast between the outer world perceived by the senses and the inner world of the imagination and emotions. It concerns a particular winter numbness, when the ego seems to retreat into itself, and the mind to take on an internal life of its own, sometimes with little connection to external reality.

Two layers of sound make up the fabric of the music. On the surface is an outer layer of virtually untreated environmental sounds, all recorded one December afternoon in Bourges and presented in almost the order in which they were recorded. These sounds are thus very specifically linked to a particular place and time. Beneath is a glowing core of abstract sounds and instrumental sounds, deliberately chosen to have as little direct resonance of a particular time or place as possible.

The relationship between the two layers is very fluid. Often they are completely independent, each following its own musical flow. Sometimes an event in one will trigger a change in the other, sometimes one might temporarily take on some characteristic of the other, sometimes they will be linked by a common articulative shaping. But there is always a barrier between them, and they are never allowed to come together to form a completely homogeneous texture.

„GoldGlow" beschäftigt sich mit dem Kontrast zwischen der äußeren, über die Sinne erfahrbaren Welt und der inneren Welt der Imagination und der Emotionen. Es geht dabei um eine besondere Form des Winterschlafs, bei dem sich das Ego scheinbar in sich selbst zurückzieht und der Geist ein Eigenleben entwickelt, welches manchmal nur geringen Bezug zur äußeren Realität hat.

Zwei Klangebenen stellen die Grundtextur der Musik dar. An der Oberfläche liegt eine äußere Schicht fast gänzlich unbearbeiteter Umweltklänge - alle an einem Dezembernachmittag in Bourges aufgenommen und nun fast in der Reihenfolge der Aufnahme wiedergegeben. Auf diese Weise sind die Klänge intensiv mit einem spezifischen Ort und einer spezifischen Zeit verknüpft. Darunter liegt ein glühender Kern abstrakter und instrumentaler Klänge, die so ausgesucht wurden, daß sie so wenig wie möglich einen definierten Ort oder eine bestimmte Zeit reflektieren.

Die Beziehung zwischen den beiden Schichten ist fließend. Oftmals sind sie völlig unabhängig voneinander – jede folgt ihrem eigenen musikalischen Fluß. Dann wieder löst ein Ereignis auf einer Ebene eine Veränderung auf der anderen aus, manchmal übernimmt eine Ebene charakteristische Elemente der anderen, bisweilen werden sie durch eine gemeinsame Ausdrucksform verknüpft. Aber es gibt immer eine Barriere zwischen den beiden, es wird ihnen niemals gestattet, zu einer komplett homogenen Textur zu verschmelzen.

Joseph Hyde (GB), born 1969, is an electronic artist specialising in sonic art and multimedia. He has studied with Jim Fulkerson, Trevor Wishart, Nigel Osborne, Vic Hoyland and Jonty Harrison, and has a PhD in Musical Composition from the University of Birmingham. He is currently working at Dartington College of Arts, as Lecturer and Academic Coordinator in Performance Technology. He is also Visiting Lecturer at Exeter University and a director of Sonic Arts Network.

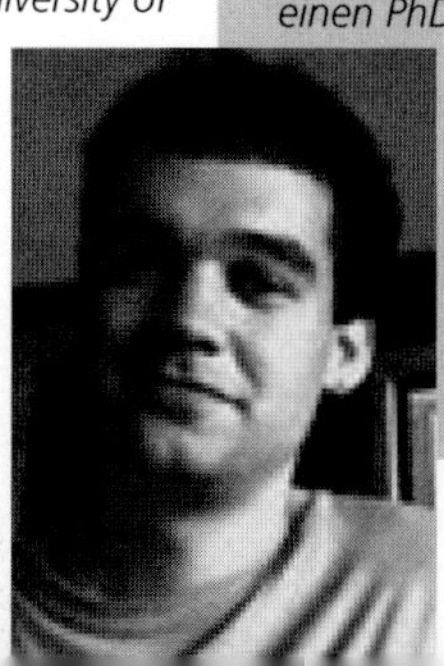

Joseph Hyde (GB), geb. 1969, ist ein auf Multimedia und Klangkunst spezialisierter Elektronikkünstler. Er hat bei Jim Fulkerson, Trevor Wishart, Nigel Osborne, Vic Hoyland und Jonty Harrison studiert und erwarb einen PhD aus Komposition an der University of Birmingham. Derzeit arbeitet er als Lektor und akademischer Koordinator für Performance-Technologie am Dartington College of Arts. Daneben ist er Visiting Lecturer an der Exeter University und einer der Leiter des Sonic Arts Network.

Songs for G

Charles Kriel

"Songs For G", written between Christmas of 1996 and May 1997, is a reflective work dealing with the character of stories I was told by a friend – an artist – who found them too difficult to tell herself within a creative work. "Songs for G" does not attempt narrative, but rather tries to create a psychological space for the listener to enter, which might be similar to both the space I occupied while composing this work, and the space habited by G throughout much of her life.

"Dottir", the first section, is rooted in daughterhood, and the antinomous life one leads as a child of the mad. The subtitles refer to sections within "Dottir", and a dialogue "directed" between child and mother.

"My eyes wet by the morning sun, he walked with his back to the west", finds its psychological source in the difficulties of creating and maintaining a mature romantic love in the wake of an embittering yet poorly questioned childhood.

"Someplace better than this place" is the second aftermath, following "My eyes ...", as "My eyes ..." follows "Dottir". The work ends ambiguously, as did G and her stories.

„Songs for G", geschrieben zwischen Weihnachten 1996 und Mai 1997, ist eine reflexive Arbeit, die sich mit einigen Erzählungen auseinandersetzt, die mir von einer Freundin – einer Künstlerin – berichtet wurden, die es selbst zu schwer fand, sie in einem kreativen Werk zu erzählen. „Songs for G" versucht nicht, den Inhalt nachzuerzählen, sondern vielmehr den Zuhörer in einen psychologischen Raum zu versetzen, der möglichst vergleichbar ist mit jenem, in dem ich mich beim Komponieren dieses Werks befand, und jenem, der von G in einem erheblichen Teil ihres Lebens bewohnt wurde.

„Dottir", der erste Teil, wurzelt im Tochter-Sein und in jenem widersprüchlichen Leben, das man als Kind einer Verrückten führt. Die Untertitel beziehen sich auf Abschnitte innerhalb „Dottirs" und auf einen „gezielten" Dialog zwischen Kind und Mutter.

„My eyes wet with the morning sun, he walked with his back to the west" (Meine Augen feucht mit der Morgensonne, ging er, dem Westen seinen Rükken kehrend) hat seine psychologische Quelle in den Schwierigkeiten, nach dem Erwachen aus einer verbitternden und dennoch nur wenig hinterfragten Kindheit eine reife romantische Liebe zu schaffen und aufrecht zu erhalten.

„Someplace better than this place" ist der zweite Nachsatz und folgt auf „My eyes...", wie „My eyes..." auf „Dottir" folgt. Das Werk endet so zwiespältig, wie auch G und ihre Geschichten sind.

Charles Kriel (GB), born 1963. Career in radio and television, parallel career as a contemporary artist. After undergraduate studies at the Atlanta College of Art, he completed his MA – Fine Arts (Combined Media) at the Chelsea College of Art in London. He was the Founding Director of Free Media Forum, a US non-profit organization for the colonization of digital media by artists. He currently lives and works in London, where he is artist-in-residence at Islington's Artec agency.

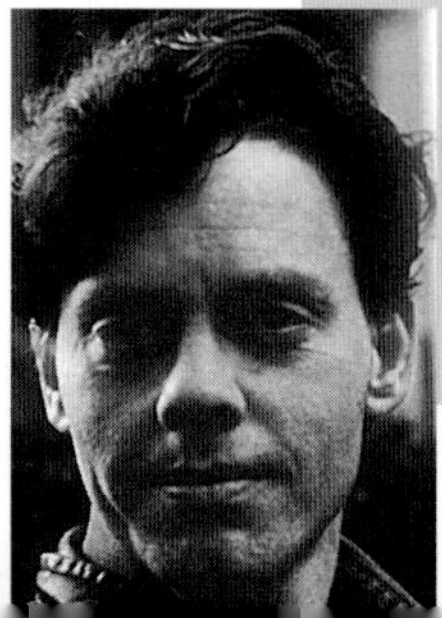

Charles Kriel (GB), geb. 1963. Karriere bei Radio und Fernsehen und als Künstler. Nach ersten Studien am Atlanta College of Art erwarb er einen MA aus angewandter Kunst (kombinierte Medien) am Chelsea Art College in London. Er war Gründungsleiter des Free Media Forum, einer US-amerikanischen Non-Profit-Organisation für die Kolonisierung der digitalen Medien durch Künstler. Derzeit lebt und arbeitet er als Artist-in-Residence an Islington's Artec Agency.

Renaissance

Åke Parmerud

Having lived and worked with electronic instruments over a period of twenty years, my attention was lately drawn to the tendency of retrospect interest in old analog synthesizers. These instruments represent to me not only a different quality of sound, but also a different mode of composing. This in turn deeply affects the musical structures coming out as the final result. Digital systems have a greater flexibility in terms of time and spectral control on a point basis. Analog systems, on the other hand, offer a way to create complex structures with a direct physical control coupled with a unique "raw" quality of sound lacking in digital instruments.

Composing with "old" analog systems is a rare phenomenon in today´s world of computers, harddisc recorders and digital synthesizers. With increasingly few exceptions, analog synthesizers are more often missing than present in modern electroacoustic workspaces. The transition from analog to digital systems as the main composing tool in electroacoustic studios has taken about ten years. The development of medieval instruments to their modern counterparts took several centuries.

"Renaissance" is based on sounds from a Serge modular synthesizer and some additional sounds from renaissance instruments like drums, lute, crumhorns and viola da gamba. My intention has been to create a virtual meeting between the sounds, composing methods and aesthetics of the analog and digital worlds, painted on a remote cavas of medieval timbres and rhythms.

Nachdem ich schon seit zwanzig Jahren mit elektronischen Instrumenten lebe und arbeite, wurde in letzter Zeit meine Aufmerksamkeit auf das aufkeimende retrospektive Interesse an den alten analogen Synthesizern gelenkt. Diese Instrumente bedeuten für mich nicht nur eine unterschiedliche Klangqualität, sondern auch einen anderen Kompositionsmodus. Dies wiederum beeinflußt stark die musikalischen Strukturen, die letztlich dabei entstehen. Digitale Systeme sind hinsichtlich der zeitlichen und spektralen Steuerung auf einer Punkt-zu-Punkt-Basis wesentlich flexibler. Analoge Systeme bieten andererseits die Möglichkeit, komplexe Strukturen in Verbindung mit direkter physikalischer Steuerung zu erzeugen, gekoppelt mit einer einzigartigen „rohen" Klangqualität, die den digitalen Instrumenten fehlt.

Mit „alten" analogen System zu komponieren ist in der heutigen Welt der Computer, Harddisk-Recorder und digitalen Synthesizer ein seltenes Phänomen geworden. Von immer weniger Ausnahmen abgesehen, wird bei modernen elektroakustischen Arbeitsräumen immer mehr auf analoge Synthesizer verzichtet. Dieser Übergang von analogen zu digitalen Systemen als Hauptwerkzeug der Komposition hat rund zehn Jahre gedauert, die Entwicklung von den mittelalterlichen Instrumenten bis zu ihrer modernen Gestalt dagegen mehrere Jahrhunderte.

„Renaissance" basiert auf Klängen aus einem modularen Serge-Synthesizer und einigen zusätzlichen Renaissanceinstrumenten wie Trommeln, Lauten, Krummhörnern und Viola da gamba. Meine Absicht war es, ein virtuelles Treffen zu arrangieren zwischen den Klängen, den Kompositionsmethoden und der Ästhetik der analogen und der digitalen Welt, gemalt auf eine ferne Leinwand aus mittelalterlichen Klangfarben und Rhythmen.

Åke Parmerud (S) has been a full time composer and artist since 1978. He has studied music but is also a trained photographer. His list of works contains instrumental music, electroacoustic music, multimedia and video. For many years he worked at EMS (The Electronic Music Studio of Stockholm) where many of his pieces were realized. He teaches computer music and composition at the Music Conservatory of Gothenburg since 1987.

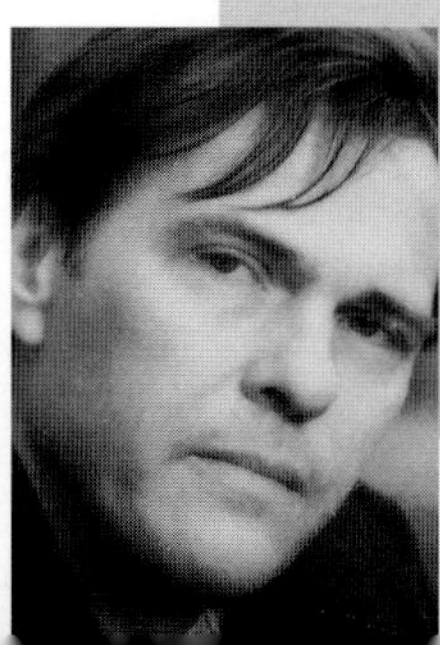

Åke Parmerud (S), Komponist und Künstler seit 1978. Er studierte Musik und ist gelernter Photograph. Sein Werk umfaßt Instrumentalmusik, elektroakustische Musik, Multimedia und Video. Mehrere Jahre arbeitete er am EMS (The Electronic Music Studio of Stockholm), wo viele seiner Stücke entstanden. Seit 1987 unterrichtet er Computermusik und Komposition am Konservatorium Göteborg.

Has/Had

Laetitia Sonami

"Has/Had" is a live performance computer piece that uses the "lady´s glove" controller of my design to map my hand and body gestures to the control of sound material and rhythmic structures.

The musical core of "Has/Had" is based on unstable rhythms; a playful wink at the stability and predictability of techno music and its variants which now dominate the popular music scene. These rhythms are on the verge of collapse and go through various transformations until they finally disintegrate, only to reappear towards the end as a vague memory passing by and accompanied by broken words ("a/avait" in French for "Has/Had", a reference to the impermanence of perceptions) and followed by remnant messages which gush in as a final and desperate attempt to say something and imprint messages when these are irrelevant and empty. The same way the mind tried to re-create stability in the rhythmic material, it tries to hold on and perceive meaning in these empty messages. The form of "Has/had" is a long structural decay. The energy at its beginning is the spring board that launches the unraveling in time of its disintegration.

„Has/Had" ist ein Live-Performance-Computerstück, in dem ich den von mir entwickelten „Lady's Glove"-Handschuh zur Abbildung meiner Hand- und Körpergesten verwende, um das Klangmaterial und die rhythmischen Strukturen zu steuern.

Der musikalische Kern von „Has/ Had" basiert auf instabilen Rhythmen - ein spielerisches Augenzwinkern gegenüber der Stabilität und Vorhersehbarkeit der Techno-Musik und ihrer Abarten, die jetzt die Popmusik-Szene beherrschen. Diese Rhythmen stehen kurz vor dem Kollaps und durchlaufen diverse Transformationen, bevor sie sich endlich auflösen, nur um gegen Ende des Stückes als vage Erinnerung vorbeizuziehen. Begleitet werden sie von gebrochenen Worten („a/avait" bedeutet im Französischen dasselbe wie „hat/hatte", ein Hinweis auf die fehlende Dauerhaftigkeit der Wahrnehmung). Auf sie folgen übriggebliebene Nachrichten, die hereinrasen als ein letzter verzweifelter Versuch, etwas zu sagen und eine Botschaft auszudrücken, wenn diese auch unwichtig und leer ist. Formal gesehen, ist „Has/Had" ist ein langer struktureller Verfall, die Energie am Anfang des Stückes ist nur das Sprungbrett, von dem das Stück in seine Desintegration in der Zeit startet.

Laetitia Sonami (F), born 1957. She studied with Joel Chadabe at the Electronic Music Studio in Albany, NY, and with Robert Ashley and David Behrman at the Center for Contemporary Music, Mills College, CA, where she received her MFA in composition in 1980. She has since been developing her own electronic musical instruments for live performance. She lives in Oakland, California, and is currently working in France.

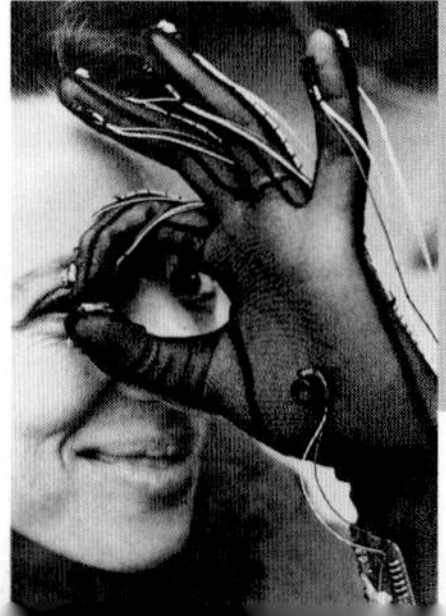

Laetitia Sonami (F), geb. 1957, studierte bei Joel Chadabe am Electronic Music Studio in Albany (New York) und bei Robert Ashley und David Behrman am Center for Contemporary Music am Mills College, Kalifornien, wo sie 1980 zum MFA aus Komposition graduierte. Seither entwickelt sie ihre eigenen Musikinstrumente für Live-Performances. Sie lebt in Oakland, Kalifornien, und arbeitet derzeit in Frankreich.

Figuras Flamencas

Mario Verandi

In this piece the source materials are from sampled flamenco musical extracts and some texts taken from the theatre play "Bodas de Sangre" ("Blood Wedding") by the Spanish writer F. G Lorca. By means of computer processing I have tried to create a surreal parallel sound world out of the recognizable flamenco sound world. The musical discourse flows as a sequence of superimposed and juxtaposed transformations and confrontations between the flamenco sound world and the surreal sound world.

The piece is clearly divided into four sections and attempts to evoke a metaphorical journey across a flamenco dreamland which is "corrupted by undefined sonic creatures". Lorca´s duende (in Spanish speaking countries, a kind of goblin which gives inspiration) had to be present in this journey and his texts were the source of inspiration behind the music as well as the core element in the organisation of the musical discourse.

The texts were recited by the young Spanish actors Javier Letran and Conchi Almeria.

"Figuras Flamencas" was composed in the Electroacoustic Music Studios at Birmingham University in 1996.

In diesem Stück stammt das Ausgangsmaterial aus gesampelten Auszügen aus Flamenco-Musik sowie einigen Texten aus dem Theaterstück „Bodas de Sangre" („Bluthochzeit") von Federico Garcia Lorca. Mittels Computerbearbeitung habe ich versucht, zu dem erkennbaren Flamenco-Klang eine surreale parallele Klangwelt zu schaffen. Der musikalische Diskurs läuft ab als Folge von überlagerten und gegeneinander gestellten Transformationen und Konfrontationen zwischen der Welt des Flamenco-Klangs und dieser surrealen Klangwelt.

Das Stück ist klar in vier Abschnitte unterteilt und versucht, eine metaphorische Reise durch ein Traumland des Flamencos heraufzubeschwören, das durch undefinierte Klangkreaturen gestört wird. Lorcas „Duende", eine Art Eingebungen verursachender Kobold, mußte bei dieser Reise natürlich dabei sein, und seine Texte waren sowohl die Quelle der Inspiration hinter der Musik wie das Kernelement zur Strukturierung des musikalischen Diskurses.

Die Texte werden von den jungen spanischen Schauspielern Javier Letran und Conchi Almeria gesprochen. „Figuras Flamencas" wurde 1996 in den Electrocoustic Music Studios der University of Birmingham komponiert.

Mario Verandi (RA), born 1960, studied Music and Computing at the University in Argentina. He continued his studies at the Phonos Electroacoustic Music Laboratory in Barcelona and later at Birmingham University in England. In addition to his acousmatic music works, he has composed music and sound environments for art installations, dance, video and theatre.

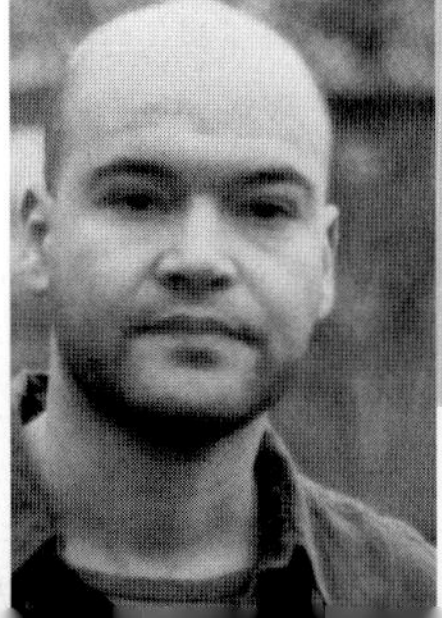

Mario Verandi (RA), geb. 1960, studierte in Argentinien Musik und Computerwissenschaften. Er setzte seine Studien am Phonos Electroacoustic Music Laboratory in Barcelona und später an der Universität Birmingham fort. Neben akusmatischen Werken hat er Musik und Klangumgebungen für Kunstinstallationen, Tanz, Video und Theater komponiert.

Klang, Kar und Melodie

Mark Wingate

"Klang, Kar und Melodie" was composed in 1996 and depicts a day in the life of a commuter in an American metropolis near the end of the millenium. This hallucinatory adventure starts innocently enough with concrete sounds of the city that gradually develop, intensify, and morph with those inside a commuter´s inner psyche. It was conceived as a concert piece, that is, a digital tape piece to be played in a concert hall with a multiple speaker sound system. This way, the sound can be further "diffused" around the audience in three-dimensional space, thus adding to the hallucinatory/cinematic narrative effect. All currently available computer hardware and software programs for digital sound manipulation were utilized. The composer also felt a need to detour from the pronounced homogeneity and conservative approach in today´s supposedly "experimental" computer music as well as its profound lack of humor. This piece further breaks the current mode by including actual pitch content, or tonalities in addition to synthetic noises, samples, digital sound processes, etc.

„Klang, Kar, und Melodie" wurde 1996 komponiert und beschreibt einen Tag im Leben eines Pendlers in einer amerikanischen Großstadt am Ende des Jahrtausends. Dieses halluzinierende Abenteuer beginnt unschuldig genug mit konkreten Klängen der Stadt, die sich langsam entwickeln, intensivieren und mit jenen im Inneren der Psyche eines Pendlers verschmelzen. Das Werk wurde als Konzertstück konzipiert, das heißt, als Stück für digitales Band, das über ein Multiple-Speaker-System in einem Konzertsaal aufgeführt wird. Auf diese Weise kann der Klang im dreidimensionalen Raum rund um das Auditorium weiter verteilt werden, was den halluzinatorisch-kinematischen Erzähleffekt noch verstärkt. Alle derzeit verfügbaren Hard- und Softwaresysteme für digitale Klangmanipulation wurden eingesetzt. Der Komponist hielt es für angebracht, bewußt aus der ausgesprochenen Homogenität und dem konservativen Ansatz in der heute als „experimentell" angesehenen Computermusik auszuscheren und von ihrer profunden Humorlosigkeit abzuweichen. Dieses Stück durchbricht die gängige Mode auch dadurch, daß es neben den synthetischen Geräuschen, Samples, digitalen Klangprozessen usw. tatsächliche tonale Inhalte (also ihrer Höhe nach definierte Töne) verwendet.

Mark Wingate (USA), born 1954, was a freelance jazz musician for a number of years in Boston, where he received the bachelor of music degree from Berklee School of Music. He is currently finishing his doctorate in composition at the University of Texas. He has been honored with fellowships from the United States Fulbright Commission and the National Endowment for the Arts among others. Wingate currently composes and teaches electroacoustic music in Austin, Texas.

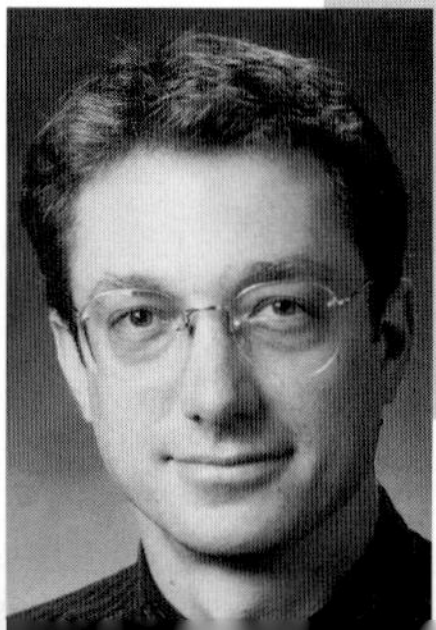

Mark Wingate (USA), geb. 1954, war jahrelang freiberuflicher Jazzmusiker in Boston, wo er zum Bachelor of Music an der Berklee School of Music graduierte. Derzeit ist er Doktorand der Komposition an der University of Texas. Er wurde unter anderem mit Stipendien der United States Fulbright Commission und des National Endowment for the Arts ausgezeichnet. Mark Wingate komponiert und lehrt gegenwärtig elektroakustische Musik in Austin, Texas.

Virtual

John Young

"Virtual" attempts to convey the idea of an illusory soundscape set in vibrant motion by the wind – an invisible and capricious source of energy for the "virtual" objects in the space, the exact nature of which we are left to imagine. This idea extended partly out of my fascination in the process of making field recordings in the natural environment, where the wind is typically regarded as a pervasive and problematic intruder. But, in fact, the wind is often a part of the sonic signature of a place (especially in the notoriously windy city in which I live), so I set out to record wind "noise" trying to take advantage of the very active but entirely natural stereo play it produces over a pair of microphones (just as it does across our ear canals). Other sound identities were designed around this idea, including controlled and turbulent air streams, tube-like resonances, and "objects" which form and flutter close to the ear, or far in the distance. Linked to this is a characteristic and consistent spatial morphology within which these sounds evolve and transform in their own ways. I was also aiming to develop a scale of sonic reality and abstraction, from the super-real sensations of direct physicality to more materially ambiguous sounds.

„Virtual" versucht, die Idee einer virtuellen Klanglandschaft zu vermitteln, die vom Wind in eine vibrierende Bewegung versetzt wird – eine unsichtbare und kapriziöse Energiequelle für die „virtuellen" Objekte im Raum, deren genaue Natur unserer Vorstellungskraft überlassen bleibt. Der Gedanke dazu entstand teilweise aus meiner Faszination, Aufnahmen der natürlichen Umgebung vor Ort zu machen, wobei der Wind normalerweise als alles überdeckender, störender Eindringling angesehen wird. Aber in Wirklichkeit ist der Wind häufig ein Teil des klanglichen Gesamtbildes eines Ortes (besonders einer so notorisch windigen Stadt wie jener, in der ich lebe). Deshalb habe ich mich aufgemacht, das Wind-„Geräusch" aufzunehmen und dabei vom sehr aktiven, aber völlig natürlichen Stereo-Spiel zu profitieren, das der Wind mit einem Paar Mikrophonen treibt (wie er das ja auch mit unseren Gehörgängen tut). Rund um diese Idee wurden andere Klangidentitäten aufgebaut, darunter gesteuerte und turbulente Luftströme, röhrenähnliche Resonanzen und Objekte, die nahe dem Ohr oder in weiter Ferne entstehen und dahinflattern. Damit verbunden ist eine charaktistische und konsistente räumliche Morphologie, innerhalb derer sich diese Klänge auf ihre eigene Weise entwickeln und verformen. Ich wollte eine Skala von klanglicher Realität und Abstraktion entstehen lassen: von den über-realen Gefühlen einer direkten Physikalität zu materiell weniger eindeutigen Klängen.

John Young (NZ) born 1962, studied composition at the University of Canterbury, where he also completed a PhD on field recordings as a sound source in electroacoustic music. In 1989, with assistance from the Queen Elizabeth II Arts Council, he travelled to the United Kingdom to further his study of electroacoustic music, returning to New Zealand the following year to join the staff of Victoria University of Wellington, where he is now Director of the Electroacoustic Music Studios.

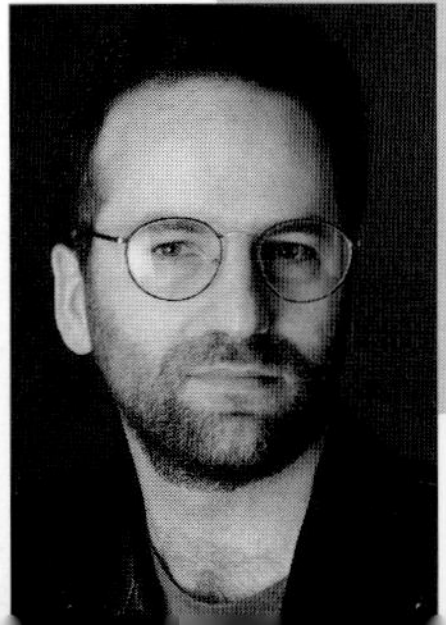

John Young (NZ), geb. 1962, studierte Komposition an der University of Canterbury, wo er über Feldaufnahmen als Klangquelle der elektroakustischen Musik dissertierte. 1989 reiste er mit Unterstützung des Queen Elizabeth II Arts Council nach Großbritannien, um seine Studien der elektroakustischen Musik fortzusetzen. Im Jahr darauf kehrte er nach Neuseeland zurück und trat in den Lehrkörper der Victoria University of Wellington ein, wo er heute Direktor der Electroacoustic Music Studios ist.

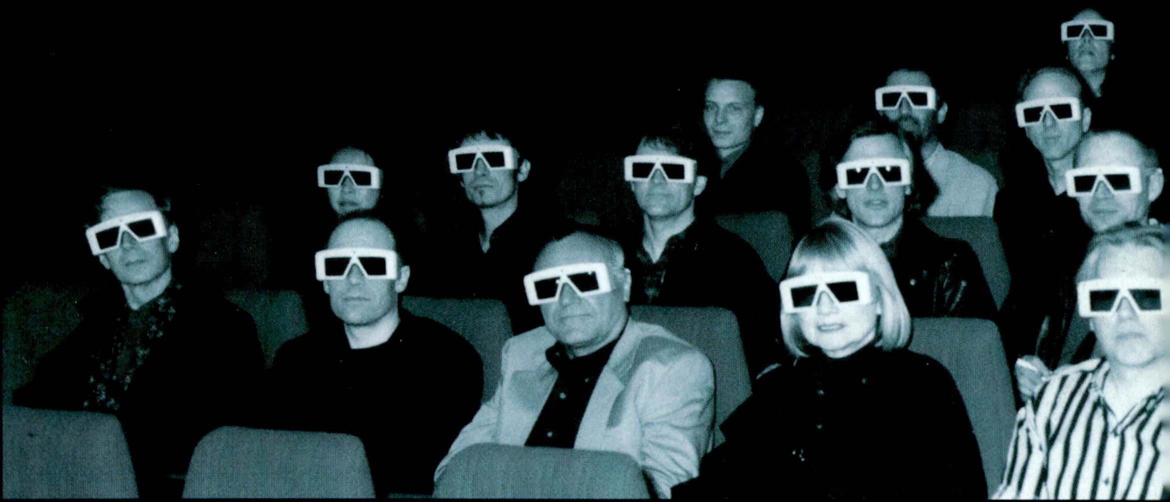

JURY

PRIX ARS ELECTRONICA 97

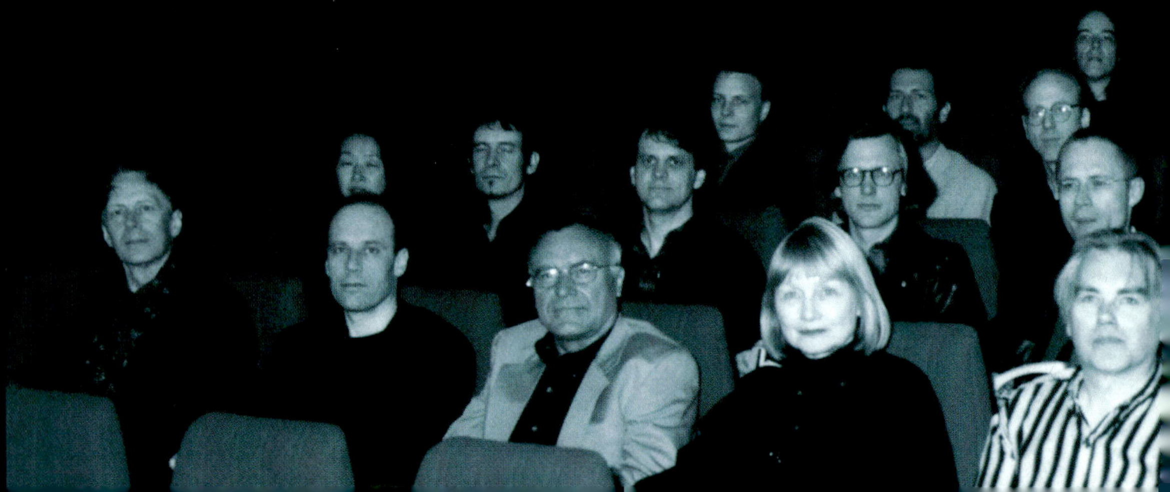

Hannes Leopoldseder
Landesintendant des ORF, Landesstudio Oberösterreich. 1979 Mitbegründer der Ars Electronica und der Linzer Klangwolke. 1987 Initiator des Prix Ars Electronica, 1991 Projektidee zum Ars Electronica Center als Museum der Zukunft in Linz.

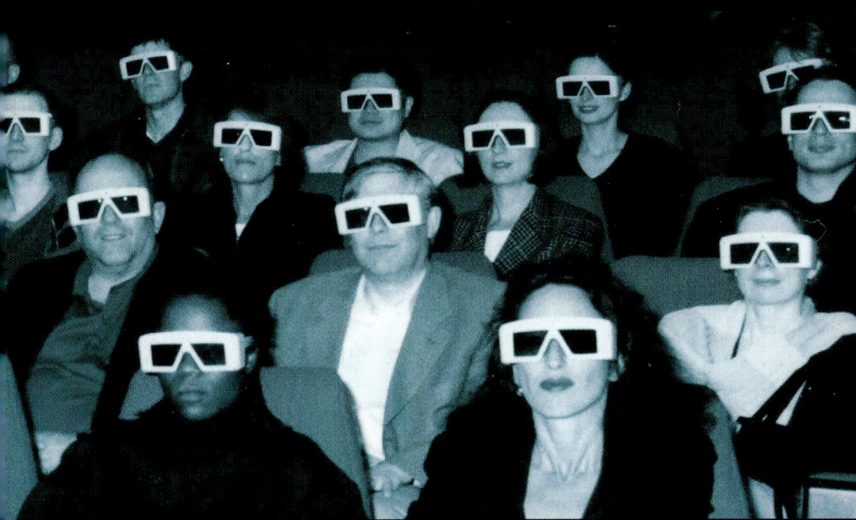

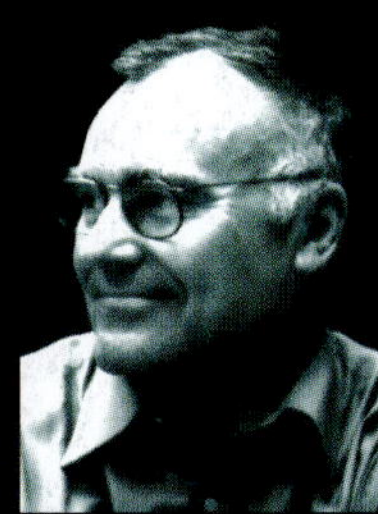

Hannes Leopoldseder
Managing Director of the Upper Austrian Region since 1974. Co-founder of the Ars Electronica Festival and the Linzer Klangwolke in 1979, initiator of the Prix Ars Electronica competition in 1987, originator of the 1991 Ars Electronica Center idea for a museum of the future in Linz. 1991 set the idea in motion of the Ars Electronica Center as a museum of the future in Linz.

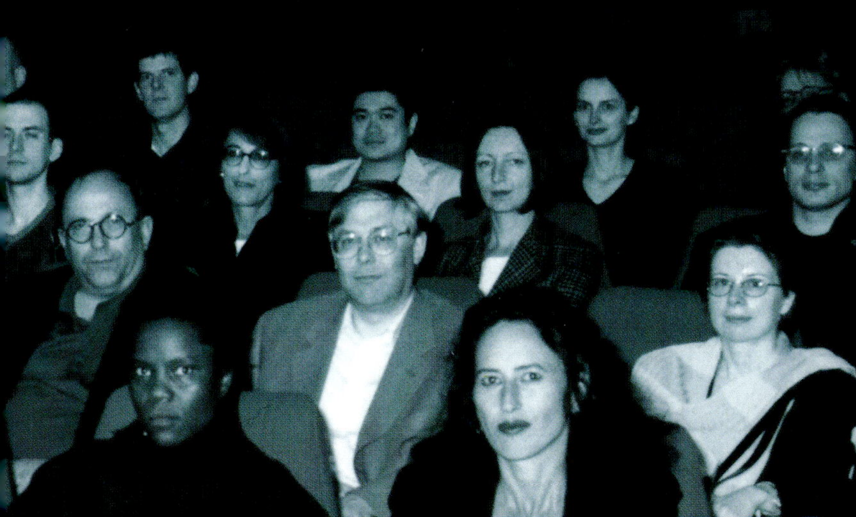

.NET JURY

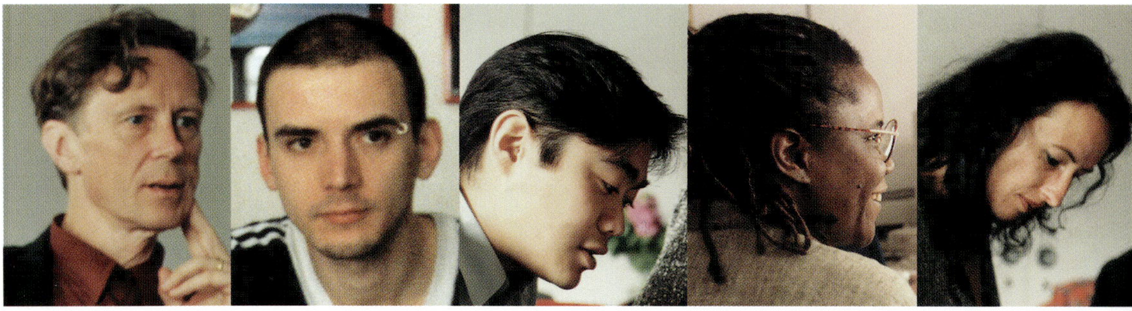

Derrick de Kerckhove

Derrick de Kerckhove (CDN), Director of the McLuhan Institute of Culture and Technology at the University of Toronto. His works on the effects of communication-media on the human nervous system include "Brainframes" and "The Alphabet and the Brain".

Derrick de Kerckhove (CDN), Direktor des McLuhan Institute for Culture and Technology, Universität Toronto. Er beschäftigt sich mit den Auswirkungen der Kommunikations-Medien auf das Nervensystem des Menschen. Wichtige Arbeiten: „Brainframes", „The Alphabet and the Brain".

Oliver Frommel

Oliver Frommel (D), born 1969; studied in Munich 1990. In 1994, Medienlabor in Munich, 1996 Ars Electronica Center, Linz. Member of FirstFloor electronix (http://www.firstfloor.org).

Oliver Frommel (D), geb. 1969. 1990 Studium in München. 1994 Medienlabor, München. 1996 Ars Electronica Center, Linz. Mitglied von FirstFloor electronix (http://www.firstfloor.org).

Joichi Ito

Joichi Ito (J/USA), developer and producer in the areas of virtual reality and multimedia. Japan correspondent for Mondo 2000, Wired and others. Numerous publications, particularly on networks.

Joichi Ito (J/USA), Entwickler und Produzent in den Bereichen Virtual Reality und Multimedia. Japan-Korrespondent für Mondo 2000, Wired u. a. Ausgedehnte Publikationstätigkeit, vor allem in bezug auf Netzwerke.

Demetria Royals

Demetria Royals (USA) is an African American Female video/filmmaker whose work has been shown throughout the United States and in Europe at Film Festivals. She has had video installations that dealt with the images of African American Women throughout American history at The Mill Valley Film Festival in California and at 651, the community art center for Brooklyn Academy Of Music. She is currently a Carnegie Mellon Fellow in the computer arts lab, where her current project is the creation of a CD ROM that will celebrate the "sheros" of the American Black Civil Rights Movement.

Demetria Royals (USA) ist eine afro-amerikanische Filmemacherin, deren Werk in den gesamten USA und bei Filmfestivals in Europa gezeigt worden ist. Ihre Videoinstallationen für das Mill Valley Film Festival (Kalifornien) und das „651", das Gemeinschafts-Kunstzentrum der Brooklyn Academy of Music, beschäftigen sich mit dem Bild afro-amerikanischer Frauen in der amerikanischen Geschichte. Als Carnegie-Mellow-Fellow arbeitet sie derzeit am Computerlaboratorium der Carnegie Mellon Universität, wo sie ein CD-ROM-Projekt betreut, das den Heldinnen der schwarzen Bürgerrechtsbewegung in Amerika gewidmet ist.

Karin Spaink

Karin Spaink (NL), born 1957, was trained as a teacher and a programmer. Works as a writer and has published seven books on various subjects, that somehow tend to be linked to that unstable object which we refer to as "the body": on sex, health, quack therapies, cyborgs etc. Spends up to eight hours per day on the Net and is currently being sued by a cult, Scientology, because of her homepage.

Karin Spaink (NL), geb. 1957, Ausbildung als Lehrerin und Programmiererin. Arbeitet als Schriftstellerin und hat sieben Bücher veröffentlicht, deren Themen alle irgendwie mit jenem unbeständigen Objekt, das wir „Körper" nennen, zu tun haben: Sexualität, Gesundheit, fragwürdige Therapien, Cyborgs usw. Sie verbringt bis zu acht Stunden täglich im Netz und wird zur Zeit wegen ihrer Homepage von der religiösen Vereinigung „Scientology" gerichtlich belangt.

INTERACTIVE ART JURY

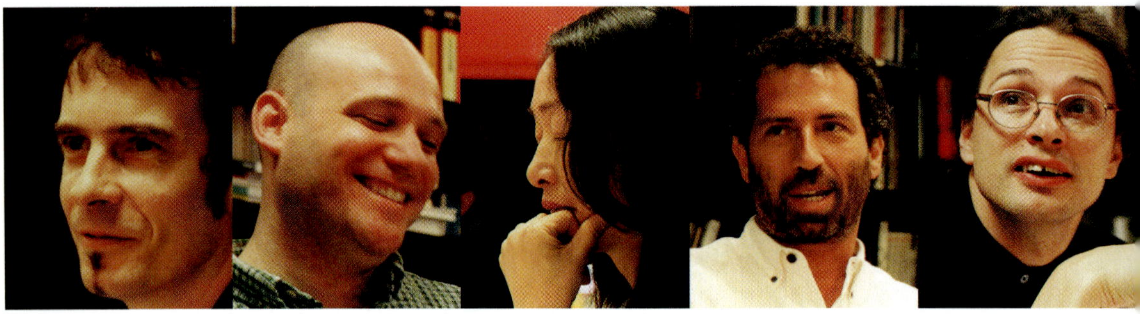

Alex Adriaansens

Alex Adriaansens (NL) studied at the Royal Academy of Arts in 's-Hertogenbosch and has worked as an artist since 1976. Together with Joke Brouwer, he initiated the V_2 Organisation as a centre for art and (media) technology in Rotterdam (Institute for the Unstable Media), of which he is now the director. In 1987, Alex Adriaansens and Joke Brouwer launched the "Manifesto for the Unstable Media". He is a member of several national advisory boards (education, art) that deal with (media) technology in a broad sense.

Alex Adriaansens (NL) studierte an der Königlichen Akademie der Künste in s-Hertogenbosch und arbeitet seit 1976 als Künstler. Gemeinsam mit Joke Brouwer gründete er die V_2-Organisation als ein Zentrum für Kunst und (Medien-)Technologie in Rotterdam (Institute for the Unstable Media), wo er derzeit Direktor ist. 1987 initiierten Alex Adriaansens und Joke Brouwer das „Manifesto for the Unstable Media". Adriaansens ist Mitglied einiger nationaler Beratungsgremien (Bildung, Kunst), die sich mit (Medien-)Technologie im weiteren Sinn befassen.

Perry Hoberman

Perry Hoberman is an installation and performance artist. His installation "Bar Code Hotel" was awarded the top prize at the 1995 Interactive Media Festival in Los Angeles, and has also been shown at Ars Electronica. Hoberman currently teaches in the graduate Computer Art Department at the School of Visual Arts in New York. He is the Art Director at Telepresence Research, a company specializing in virtual reality and telepresence installations for arts and industry.

Perry Hoberman ist Installations- und Performance-Künstler. Für „Bar Code Hotel" bekam er den ersten Preis des 1995 Interactive Media Festival in Los Angeles; diese Installation wurde auch beim Ars Electronica Festival gezeigt. Hoberman unterrichtet an der Graduate Computer Art Department der School of Visual Arts in New York. Er ist künstlerischer Direktor bei Telepresence Research, einer Firma, die sich auf Virtual Reality und Telepräsenz-Installation für Kunst und Industrie spezialisiert hat.

Machiko Kusahara

Machiko Kusahara (J) studied mathematics and history of science at Tokyo's International Christian University and is an Associate Professor of Media Art at the Faculty of Arts, Tokyo Institute of Polytechnics. She has been teaching computer graphics theory and media since 1986 and has published several books on computer graphics and A-Life. Her recent research has been centered around the transition of the nature of artistic creativity in interactive art, especially in relation to the concept of networking and A-Life.

Machiko Kusahara (J) studierte Mathematik und Wissenschaftsgeschichte an der Internationalen Christlichen Universität Tokio und ist Associate Professor für Medienkunst an der Kunstfakultät des Tokyo Institute of Polytechnics. Seit 1986 unterrichtet sie Theorie der Computergraphik und Medien und hat zahlreiche Bücher über Computergraphik und A-Life veröffentlicht. Ihre derzeitige Forschungstätigkeit konzentriert sich auf die Frage des Übergangs der künstlerischen Kreativität in die interaktive Kunst, insbesondere in Hinblick auf Vernetzungskonzepte und A-Life.

Michael Naimark

Michael Naimark (USA) was instrumental in making the first interactive laserdiscs in the late 1970s at MIT and has worked with projection and immersive virtual environments. Naimark has held faculty appointments at the San Francisco Art Institute, San Francisco State University, California Institute of the Arts, MIT, the University of Michigan, and is on the Editorial Boards of Presence and Leonardo Electronic Almanac. He created a BS in Cybernetic Systems as an independent major from the University of Michigan in 1974 and received an MS in Visual Studies and Environmental Art from MIT in 1979.

Michael Naimark (USA) war federführend an der Erstellung der ersten interaktiven Laserdisks am MIT in den späten 70er Jahren beteiligt und hat sich mit Projektion und immersiven virtuellen Environments beschäftigt. Er hat Lehraufträge am San Francisco Art Institute, an der San Francisco State University, am California Institute of the Arts, am MIT, an der University of Michigan und gehört zu den Herausgebern von „Presence" und des „Leonardo Electronic Almanac". 1974 erhielt er einen BS aus Kybernetik an der University of Michigan und graduierte 1979 am MIT zum MS aus Visuellen Studien und Umweltkunst.

Gerfried Stocker

Gerfried Stocker (A), born 1964. In 1991, he founded x-space, a group of artists-technicians who are involved in interdisciplinary projects. Up to 1995, numerous x-space projects have been carried out, including EXPO Sevilla '92, Venice Biennale '93, FISEA '93/94/Minneapolis/Helsinki, SIGGRAPH '94/95-Orlando/Los Angeles, ISEA '95 Montreal. In 1992/93 Stocker was the artistic director of the "Steirischen Kulturinitiative" and was also involved in setting up the artists network ZEROnet. Since 1995, Gerfried Stocker has been the artistic director of the Ars Electronica Festival and the managing director of the Ars Electronica Center.

Gerfried Stocker (A), geb. 1964, gründete 1991 x-space, eine Künstler-Techniker-Gruppe zur Realisierung interdisziplinärer Projekte. Bis 1995 wurden zahlreiche x-space-Projekte realisiert, u. a. EXPO-Sevilla '92, Biennale Venedig '93, FISEA-'93/Minneapolis/ Helsinki, SIGGRAPH '94/95-Orlando/ Los Angeles, ISEA '95 Montreal. In den Jahren 1992/93 war Stocker künstlerischer Leiter der „Steirischen Kulturinitiative" und widmete sich u. a. dem Aufbau des Künstler-Netzwerkes ZEROnet. Seit 1995 ist Gerfried Stocker künstlerischer Leiter des Ars Electronica Festival und Geschäftsführer des Ars Electronica Center.

COMPUTER ANIMATION JURY

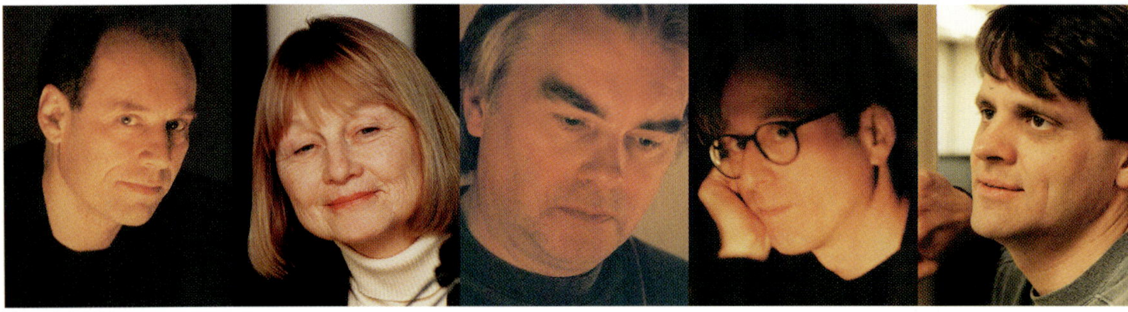

Rolf Herken

Rolf Herken (D), born 1954, studied Theoretical Physics and Mathematics at the Freie Universität Berlin. In 1986 he founded the company "mental images" in Berlin, which specializes in the development of high image quality visualization software. His interests are in computer graphics with special emphasis on image synthesis, and in artifical intelligence, specifically mental imagery and vision.

Rolf Herken (D), geb. 1954, studierte Theoretische Physik und Mathematik an der Freien Universität Berlin. 1986 gründete er in Berlin die Firma „mental images", die auf die Entwicklung von Hochleistungs-Visualisierungssoftware spezialisiert ist. Sein Hauptinteresse gilt der Computergraphik, vor allem der Bildsynthese, und der künstlichen Intelligenz, vor allem der Rolle von Vorstellungsbildern.

Mickey McGovern

Mickey McGovern (USA) has been working in the visual effects film industry since 1987. During this time she has worked on staff for George Lucas' Industrial Light and Magic, Richard Edlund's Boss Film, and Sony Pictures Imageworks. Mickey has produced blue / greenscreen, element, model, plate photography, digital compositing and effects animation for several Hollywood films. Among these films are "Michael", "James and the Giant Peach", and "Speed". Mickey is currently the Executive Producer of MillFilm, Ltd. in London, England, which is a film division that has grown out of The Mill and is associated with Tony and Ridley Scott.

Mickey McGovern (USA) arbeitet seit 1987 im Visual-Effects-Bereich der Filmindustrie. In diesen Jahren war sie Mitarbeiterin bei George Lucas' Industrial Light & Magic, bei Richard Edlunds Boss Film und bei Sony Pictures Imageworks. Sie hat die Bluescreen-, Element-, Modell-, Einzelbildphotographie, die digitale Montage- und Effektanimation für zahlreiche Hollywoodfilme produziert, darunter für „Michael", „James and the Giant Peach" und „Speed". Derzeit ist McGovern Executive Producer für MillFilm, Ltd., London, der aus The Mill in Partnerschaft mit Tony und Ridley Scott entstandenen Filmabteilung.

A. J. Mitchell

A. J. Mitchell (GB), born 1947; cameraman at BBC-TV; 1976 video effects supervisor; 1980 freelance als lighting cameraman mainly on effects and pop promotions; 1981 The Moving Picture Company as a commercial director, cameraman and effects supervisor; 1987 director of special effects; 1990 director of Printed Picture Company.

A. J. Mitchell (GB), geb. 1947; Kameramann bei BBC-TV; 1976 Leiter Video-Effects; 1980 freischaffender Beleuchtungskameramann hauptsächlich für Special-Effects und Pop-Werbung; 1981 Geschäftsführer, Kameramann und Special-Effects-Leiter der Moving Picture Company; 1987 Direktor für Special Effects; 1990 Direktor der Printed Picture Company.

Michael Wahrman

Michael Wahrman (USA) is an independent designer and supervisor of digital effects and production technology. He has contributed to the design, planning and supervision of dozens of feature films and special venue productions. Currently he is working as an independent digital effects supervisor, a consultant to Viacom International, a designer of creative technology for Post Perfect (New York Media Group), and designing a series of production technologies for the independent computer animator.

Michael Wahrmann (USA) ist unabhängiger Designer und Supervisor für Digital Effects und Produktionstechnologie. Er war bei Dutzenden von Spielfilmen und Auftragsarbeiten für Design, Planung und Supervision verantwortlich. Zur Zeit arbeitet er als unabhängiger Digital-Effects-Supervisor, als Konsulent für Viacom International, als Designer kreativer Technologie für Post Perfect (New York Media Group) und entwirft eine Serie von Produktionstechnologien für unabhängige Computeranimatoren.

Chris Wedge

Chris Wedge (USA) has a BFA from the State University of New York and an MA in Computer Graphics and Art Education from Ohio State University. He is Vice President and Creative Director of Blue Sky Studios, Inc. and is considered one of the top computer animation directors in the world. He was director of animation at MAGI/ SynthaVision, where he directed and/or animated projects for television commercials and sequences for the Walt Disney feature film "Tron".

Chris Wedge (USA) graduierte an der State University of New York zum BFA und an der Ohio State University zum MA aus Computergraphik und Kunsterziehung. Er ist Vizepräsident und Kreativdirektor der Blue Sky Studios, Inc., und gilt als einer der führenden Regisseure für Computeranimation weltweit. Er arbeitete als Director of Animation bei MAGI/ SynthaVision, wo er für Regie bzw. Animation von TV-Spots sowie Sequenzen des Walt-Disney-Films „Tron" verantwortlich war.

COMPUTER MUSIC JURY

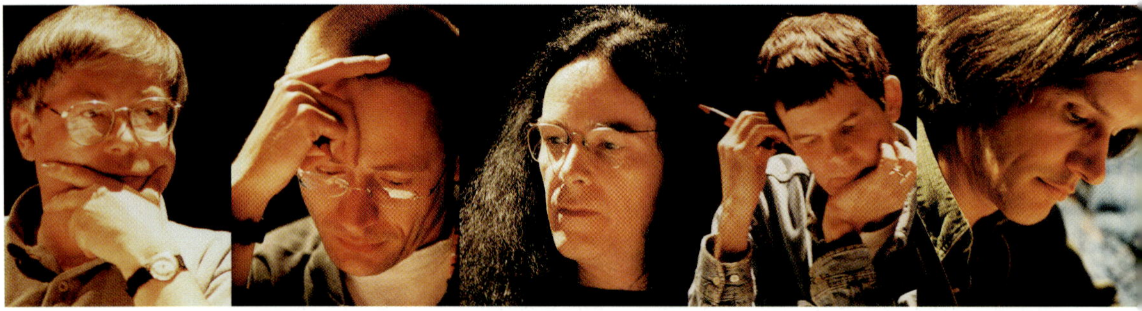

Stephen Arnold

Stephen Arnold (GB) studied with Peter Maxwell Davies, Jonathan Harvey and Alexander Goehr, and at the Universities of Southampton and Nottingham, gaining a PhD for a study of Milton Babbitt. He is Director of the Computer Music Studios at the University of Glasgow, where he heads a team engaged in composition and research relating to Music Technology. His present research relates to the development of audio applications designed to exploit the high performance of ATM networks. He is Vice-President for Conferences for the International Computer Music Association.

Stephen Arnold (GB) studierte bei Peter Maxwell Davies, Jonathan Harvey und Alexander Goehr sowie an den Universitäten von Southampton und Nottingham. Für sein Doktorat schrieb er eine Studie über den amerikanischen Komponisten Milton Babbitt. Er ist Direktor der Computer Music Studios an der University of Glasgow, wo er ein Team leitet, das sich der Komposition, Lehre und Forschung im Bereich der Musiktechnologie widmet. Sein derzeitiges Forschungsgebiet ist die Entwicklung von Musik- und Audioanwendungen zur Nutzung der Geschwindigkeit in ATM-Netzwerken. Er ist Vizepräsident für Konferenzen der International Computer Music Association.

Sam Auinger

Sam Auinger (A), born 1956, lives in Linz. Intensive involvement with questions of composition, computer music, sound design and psychoacoustics since the early 1980s. Work in film, theater, radio, video ... Exhibitions and festivals ... in Europe and the US ... Since 1989, regular cooperation on sound installations with Bruce Odland: Garten der Träume (Ars Electronica 90), TrafficMantra (Forum Trojanum, Rome 1991). They were also responsible for the sound design for Peter Sellar's "Die Perser" (1993). 1997 Artist in Residence, DAAD.

Sam Auinger (A), geb. 1956, lebt in Linz. Seit den frühen 80er Jahren intensive Beschäftigung mit Fragen der Komposition, Computermusik, Psychoakustik und des Sound-Design. Arbeit in Film, Theater, Radio, Video … Ausstellungen und Festivals … in Europa und den USA … Seit 1989 regelmäßige Zusammenarbeit bei Sound-Installationen mit Bruce Odland: Garten der Träume (Ars Electronica 90), TrafficMantra (Forum Trojanum, Rom 1991). Sie zeichneten auch verantwortlich für das Sound-Design von Peter Sellars „Die Perser" (1993). 1997 Artist-in-Residence, DAAD.

Naut Humon

Naut Humon (USA) is the conductor, curator and one of the composers in Sound Traffic Control, a San Francisco-based media consortium that features a real-time spatial 3D audio network through which musicians, sound sculptors, DJ's and film/video artists create an interwoven surround-sound orchestra. He is director of operations for Recombinant – a massive "Diffusion Jockey" summit that crossfades the junction between electroacoustic orchestras, visualistic ballistics and audio DJ subcultures.

Naut Humon (USA) ist Dirigent, Kurator und einer der Komponisten von Sound Traffic Control, einem in San Francisco ansässigen Medienkonsortium, das ein dreidimensionales räumliches Echtzeit-Audio-Netzwerk umfaßt, das es Musikern, Klangbildhauern, Diskjockeys und Film- bzw. Videokünstlern erlaubt, ein dicht verwobenes Sound-Surround-Orchester zu schaffen. Er ist operativer Leiter von Recombinant, einem großen „DiffusionJockey"-Gipfel, der einen fließenden Übergang herstellt zwischen elektroakustischen Orchestern, „visualistic ballistics" und der Audio-DJ-Subkultur.

Ben Neill

Ben Neill (USA) is a composer and performer based in New York City. Much of his recent music is written for the mutantrumpet, an electroacoustic instrument of his own design. Neill studied composition with La Monte Young and holds a Doctorate degree from Manhattan School of Music. He has been an Artist in Residence at Steim in Amsterdam. Since 1992 he has been the Music Curator of The Kitchen in New York.

Ben Neill (USA) ist ein in New York lebender Komponist und Musiker. Ein erheblicher Teil seiner jüngsten Kompositionen wurde für „Mutantrumpet" geschrieben, ein von ihm selbst entwickeltes elektroakustisches Instrument. Neill studierte Komposition bei La Monte Young und hat an der Manhattan School of Music promoviert. Er war Artist-in-Residence am Steim in Amsterdam. Seit 1992 ist er der Musik-Kurator von The Kitchen, New York.

Robert Normandeau

Robert Normandeau (CDN), born 1955, founding a member of the Canadian Electroacoustic Community (CEC) and of Réseaux, a group which promotes creative collaborations of art and the media. 1986–1993 member of the Association pour la création et la recherche électroacoustique du Québec (ACREQ), since 1988 lectures in acoustics and electroacoustics at the Faculty of Music of the University of Montreal. Current endeavours focus on acousmatic music.

Robert Normandeau (CDN), geb. 1955, Gründungsmitglied der Canada Electroacoustic Community (CEC) und von Réseaux, einer Gruppe zur Förderung der kreativen Zusammenarbeit zwischen Kunst und Medien. 1986–1993 Mitglied der Association pour la création et la recherche électroacoustique du Québec (ACREQ), seit 1988 Vorlesungen über Akustik und Elektroakustik an der Musikfakultät der University of Montreal. Derzeitiger Arbeitsschwerpunkt ist akusmatische Musik.

Participants .NET

A & G
Ecole Nationale Supérieure
des Beaux Arts
14, rue Bonaparte
75006 Paris, F
mbenayoun@compuserve.
com
http://artandglory.ensba.fr

Sabin Aell
Semmelweisstr. 2
4020 Linz, A
sabin.aell@silverserver.co.at
http://www.rawpower.co.at/r
awpower/hannibal

Amy Alexander
California Institute of Arts
24700 McBean Pkwy.
Valencia, CA 91355, USA
amy@emsh.calarts.edu
http://shoko.calarts.edu/~al
ex/recycler.html

Animal With No Name
130 West 3rd Street #4
New York, NY 10012, USA
animal@flyvision.org
http://www.flyvision.org/SI/A
WNN/SS/index.html

Yves Arikoglu
Project 03
Beymestr. 10a
12167 Berlin, D
yves@project03.com
http://www.itw.fta-
berlin.de/studio

Allan Hunt Badiner
Deer Point - Lime Creek
Big Sur, CA 93920, USA
ahbadiner@igc.org
http://www.odwalla.com

Thomas Balzer
Platz der Vereinten
Nationen 2
10249 Berlin, D
balzer@berlin.snafu.de
http://www.azwei.de/balzer/
nasenroman.html

Alfred Banze
Hans-Sachs-Str. 19
50931 Köln, D
alfred@khm.de
http://www.khm.de/~alfred

Bill Barminski
DE-LUX'O
P.O. Box 661635
Los Angeles, CA 90066, USA
deluxo1@aol.com
http://www.cyclopsboy.com

Scott Becker
P.O.Box 578956
Chicago, IL 60657, USA
artscb@interaccess.com
http://www.artnetweb.com/p
rojects/sbeck2/sens.html

Maurice Benayoun
Z.A Production
64, rue de la Folie Méricourt
75011 Paris, F
mbenayoun@compuserve.com
http://panoramix.univ-
paris1.fr/UFR04/benayoun/Et
Moi/index.html

Joachim Blank
Internationale Stadt e.V.
Prinzessinnenstr. 19/20
10569 Berlin, D
blank@icf.de
http://www.documenta.de/wi
thout_addresses

Mark Blaschitz
Technische Universität Graz
Rechbauerstr. 12
8010 Graz, A
wb@xarch.tu-graz.ac.at
http://xarch.tu-
graz.ac.at/proj/wb/96

Christoph Blase
Wiener Weg 3b
50858 Köln, D
christophblase@compuserve
.com
http://blitzreview.thing.at

Blue Klein
Via Puccionotti, 19
50129 Firenze, I
bluklein@dada.it
http://www.dada.it/dadart/lijn

Göran Boardy
Elvsborgsgatan 33
414 72 Göteborg, S
goran.boardy@valand.gu.se
http://www.valand.gu.se/Fin
eArts/staff/GBoardy/gravity/i
ndex.html

Paul Bonaventura
Ruskin School of Drawing
and Fine Art
74 High Street
Oxford OX1 4BG, GB
paul.bonaventura@ruskin-
school.ox.ac.uk
http://www.ruskin-
sch.ox.ac.uk/lab

Sylke Booms
P+BOOMS
Höhenweg 78
52072 Aachen, D
infoboard@P-Booms.com
http://www.inphutation.com

Charles Boone
1122 South Campbell
Avenue
Wheaton, IL 60187, USA
http://www.netcom.com/~eu
biedad/stations.htm

Lee Boot
Goethe-Institut Washington
814 Seventh Street, NW
Washington,
DC 20001-3718, USA
goethedc@artswire.org
http://www.glows.com/cyber
clipper/

Brad Brace
1005 Market St. #411
San Francisco, CA 94103, USA
bbrace@netcom.com
http://www.teleport.com/~b
brace/12hr.html

Matthew Brand
MIT Media Lab
20 Ames Street
Cambridge,
MA 02139-4307, USA
brand@media.mit.edu
http://www.media.mit.edu/~
brand/artist.html

Marion Breitschopf
BLACK BOX SYSTEMS
Bennogasse 8/6
1080 Wien, A
lovion@blackbox.at
palace://palazzo.blackbox.at

Michael Bretterbauer
Computer-Technik
Bretterbauer & Kothgassner
GesmbH
Tienenweg 361
8911 Admont, A
mike@schurke.at
http://www.schurke.at

Sawad Brooks
20 Ames Street E 15-443
Cambridge, MA 02139, USA
sawad@mediamit.edu
http://www.media.mit.edu/~
sawad/graffiti/

Paul Brown
2 Susan Court
Cornubia, QLD 4130
AUS
paul-brown@siggraph.org
http://www.msstate.edu/fine
art_online/home.html

Christian Bruckmayr
Piringerhofstr. 11
4020 Linz, A
cbruckmayr@silverserver.co.
at
http://www.rawpower.co.at/r
awpower/cmc

Eva Brunner-Szabo
Währinger Gürtel 51/9
1180 Wien, A
e.b.szabo@xpoint.at
http://ourworld.compuserve.
com/homepages/erinnerun
gen

Helena Bulaja-Madunnic
Radnicki dol 8
10000 Zagreb, HR
http://www.vidi.hr/helena/loo
k.html

Heath Bunting
Top Flat, 2 Neals Yard
London, WC2H 9DP, GB
heath@cybercafe.org
http://www.irational.org/cctv/

Marcy Burt Butz
Keltenstraße 23
8044 Zürich, CH
burtz@burtz.ch
http://www.whs.ch

Christophe Chamerat
Skydebanegade 3, 4 tv
1709 Kopenhagen V, DK
chrischamerat@dk-online.dk
http://www2.dk-
online.dk/users/chistophe

Helen Cho
Quiet Time
6056 Romany Road
Oakland, CA 94618, USA
ifrancis@quiet-time.com
http://www.quiet-time.com

Murry-C. Christensen
Knowledge Transfer
International
747 Third Avenue
New York, NY 10017, USA
mc2@mchristensen.com
http://www.mchristensen.co
m/mental/mental.htm

Paul Clifford
Chichester Institute HE
College Lane
Chichester PO19 4PE, GB
jupiter@netcomuk.co.uk
http://www.rmplc.co.uk/edu
web/sites/kentlea/rsg/Rainha
m/index.html

Jean-François Colonna
CMAP
Ecole Polytechnique
91128 Palaiseau Cédex, F
colonna@cmapx.polytechniq
ue.fr
http://blanche.polytechnique
.fr/lactamme/Mosaic/descript
eurs/infinity.03.html

Holly Crawford
3701 Sacramento St. #170
San Francisco, CA 94118, USA
holly@netwiz.net
http://www-
leland.stanford.edu/~geo/ho
lly_crawford.html

Lyndall Culbertson
45 Tudor City Place #2201
New York, NY 10017, USA
linoleum@interport.net
http://www.users.interport.net/
~linoleum/Expo/experiment.ht
ml

Alexis Da Silva
4403 Saint-Jacques Street
Montreal, PQ H4C 1K1, CDN
d172321@er.uqam.ca
http://www.er.uqam.ca/nobel/d
172321/index.html

Diana Danelli
Milan University
Via Comelico 39
20134 Milano, I
dianadnl@ipo.tesi.dsi.unimi.it
http://www.dsi.unimi.it/~ars/La
nimae.html

Sharon Daniel
University of California, Santa
Cruz
Baskin Visual Arts Center
Santa Cruz, CA 94114, USA
sdaniel@cats.ucsc.edu
http://www.decordova.org/

Dominique de Bardonnèche
Editions IDEM
49, rue Louis de Savoie
1110 Morges, CH
debab@pingnet.ch
http://www.museum.ch/Lausan
ne Antique.htm

Pierre de Kerangal
130 W 3rd Street #4
New York, NY 10012, USA
pierre@flyvision.org
http://www.nyu.edu/pages/ngc/
projects/JetLog/index.html

Petra De Nijs
2e Jan Steenstraat 62-1
1074 CR Amsterdam, NL
pdenijs@desk.nl
http://www.desk.nl/~pdenijs

Ryoichiro Debuchi
Atom Co., Ltd., Asaki - Seimei
Bldg.
1-7-8 Shibuya, Shibuya-ku
Tokyo 150, J
debuchi@atom.co.jp
http://www.atom.co.jp/vrml2/o
cean/index.html

Lora Dibner
225 East 4th Street #17
New York, NY 10009, USA
lora@sva.edu
http://www2.sva.edu/hereandt
here/lora/

Tim Diggins
RED56
91 Brick Lane
London E1 6QN, GB
http://www.red56.co.uk
/advent

David Diggles
7 Sankey St,
Highgate Hill
Brisbane 4101, AUS
aladrin@ice-t.com.au
http://www.aladrin.org

Gisela Domschke
35-C Glenloch Road
London NW3 4DJ, GB
fitapura@pratique.fr
http://www.cicv.fr/chez/

Karel Dudesek
VAN GOGH TV
Glüsingerstr. 40c
21217 Seevetal, D
100542.1571@compuserve.com

Uwe Eßer
Kürtenstr. 91
40472 Düsseldorf, D
http://www.zakk.de/einstieg13g

Karlheinz Essl
Am Ölberg 26-30
3400 Klosterneuburg, A
essl@ping.at
http://www.ping.at/users/essl/works/mindshipmind.html

Dmitry Feodoroff
Ul, Moldagulovoy 28/4
Art. 40
111538 Moscow, RU
info@cronyx.ru
http://www.mosdesign.ru/concept

Robert Andreas Fischer
Doeltschihalde 24
8055 Zürich, CH
robfi@vmz.unizh.ch
http://www.to.or.at/~riviera

Carol Flax
5690A Casitas Pass Rd.
Carpinteria, CA 93013, USA
cflax@west.net
http://www.cmp.ucr.edu/flax/2poems.html

Floating Point Unit
60 Berry Street
Brooklyn, NY 11211, USA
mimic@inch.com
http://www.thing.net/~floating

Amy Franceschini
Atlas Web Design
1201 B Howard St. 2nd Fl.
San Francisco, CA 94103
USA
ame@sirius.com
http://www.atlasmagazine.com
Cornelia Franke
Künstlerhaus Mousonturm
Daimlerstraße 32
60314 Frankfurt, D
nanna@inm.de

Kay Fricke
Schöffelgass 6
3002 Purkersdorf, A
kay@vis-med.ac.at
http://judge.vis-med.ac.at:8080/

Paul Friedlander
43 Narcissus Road
London NW6 1TL, GB
praskovi@clara.net
http://home.clara.net/praskovi

Christian Fröhlich
Heidenreichring 48/7
8054 Graz-Seiersberg, A
froeh@sbox.tu-graz.ac.at
http://xarch.tu-graz.ac.at/home/w/arch/space.music/

Matthew Fuller
121 Bethnal Green Road
London E2 7DG, GB
escapecommitee@compuserve.com
http://www.phreak.co.uk/i_o_d/home.html

György Galantai
Artpool Art Research Center
Liszt Ferenc tTr 10, I.1.
1061 Budapest, H
artpool@artpool.hu
http://www.artpool.hu/RayJohnson.html

Gerald Ganglbauer
GANGAN PUBLISHING
P.O.Box 522
Strawberry Hills,
NSW 2012, AUS
gangan@matra.com.au
http://www.matra.com.au/~gangan

Paul Garrin
name.space
11E. 4th Street 2F
New York, NY 10003, USA
info@autono.net
http://namespace.autono.net

Gregory Garvey
Concordia University
1455 de Maisonneuve Blvd.
West
Montreal, QC H3G 1M8, CDN
ggarvey@vax2.concordia.ca
http://www.digearth.bcit.bc.ca/dedocs/ggarvey/GBscreen.html

Knut Gerwers
Forster Str. 50
10999 Berlin, D
kg@mediopolis.de
http://www.thing.de/projekte/future/
Copper Giloth
University of Massachusetts
125 LGRC
Amherst, MA 01003, USA
giloth@art.umass.edu
http://www.umass.edu/thirdperson

Valéry Grancher
53, Rue de Seine
75006 Paris, F
vgranger@imaginet.fr
http://www.imaginet.fr/nomemory

Hermann-Josef Hack
Hangweg 11
53757 Sankt Augustin, D
hack@gekko.de
http://www.hack-roof.de/

Emily Hartzell
New York University Center
for Advanced Technology
719 Broadway,
12th Floor
New York, NY 10003, USA
parkbnch@cat.nyu.edu
http://www.cat.nyu.edu/parkbench

Auriea Harvey
ENTROPY8 DIGITAL ARTS
114 East 1st St. #20
New York, NY 10009, USA
chaos@entropy8.com
http://www.entropy8.com

Bunting Heath
Top Flat, 2 Neals Yard
London WC2H 9DP, GB
heath@cybercafe.org
http://www.irational.org/cctv/

Lynn Hershman-Leeson
HOTWIRE PRODUCTIONS
327 Ritch Street
San Francisco, CA 94108
USA
lynn2@well.com
http://www.construct.net/

June Houston
127 Mac Dougal Street #953
New York, NY 10012, USA
june@flyvision.org
http://www.flyvision.org/sitelite/Houston/GhostWatcher/

G.H. Hovagimyan
PSEUDO PROGRAMS
600 Broadway
New York, NY 10012, USA
gh@thing.net
http://pseudo.com/netcast/shows/adirt/index.html

Lin Hsin Hsin
2 Jalan Daliah #05-02
Singapore 438516, SGP
linhh@pacific.net.sg
http://www.lhham.com.sg/exhibit

Markus Huemer
Gereonswall 23 b
50677 Köln, D
huemer@khm.de
http://www.softa.com/mh
Leslie Huppert
Hilschbacherstr. 31
66292 Riegelsberg, D
fekoooo@stud.uni-sb.de
http://www-wj.cs.uni-sb.de/art/leslie/p2

Lisa Hutton
4606 Castelar St.
San Diego, CA 92107-1412
USA
lhutton@sgva-serv1.ucsd.edu
http://artslab.ucsd.edu/ARTSLAB/LisaHutton/LLHpage.html

Internationale Stadt e.V.
Prinzessinnenstr. 19/20
10969 Berlin, D
is@icf.de
http://www.icf.de

Mogens Jacobsen
artnode.dk
Udstfllingsbygningen
Nikolay P
1153 Copenhagen K, DK
jacobsen@artnode.dk;
mogens@mondo.dk
http://www.artnode.dk

Geismar Jarg
INTERNET DESIGN
DORFSCHENKE
Am Damm 10
47229 Duisburg, D
geismar@pop.gun.de
http://www.uni-duisburg.de/HRZ/Art/GEISMAR/DOG

Manuela Jungmann
Box 293,
2440-16th Street
San Francisco, CA 94103
USA
http://www.dreamstream.org

Markus Käch
Dufourstr.15
6003 Luzern, CH
m.kaech@mail.tic.ch
http://www.moving-art-studio.com/september/media_diseases.html

Daniel Kelly
4077 Crystal Drive
Beulah, MI 49617, USA
dankelly@artisthouse.com
http://www.artisthouse.com

Raivo Kelomees
Estonian Academy Ofarts
TARTU MNT 1
0001 Tallinn, EW
raivo@artun.ee
http://www.artun.ee/center/emedia/emedia.html
Andruid Kerne
NYU Media Research Lab
719 Broadway, 12th Floor
New York, NY 10003, USA
andruid@cat.nyu.edu
http://www.cat.nyu.edu/ecology

Omar Khan
66 Cumberland St.
San Francisco, CA 94110,
USA
omar@harappa.com
http://www.harappa.com

Paul Kim
Avalanche Systems Inc.
304 Hudson St. 6th floor
New York, NY 10013, USA
seidler@avsi.com
http://www.misterboy.com

Rudolf Kinsky
P.O.Box 750355
Forest Hills, NY 11375, USA
rudolfl@idt.net
http://www.art.net/~rudolf/hpage2.html

Ken Kobland
306 Bowery
New York, NY 10012, USA
vonkob@aol.com

Asano Kohei
Tokyo Kougei Daigaku
Honcho 2-9-5
164 Nakano Tokyo, J
im94b002@t-kougei.ac.jp

Stef Kolman
KKEP
van Hallstraat 679
1051 HG Amsterdam, NL
info@kkep.com
http://www.kkep.com

Franz Krahberger
ELECTRONIC JOURNAL
LITERATUR PRIMÄR
Schönburgstr. 44/9
1040 Wien, A
crow@thing.or.at
http://ejournal.thing.at/ejournal

Klaas Krüger
Nernstweg 27
22765 Hamburg, D
klaas_krueger@magicvillage.de
http://www.klaas.ibp.de

Harry Kurz
Weidesgrün 18
95152 Selbitz, D
harry_kurz@hof.baynet.de
http://www.hof.baynet.de/~harry_kurz/

Dooley Le Cappelaine
284 Mott Street #9K
New York, NY 10012
USA
dooley@thing.net
http://www.thing.net/dooley

Iury Lech
INSOLIT PROD.
Espolsasacs 4-4-1a
08002 Barcelona, E
insolit@redestb.es
http://www.geocities.com/SoHo/4215/

Thomas Lehner
servus.at- OÖ Kunst und
Kultur im Netz
Kirchengasse 4
4040 Linz, A
servus@servus.at
http://www.servus.at

Marketta Leino
Yläaitankatu 4 A 23
40320 Jyväskylä, SF
marketta@art.net,
marleino@mleino.pp.fi
http://www.art.net/Studios/Visual/Marketta/fear.html

Sabrina Liao
430 W. 34 St. #12J
New York, NY 10001, USA
liaos@cat.nyu.edu
http://www.acm.org/sigchi/chi97/chikids

Andreas Lindstrom
Avalanche Systems
304 Hudson Street
New York, NY 10013, USA
info@avsi.com
http://www.cosmomag.com/

Andreas Löschner-Gornau
Ludwig-Wucherer Str. 69
06108 Halle, D
loeschner@andreas.hal.eunet.de
http://www.kunst-halle.mda.de

Annette Loudon
Construct Internet Design
448 Bryant Street
San Francisco,CA 94107
USA
webmstrs@construct.net
http://www.construct.net/stratus

Margaret Crane
Xerox Parc
3333 Coyote Hill Road
Palo Alto, CA 94304, USA
conwis@parc.xerox.com
http://www.pair.xerox.com/cw/conventional_wisdom

Lajos Major
Eszterhazy Karoly Teachers' Training College
Szechenyi Ter 1
3300 Eger, H
ali@gemini.ektf.hu
http://www.ektf.hu/ra/za

Steve Mann
M.I.T. Building
20 Ames Street
Cambridge, MA 02139, USA
steve@media.mit.edu
http://www.n1nlf-1.media.mit.edu/shootingback.html

Thomas Markert
Lunatic Interactive GmbH
Brunnenstr.181
10119 Berlin, D
markert@lunatic.de
http://www.lunatic-berlin.com/index.htm

Sebastian Marquez
Norrtullsgatan 22
Stockholm, S
seant@remotecom.se
http://www.remotecom.se/users/snmz/chano2.htm

Juliet Martin
450 6th Avenue #2F
New York, NY 10011, USA
juliet@sva.edu
http://www2.sva.edu/threads/juliet/seeme

Andrea Mayr
Ramperstorffergasse 41/10
1050 Wien, A
194152164137@[194.152.164.137]
http://194.152.164.137

Antonio Mazzitelli
MGW – MediaGroup Werdenberg
Torstrasse 1
9470 Buchs, CH
rondo@rondo.rol.ch
http://www.obscure.ch

Mark-S. Meadows
Construct Internet Design
448 Bryant Street
San Francisco, CA 94107
USA
pighed@construct.net
http://www.construct.net/projects/crutch

Pedro Meyer
ZoneZero
1333 S. Beverly Glen, #1004
Los Angeles, CA 90024
USA
pedro@zonezero.com
http://Zonezero.com

Merel Mirage
c/o KHM, Peter Welter
Platz 2
50676 Köln, D
merel@khm.de
http://art.cwi.nl/stedelijk/capricorn/mirage/start.html

Kostya Mitenev
ul. Doblesty 18 kv. 318
198331 St. Petersburg, RU
tenev@cyberun.spb.su
http://www.dux.ru/virtual/digbody/undina/!undina.htm

Georg-Th. Mnich
Lunatic Interactive GmbH
Brunnerstr. 181
10119 Berlin, D
mnich@lunatic.de
http://www.traum-vom-sehen.de/ausstellung/ausstellung.html
http://www.traum-vom-sehen.de/ausstellung/galerie/galerie.html

Ahuva Mu'alem
63/53Stern St.,Kiryat Hayovel
96750 Jerusalem, IL
ahumu@cs.huji.ac.il
http://www.cs.huji.ac.il/~ahumu

Robbin Murphy
artnetweb
426 Broome Street
New York, NY 10013, USA
murph@artnetweb.com
http://artnetweb.com/port

Vladimir Muzhesky
Fürstenwall 192
40215 Düsseldorf, D
106352.243@compuserve.com
http://heimat.de/nettropics/nettropics.html

Julie Myers
66D Crowndale road, Camden
London NWI ITP, GB
Julie9@QP1.mdx.ac.uk
http://www.backspace.org/noseyparker

Wolfgang Nagy
Virtuality Club Austria
Laxenburgerstr. 24
1100 Wien, A
virtwolf@netway.at
http://www.vr.vip.at/

Tetsushi Nakamura
EXE140 #308 Shitami-Nishigawa
739 Higashi-Hirishoma, J
tes@alifegarden.com
http://alifegarden.com/ale.html

Rienko Namakura
Kaidori 1-34-1#708
206 Tama, Tokyo, J
rieko@renga.nttic.or.jp
http://www.renga.com/rengeiza_e/index.htm

Mark Napier
45/E14 St #12H
New York, NY 10009, USA
NAPIER@INTERPORT.NET
http://www.interport.net/~napier/stolen

Timothy Nohe
University of Massachusetts Boston
100 Morrissey Boulevard
Boston, MA 02125-3393
USA
nohe@umbsky.cc.umb.edu
http://omega.cc.umb.edu/~nohe/OTIOSE

Gordana Novakovic
Radio B92 - Opennet
Makedonska 22
11000 Belgrad. YU
shile@opennet.org
http://www.shirt.opennet.org

Gerwald Oberleitner
CCC - Computer Communications
Fernkornstr.17/1/6
1100 Wien, A
oberleit@ccc.or.at
http://www.ccc.or.at

Dirk Paesmans
Franshalsstr. 61
2162 CL Lisse, NL
jodi@jodi.org
http://www.jodi.org

Rebeka Pal
Magyar Kepzömüveszeti Föiskola
Andrassy Ut 69-71
1062 Budapest, H
akeben@c3.hu.
http://www.c3.hu/InterMedia/Students/Pal_ZsR/

Melentie Pandilovski
Soros Center for Contemporary Arts - Skopje
Ruzveltova 34
91000 Skopje, MK
mpandil@soros.org.mk
http://www.soros.org.mk/scca/empire

Chris Peregoy
1201 West Ostend Street
Baltimore, MD 21230, USA
peregoy@umbc.edu
http://www.gl.umbc.edu/~peregoy/story/story.html

M.R. Petit
104 Suffolk Street,
Apt. #3
New York, NY 10002, USA
petit@echonyc.com
http://www.turbulence.org

Clifford Pickover
IBM Thomas J. Watson Research Center
Yorktown Heights, NY 10598
USA
cliff@watson.ibm.com
http://sprott.physics.wisc.edu/pickover/home.htm

Emanuel Pimenta
Rua Tierno Galvan,
4-11B
1070 Lisbon, P
edmp@mail.telepac.pt
http://www.asa-art.com/woiksed/w.htm

Ine Poppe
WOMEN WITH BEARDS
Postbus 3970
1001 AT Amsterdam, NL
beards@dds.nl
http://www.dds.nl/~beards

Jeremy Quinn
4 Heathfield,
Rushcroft Road
London SW2 1LE, GB
jermq@hrc.wm1n.ac.uk
http://raft.vmg.co.uk/fsol/

Jean Ranger
5175 Drolet
Montréal,
Québec H2T 2H3, CDN
jranger@mediarenaissance.com
http://www.fivaonline.com/mindroots.htm

Sonya Rapoport
6 Hillcrest Court
Berkeley,
CA 94705-2805, USA
rapop@socrates.berkeley.edu
http://www.lanminds.com/local/sr/brutal

Erwin Redl
16 Clifford Place Apt. 2
Brooklyn, NY 11222, USA
parallel@thing.net
http://www.thing.net/~parallel/

Wolfgang Reinisch
Idlhofgasse 52
8020 Graz, A
w@xarch.tu-graz.ac.at
http://xarch.tu-graz.ac.at/home/w/proj/weiter

Alan Rhodes
Illumination Graphics
4641 Marquette
Montréal,
Québec H2J 3Y3, CDN
alan@cam.org
http://www.cam.org/~alan

Winfried Ritsch
Klangatelier Algorythmics
Leitnergasse 7A
8010 Graz, A
ritsch@iem.mhsg.ac.at
http://iem.mhsg.ac.at/

Miroslaw Rogala
329 W. 18th Street, Suite 900
Chicago, IL 60616, USA
rogala@mcs.com
http://www.mcs.net/~rogala/eGarden

Sinisa Rogic
Radio B92-Opennet
Makedonska 22/V
11000 Belgrade, YU
shile@opennet.org
http://www.rex.opennet.org

Nelli Rohtvee
KÄRBERI 28-191
0038 Tallinn, EW
xnellix@hotmail.com
http://www.artun.ee/homepages/nelli/nelli.html

Gustavo Romano
Güemes 4147C
1425 Buenos Aires, RA
gusrom@webtv.com.ar
http://www.fadu.uba.ar/findelmundo

Avi Rosen
Technion I.I.T.
Faculty of Electrical Engineer
32000 Haifa, IL
avi@savyon.technion.ac.il
http://www.technion.ac.il/~ravi/vrml/vrml.html

Florian Rötzer
Telepolis - Verlag Heinz Heise
Kühbachstr.11
81543 München, D
fr@tp.heise.de
http://www.heise.de/tp

Joachim Sauter
ART + COM
Budapester Str. 44
10787 Berlin, D
js@artcom.de
http://www.artcom.de/projects/invisible_shape/welcome.html/

Herbert Schager
Im Tal 11
4040 Linz, A
hschager@fishnet.co.at
http://www.servus.at/~schager/

Rolf Schmidt
rolf@clark.net
http://www.clark.net/pub/rolf/mmf/

Steve Schneider
Avalanche Systems
304 Hudson Street
New York, NY 10013, USA
info@avsi.com
http://www.borderequalszero.com

Christian Schrenk
Kuddelmuddel Ich & Du
Langgasse 13
4020 Linz, A
schrenkc@ping.at
http://www.rito.com/kuddel

Markus Schulthess
Hirschengraben 13
6003 Luzern, CH
mas@centralnet.ch
http://www.mesch.ch/container-city

Björn Seeger
Magic & Less
Hufelandstr.12
10407 Berlin, D
bjorn@smallest.com
http://www.smallest.com/netman

Peter Seidler
Avalanche Systems
304 Hudson Street
New York, NY 10013, USA
info@avsi.com
http://www.carnegiehall.org

Nikolai Selivanov
Bolotnikovskaja str., 38/1/58
113209 Moscow, RU
nfuture@rsuh.ru
http://www.geocities.com/Paris/LeftBank/5510

SENSORIUM Team
Project TAOS Inc.
7-5-2-101 Minami-Aoyama, Minat
107 Tokyo, J
nish@kt.rim.or.jp
http://www.sensorium.org

Jacques Servin
POB 14872
San Fransisco, CA 94114
USA
jacq@quake.net
http://www.quake.net/~jacq/Beast

Elisha Sessions
Avalanche Systems
304 Hudson Street
New York, NY 10013, USA
info@avsi.com
http://www.faoschwarz.com

Eric Shoppen
Scan/Media-GN
Hoendiepskade 23a
9718 BG Groningen, NL
kick@scan.media-gn.nl
http://www.media-gn.nl/people/kick/kick.html

Alexei Shulgin
Polotskaya 29-1-39
121355 Moscow, RU
easylife@glasnet.ru
http://www.c3.hu/hyper3/form/

Silver
U Akademie 4
17000 Prague, CZ
silver@avu.cz
http://www.avu.cz/~silver/

John Simon
235 W. 22nd Street #7k
New York, NY 10011, USA
jfsjr@interport.net
http://www.interport.net/~jfsjr/eicon.html

E. Jay Sims
Sims Art, Inc.
306 Bowery
New York, NY 10012
USA
EJSims@aol.com
http://www.dalton.org/faculty/7268

Teo Spiller
HMZ
Vojkova 1B
1000 Ljubljana, SLO
TEO.SPILLER@FRACTAL.SI
http://WWW.FRACTAL.SI/TEO/ESMERALDA

Wolfgang Staehle
The Thing
601 W 26th Street
New York, NY 10001, USA
wolfgang@thing.net
http://www.thing.net/thingnyc/html/palace.html

Ed Stastny
PO Box 241113
Omaha, NE 68124-5113, USA
ed@sito.org
http://www.sito.org/synergy/gridcosm

Rose Stasuk
1210 Ryan Street
Clermont, FL 34711, USA
roric@gate.net
http://www.ucet.ufl.edu/~rstasuk/figure.html

Achim Stoesser
Hagenbach Str.6
76187 Karlsruhe, D
stoesser@ira.uka.de
http://i31www.ira.uka.de/~stoesser/Vegan/step.html

Igor Stromajer
INTIMA Virtual Base /
Creative Intimate Lab
Jakceva 2
1000 Ljubljana, SLO
igor.stromajer@guest.arnes.si
http://www2.arnes.si/~ljintima2/intima1.html

Zoltan Szegedy-Maszak
Hungarian Academy of fine Arts
Andrassy Ut 69-71
1061 Budapest, H
SZMZ@C3.HU
http://www.c3.hu/cryptogram

Athena Tacha
291 Forest St.
Oberlin, OH 44074, USA
ftacha@oberlin.edu
http://www.oberlin.edu/~atacha/death.html

TechnoSphere Team
Gordon Selley/Rycharde Hawkes/Jane Prophet
techno@lcpdt.linst.ac.uk
http://www.technosphere.org.uk

Oliver Teuber
Neviandtstr.42
42117 Wuppertal, D
ot@rsk.com
http://www.punktsubstanz.de

Henning Timcke
Stadtturmstr. 5
5400 Baden, CH
ideen@kunst.ch
http://www.kunst.ch

John Tonkin
P.O.Box 1476 Potts Point
Adelaide,
South Australia 5000, AUS
john.tonkin@bigpond.com
http://www.merlin.com.au/johnt

Andreas Troeger
160 1st Ave. 2D
New York, NY 10009, USA
sysop@zero.tolerance.org
http://zero.tolerance.org

Pascale Trudel
Studio Dragon
7009 Hamilton
Montréal,
Québec H4E 3C8, CDN
sdragon@cam.org
http://www.cam.org/~sdragon/ssde.html

Jerlian Tsao
Sinologic Software
P.O. Box 5242
Berkeley, CA 94705, USA
jerlian@sinologic.com
http://www.sinologic.com/persimmon

Michele Turre
P.O. Box 338
Conway, MA 01341, USA
mturre@art.umass.edu
http://www-unix.oit.umass.edu/~mturre/treefix

Mathilde uP (muPe)
Tuinstraat 11
1015 NX Amsterdam, NL
mupe@desk.nl
http://www.desk.org/mupe/bio/index.htm

Anna Ursyn
University of Northern Colorado
Department of Visual Arts
Greeley, CO 80639, USA
azursyn@Bentley.UnivNorthCo.edu
http://www.arts.unco.edu/vagallery/

Giorgio Vaccarino
Arslab Project Committee
Via Vanchiglia 24
10124 Torino, I
arslab@arpnet.it
http://netville.zhort.it

Johann Van Der Schiff
SCAN / media-gn
Hoendiepskade 23 A
9718 BG Groningen, NL
schijff@scan.media-gn.nl
http://www.media-gn.nl/people/johannvds/tvhcover.htm

Paul Vanouse
507 East End Avenue, Apt. 1
Pittsburgh,PA 15221, USA
pv28@andrew.cmu.edu
http://www-crca.ucsd.edu/~pdc

Fernando Varela Cisneros
Avandaro 54 Col Vergel de coyo
14390 Mexico, D.F., MEX
itziar@davinci.cnart.mx
http://www.netxpression.com/artew3

Konstantin Vassilev
P.O.Box 20
1612 Sofia, BG
konart@geocities.com
http://www.geocities.com/Athens/Forum/3132

Roman Verostko
5535 Clinton Ave South
Minneapolis, MN 55419, USA
roman@mcad.edu
http://www.concentric.net/~verostko/u.html

Mark von Rahden
die Farm
St.-Jürgen-Str. 1
28205 Bremen, D
markotic@farm.de
http://www.farm.de/x/

Akke Wagenaar
S.C.A.N.
Hoendiepskade 23A
9718 BG Groningen, NL
akke@khm.uni-koeln.de
http://err.org/akke

Michael Wagner
Fuhrmanngasse 9
3100 St. Pölten, A
wagner@geometrie.tuwien.ac.at
http://picasso.geometrie.tuwien.ac.at/virtual.gallery/

Andràs Wahorn
iNteRNeTTo, IDG Hungary LTD.
Màrvàny U. 17
1012 Budapest, H
nyiro@internetto.hu
http://www.internetto.hu/este/wahorn

Pamela Walatka
MRJ at NASA Ames Research Center
T 27A-2 NASA Ames
Moffett Field, CA 94035, USA
walatka@nas.nasa.gov
http://science.nas.nasa.gov/~walatka/WebToons/toon.html

Kerstin Weiberg
Brunnenstraße 143
10115 Berlin, D
kweiberg@is.in-berlin.de
http://www.icf.de/future-perfect

Emmett Williams
VIRTUALITAS Fine Arts Online
Lützenstr.10
10711 Berlin
D
http://www.virtualitas.com/finearts/doubt

Dajuin Yao
6429 Hillegass Avenue
Oakland, CA 94618
USA
dajuin@sinologic.com
http://www.sinologic.com/webart

Kominos-Kostant Zervos
477 Milton Rd, Auchenflower
Brisbane 4066
AUS
S271502@Student.ug.edu.au
http://student.uq.edu.au/~s271502

Sabine Zimmerhansl
3032 Bunker Hill Lane, Sui.207
Santa Clara, CA 95054
USA
zimmerhansl@hotmail.com
http://www.remedy.com/career

Nara Zoyd
SCAN
Hoendiepskade 23A
9718 BG Groningen
NL
n.zoyd@scan.media-gn.nl
http://www.media-gn.nl/pataverse

Participants Interactive Art

Frederick Abrams
ul. Studzienna 5/9
53-304 Wroclaw, , L
100743.1717@compuserve.com

Jean-Louis Accettone
Monac.1
164 Rue Abelard
59000 Lille
F

Tania Aedo
Multimedia Center
Churubusco y Tlapan s/n
Mexico City, MEX
ladron@mpsnet.com.mx

Josu Aguinaga
Universidad del Pais Vasco-Fac. de Bellas Artes
Campus de Leioa
48940 Leioa-Bizkaia, E

Kyo Akabane
3-95, Ryoke-cho, IAMAS
503 Ogaki City, J

Metthew Akers
2325 W. North AV.
Chicago, IL 60647, USA
Makers@artic.edu

Max Almy
2959 Seabreeze Drive
Malibu, CA 90265, USA
magika@earthlink.net

Laurence Arcadias
1231 #A Oxford Str.
Berkeley, CA 94709
USA
arcadias@slip.net

Björn Askefoss
Nedrevei 8
3155 Horten, N
askefoss@notam.uio.no

Jim Avignon
Media art group eyeyou
161 A Herne Hill
London SE24 9LR, GB

Ingrid Bachmann
3255 West Belden #1W
Chicago, IL 60647, USA
ibachmann@artic.edu

Edouard Bannwart
echtzeit GmbH
Kantstr. 165
10623 Berlin, D

Perry Bard
313 Church Street
New York, NY 10013
USA

Bill Barminski
DE-LUX'O
PO Box 661635
Los Angeles, CA 90066
USA

Marcus Bastel
Niuwezijds Voorburgwal 171
1012 RK Amsterdam, NL
bastel@xs4all.nl

Trevor Batten
Kanaalstraat 15
1054 WX Amsterdam, NL

Scott Becker
PO Box 578956
Chicago, IL 60657, USA
artscb@interaccess.com

Armand Behar
ART 3000
Château de l'Eglantine
78350 Jouy en Josas, F
art3000@calvanet.calvacom.fr

Zoe Beloff
153 Norfolk Street #5H
New York, NY 10002
USA
zoe@interport.net

Jon Berge
PARTNERS MULTIMEDIA SERVICES
755 Oak Street
Columbus, Ohio 43205
USA
jberge@osu.edu

Michael Bielicky
Kubelikova 38
13000 Praha 3, CZ
x@avu.cz

BINO & COOL
Poste Restante
40014 Gothenbourg, S
colbengtsson@hotmail.com

Natalie Bookchin
2222 Carmel Valley Rd. #11
Del Mar, CA 92014, USA
nbookchin@ucsd.edu

Charles Boone
1122 South Campbell Avenue
Wheaton, IL 60187, USA

Svetek Bozidar
Splitska 6
1000 Ljubljana, SLO

Kevin-Sawad Brooks
Fine Arts 123, Indiana University
Bloomington, IN 47405
USA

Richard Brown
Royal College of Art
Kensington Gore
London SW7 2 EU, GB
r.brown@rca.ac.uk

Michael Buchwald
PIXFIXION
19 Heisesgade
2100 Kopenhagen, DK
michbuch@centrum.dk

Helena Bulaja-Madunnic
Radnicki dol 8
10000 Zagreb, HR

Mark Carroll
Holocene
1216 Vintage Place
Nashville, TN 37215
USA
mc@edge.net

Isabelle Chemin
43, Quai de Juillet
14000 Caen, F

Motoshi Chikamori
University of Tsukuba,
Media Department
1-1-1 Tennoudai
Tsukuba 305, J
motoc@geijutsu.tsukuba.ac.jp

Helen Cho
6056 Romany Road
Oakland, CA 94618, USA

Insook Choi
405 N. Mathews
Urbana, IL 61801, USA

Arno Coenen
Hoendiepskade 23A
9718 BG Groningen, NL

Eric Coignoux
PARTIZAN MIDI MINUIT
71, rue Fondary
75015 Paris, F

Alison Cornyn
142 East 13th Street
New York, NY 10003
USA
acorn@picture-projects.com

Alba D'Urbano
Heinrich-Budde-Str. 3
04157 Leipzig, D

Jean-Marie Dallet
3, rue Saint-Martin
95240 Cormeilles en Parisis, F
dallet@club-internet.fr

Gary Danner
STATION ROSE
Rothschildallee 18
60389 Frankfurt, D

Labrador Ruiz De La Hermosa
...dijo el monje
C/dos Hermanas 7,3
28012 Madrid, E

Joseph DeLappe
University of Nevada,
Department of Art 224

Reno, NV 89557, USA
delappe@scs.unr.edu

Chris Dodge
148 Fifth Street, Apt. #1
Cambridge, MA 02141
USA
cdodge@media.mit.edu

Maria Dundakova
Atelier für Gestaltung
Bachstr. 117B
5000 Aarau, CH

Paul Earls
MIT / CAVS
265 Massachusetts Ave.
Cambridge, MA 02139
USA
earls@mit.edu

Christoph Ebener
Kaiser-Wilhelm-Str. 73
20355 Hamburg, D

Arthur Elsenaar/
Remko Sha
Keizersgracht 333
1016 EG Amsterdam, NL
ARTELSE@MEDIA-GN.NL

eyeyou
Media art group eyeyou
161 A herne Hill
London SE24 9LR, GB
eyeyou@backspace.org

Carlos Fadon Vicente
Rua Livreiro Saraiva 236
01237-020 Sao Paulo
SP, BR

Richard Fajnor
Lelkova 63
63700 Brno, CZ
richard@vmp.vutbr.cz

Matt Fanuele
Avalanche Systems, Inc
304 Hudson Street,
7th floor
New York, NY 10013
USA

Sidney Fels
ATR MI & C Research Laboratories
Seika-cho, Soraku-gun
Kyoto, 619-02, J
fels@mic.atr.co.jp

Sandra Fischer
Knittlingerstraße 32
75417 Mühlacker, D

Franz Fischnaller
F.A.B.R.I.CATORS
Via Fratelli Bronzetti 6
20129 Milano, I
fabricat@galactica.it

Floating Point Unit
60 Berry Street
Brooklyn, NY 11211, USA
mimic@inch.com

Friedrich Förster
Hechinger Straße 203
72072 Tübingen, D
f.foerster@proaudio.de

Masaki Fujihata
Keio University, Environmental Information Faculty
5322 Endo
252 Fujisawa, J
masaki@sfc.keio.ac.jp

Sera Furneaux
22 Courtland Road
Liverpool L18 2EQ, GB
furneaux@sera.u-net.com

Paul Garrin
11E. 4th Street 2F
New York, NY 10003, USA

Gregory Garvey
Concordia University
1455 de Maisonneuve Blvd. West
Montreal, QC H3G 1M8
CDN
ggarvey@vax2.concordia.ca

Thomas Geoffrey
24700 Mc Bean Parkway, RL-23
Valencia, CA 91355, USA
gpthomas@muse.calarts.edu

Walter Giers
Badmauer 14
73525 Schwäbisch-Gmünd
D

Piero Gilardi
Corso Casale 121
10132 Torino, I
piero.gilardi@rarino.alpcom.it

Anna Görner
Liebermannstr. 120
13088 Berlin, D

Bogdan Grabuloski
ul. Borka Taleski 103
97500 Prilep, YU

Christian Greuel
FAKESPACE Inc.
241 Polaris Ave.
Mountain View,
CA 94943, USA

gruppe 42
Rektor-Klaus-Str. 100
73525 Schwäbisch Gmünd
D
gruppe-42@hfg-gmuend.de

Hanna Haaslahti
Siltasaarenkärki 3 c 28
00530 Helsinki, SF
hhaas@mlab.niah.fi

Kazuhiko Hachiya
5-31-6 Nishi Gotanda
Shinagawa-ku,
Tokyo 141, J
QZE17170@niftyserve.or.jp

Perry Hall
Ocurix Films
309 West 106th Street #5B
New York, NY 10025
USA

Gusztav Hamos
Pestalozzistr. 81
10627 Berlin, D
guszt@berlin.snafn.de

Emily Hartzell
New York University Center
for Advanced Technology
719 Broadway,
12th Floor
New York, NY 10003
USA

Björn Hausner
Baumgartnerstraße 17
81373 München, D

Hanlin Heath
Ohio State University/ACCAD
1224 Kinncar Road
Columbus, OH 43204
USA
hhanlin@cgrg.ohio-state.edu

Suvi Hein
Eisenacherstr. 63
10823 Berlin, D
sthieke@aol.com

Claudia Herbst
1000 Oberlin Dr.
Glendale, CA 91205
USA
cobalt3@earthlink.net

Lynn Hershman-Leeson
HOTWIRE PRODUCTIONS
327 Ritch Street
San Francisco, CA 94108
USA
lynn2@well.com

Megan Heyward
76 Wemyss St., Marrickville
Sydney, NSW 2204, AUS
m.heyward@uts.edu.au

Paul Higham
1331 Tyler Street #203
Minneapolis, MN 55413
USA
highaoo1@gold.tc.umn.edu

Wolfgang Hilbert
Hauptstr. 1
2102 Bisamberg, A

Elizabeth Hinkle-Turner
301 West McGee Street
Sherman, Texas 75092
USA
ehturner@serial.music.uiowa
.edu

Ishibashi Hiroaki
Int. Academy of Media Arts
and Science
3-95 Ryoke-cho
503 Ogaki-Gifu, J
aki@iamas.ac.jp

Matjaz Hmeljak
Università di Trieste - DEEI
Via Valerio 10
34127 Trieste, I
hmeljakm@univ.trieste.it

Ulrich Höhne
Rudolfstraße 26
76131 Karlsruhe, D
eao2@rz.uni-karlsruhe.de

Carol-Ann Holzberger
32, rue du Javelot
75013 Paris, F

Markus Huemer
Gereonswall 23 b
50677 Köln, D
huemer@khm.de

Tjark Ihmels
Postfach 64
04161 Leipzig, D

Catherine Ikam
8, rue des Haies Fleuries
93100 Montreuil, F

Haruo Ishii
Nagoya City Youth Cultural
Center / HAKUHODO
3-18-1 Naka-ku
Nagoya-shi 460, J
MXC00275@niftyserve.or.jp

Setsuko Ishii
1-23-26-404, Kohinata
Bunkyo-ku, Tokyo 112
J

Toshio Iwai
International Academy of
Media Arts and Sciences
3-95 Ryouke-chou
Ogaki-City 503, J

Hiroo Iwata
Institute of Engineering
Mechanics
University of Tsukuba
Tsukuba 305, J
iwata@kz.tsukuba.ac.jp

Pamela Jennings
Mind Field Productions
333 West Charleston Rd.
Palo Alto, CA 94306
USA
mindfield@earthlink.com

KP-Ludwig John
Guerickestr. 21A
80805 München, D

Andrew Johnson
Electronic Visualization Lab.,
Univ. of Illinois
851 South Morgan Street
Chicago, IL 60607-7053
USA
ajohnson@eecs.uic.edu

Sue Johnson
164 5th Avenue #2
Brooklyn, NY 11217, USA
suej@picture-projects.com

Eduardo Kac
2817 Point Court
Lexington, KY 40517
USA
ekac1@pop.uky.edu

Daniel Kelly
4077 Crystal Drive
Beulah, MI 49617, USA
dankelly@artisthouse.
com

Tamio Kihara
Nippon Telegraph and
Telephone Corporation
1-1 Hikarinooka, Yokosuka-
shi
Kanagawa 239, J

Paul Kim
109 Fillmore St.
San Fransisco,
CA 94117, USA

Seiko Kinosita
KS & Partners Co. Ltd.
Senriyama-nisi 5-43-1
Suita Osaka 565, J
ksp@yo.rim.or.jp

Jutta Kirchgeorg
School of Visual Arts
141 W 21 Street
New York, NY 10010
USA
jutta@interport.net

Claudio Knöpli Tamaroglio
Simonstr. 2
3012 Bern, CH

Knowbotic Research
Schillingstr. 32
50670 Köln, D
kr+cf@khm.de

Keiko Kobayashi
101-4-5-6 Kawanahonmachi
Showaku, Nagoya 466, J
koba@iamas.ac.jp

Marko Kosnik
Egon March Institute
Luznarjeva 19
4000 Kranj, SLO

Mihoko Kosugi
53 Shimokanegura Enmyoji
618 Oyamazaki-cho, Kyoto-
Fu, J

Manfred Kroboth
Jan van Eyck Academy
Academieplein 1
6211 KM Maastricht, NL

Volker Kuchelmeister
Welfenstr. 13
76137 Karlsruhe, D

Martin Kusch
Wolfgang-Schmälzl-Gasse
22/12
1020 Wien, A
kusch@vis-med.ac.at

Dooley Le Cappellaine
284 Mott Street #9K
New York, NY 10012
USA
dooley@thing.net/dooley

Bonnie Lebesch
2226 Eastlake Avenue East
#61
Seattle, WA 98102, USA
bonnie@bohem-int.com

George Legrady
Merz Akademie
Teck Str. 58
70190 Stuttgart, D
Legrady@sfsu.edu

Golan Levin
Interval Research
Corporation
1801-C Page Mill Road
Palo Alto, CA 94304
USA

Marita Liulia
Uudenmaank 2 K
00120 Helsinki, SF

Peter Löderbusch
media project GmbH
Glashütter Str. 100
01277 Dresden, D
media05@ibm.net

Emilio López-Galiacho
Sangenjo 21, 8-D
28034 Madrid, E

Michael Lutz
Domagkstr. 33 /
Haus 38
80807 München, D

Mark Madel
De Ruyterkade 149
1011 AC Amsterdam, NL
madel@compuserve.com

Akitsugu Maebayashi
603, 4-15-40 Shimorenjaku
Mita
Tokyo 181, J
mae884@tama.or.jp

Yesi Maharaj Singh
F.A.B.R.I.C.A.T.O.R.S
Via Fratelli Bronzetto 6
20129 Milano, I

Peter Maloney
Chelsea College of Art and
Design
Manresa Road
London, SW3 6LS, GB
pbs@chel-vr.demon.co.uk

Steve Mann
MIT Building
20 Ames Street
Cambridge, MA 02139
USA
steve@media.mit.edu

Horst Martens
OUTER SPACE GROCERY
Schlangenbacherstr. 16c
14197 Berlin, D
hmartens@cybernet-ag.de

Jacquelyn Martino
Philips Research Labs
345 Scarborough Rd.
Briarcliff, NY 10510, USA
jam67@columbia.edu

Benoît Maubrey
Die Audio Gruppe
Bahnhofstraße 47
14806 Baitz, D
maubrey@berlin.snafv.de

Mladen Milicevic
6543 Haley Drive
Columbia, SC 29206, USA
mmladen@sc.edu

Brad Miller
22/21 Saint Neots Avenue
Potts Pt.,
Sydney NSW 2011, AUS
brad.miller@unsw.edu.au

Merel Mirage
c/o KHM
Peter-Welter-Platz 2
50676 Köln, D
merel@khm.de

Kostya Mitenev
ul. Doblesty 18 kv. 318
198331 St. Petersburg
RU
tenev@cyberun.spb.ru

Joanna Moraszczyk
Sekocinska 17 m. 17
02-313 Warsaw, PL

Knut Mork
Svenstuvn. 6A
0389 Oslo, N

Tomoko Mukai
Taubengasse 1
50676 Köln, D
Tomoko@khm.de

Patricia Nelson
Ball State University, Dept. of Art
Muncie, Indiana 47306
USA

Norbert Nestler
Brockmanngasse 87/II
8010 Graz, A

Norie Neumark
2/15 Baden St. Coogee
2034 Sydney
AUS
n.neumark@uts.edu.au

Chau Nguyen
3021 Huff Ave. #202
San Jose, CA 95128
USA
csnguyen@ucdavis.edu

Christian Paraschiv
152 Rue Rateau Atelier no.8
93120 La Courneuve, F

M.R. Petit
104 Suffolk Street,
Apt. #3
New York, NY 10002
USA

Rosen Petkov
Computer Art Centre - Sofia
10, Narodno Sabranie SQR
1000 Sofia, BG

Nick Philip
h20 Media
615 Cole St, #17
San Francisco, CA 94107
USA

Emanuel Pimenta
Rua Tierno Galvan,
4-11B
1070 Lisbon, P
edmp@mail.telepac.pt

The KidsRoom Group
MIT Media Laboratory
20 Ames Street Room E15-368C
Cambridge, MA 02139
USA

Pixelpeppy
80 Greene
New York, NY 10012
USA
peppy@thing.net

Sally Pryor
PO Box 1001
Darlinghursr 2010, AUS
S.Pryor@hum.uts.edu.au

Ramsden Stuart
Australian Centre for Arts and Technology
PO Box 804
Canberra, ACT 2601
AUS

Gerald Raunig
POLYTROPOI
Sechsschimmelg. 17/17
1090 Wien, A
ptt@tO.or.at

Joe Reitzer
University of Illinois at Chicago
851 S. Morgan St.,
Rm. 1120 S
Chicago, IL 60607-7053
USA
reitzer@evl.uic.edu

Don Ritter
204, 15 Street, 3rd Floor
Brooklyn, NY 11215, USA
ritter@interport.net

Jean-Pierre Rodella
31, rue Chapon
75003 Paris, F

Miroslaw Rogala
329 W. 18th Street, Suite 900
Chicago, IL 60616, USA
rogala@mcs.com

Giles Rollestone
Computer Related Design
Royal College of Kensington
London SW7 2EU, GB

Avi Rosen
Technion I.I.T.
Faculty of Electrical Engineer
32000 Haifa, IL
avi@savyon.technion.ac.il

Gerald Rossbacher
Hochschule für Gestaltung Linz
Sonnsteinstr. 16-18
4040 Linz, A

Alyssa Rothwell
9 Bennett street Newtown
Sydney 2042, AUS
alyssar@ozemail.com.au

Hiroshi Sakamoto
Fuji Television Network
2-4-8 Daiba, Minato-ku
137-88 Tokio, J
h-akamo@po.iijnet.or.jp

Michael Saup
Fuerstenbergerstr. 167
60322 Frankfurt/M., D
saup@em.uni-frankfurt.de

Joachim Sauter
ART + COM
Budapester Str. 44
10787 Berlin, D

Sigbert Schindler
Kaiserin-Augusta-Str. 80
12099 Berlin, D

rich.art Schneider
Wilhelm Kreß-Platz 29/56/12
1110 Wien, A
richart@vis-med.ac.at

Bill Seaman
117 N.Beaumont Ave.
Catonsville, MD 21228
USA
seaman@umbc.edu

Serog
54 White St.
Lilyfield 2040, AUS
serog@netspace.net.au

sha
Lederergasse 22/7
1080 Wien, A

Renate Sturmer
203 Chrystie Street
New York, NY 10002
USA
rsturmer@dti.net

Janos Sugàr
Varoshaz 4
1052 Budapest, H
sj@dial.isys.hu

Nobuya Suzuki
Academy of Media Arts
3-95 Ryokechon
Oogaki-city 503, J
zuckey@iamas.ac.jp

Athena Tacha
291 Forest St.
Oberlin, Ohio 44074
USA
ftacha@oberlin.edu

Leonello Tarabella
Computer Music Lab of CNUCE/C.N.R.
Via S.Maria,36
56126 Pisa, I
L.Tarabella@cnuce.cnr.it

Nell Tenhaaf
5650-1 Forbes Ave. Apt. 2
Pittsburgh, PA 15217
USA
tenhaaf@andrew.cmu.edu

Rice Dixon Tennessee
Scrutiny Associates
270 Riverside Drive #12A
New York, NY 10025
USA

Henning Timcke
Stadtturmstr. 5
5400 Baden, CH

Naoko Tosa
ATR Media Integration Labs.
Seika-cho Soraku-gun
Kyoto 619-02, J

Marie-Hélène Tramus
Université Paris, Arts et Technologies de l'image
2, rue de la Liberte
93526 Saint-Denis, F

John Underkoffler
MIT Media Laboratory
20 Ames Street
Cambridge, MA 02139
USA
jh@media.mit.edu

Dirk Urbaniak
Stereolize Film + TV Design GmbH
Elisabethstr. 44
80769 München, D

Jasper Van den Brink
P.A.R.K. 4D TV
Korte Leidsedwarsstraat 12
1017 RC Amsterdam, NL

Paul Vanouse
507 East End Avenue, Apt. 1
Pittsburgh, PA 15221
USA
pv28+@andrew.cmu.edu

Konstantin Vassilev
PO Box 20
1612 Sofia, BG
konart@geocities.com

Victoria Vesna
UC Santa Barbara
Department of Art Studio
Santa Barbara,
CA 93106, USA

Vincent John Vincent
VIVID GROUP
317 Adelaide St. W. #302
Toronto, Ontario
M5V 1P9, CDN
vincent@vividgroup.com

Bill Vorn
5565 Clark
Montréal, QC H2T 2V5
CDN

Yntse Vugts
Postfach 16612
1001 RC Amsterdam, NL

Bernhard Wagner
Universität Zürich, Institut für Informatik
Winterthurerstr. 190
8057 Zürich, CH
wagner@echtzeit.ch

Eku Wand
SCREENDESIGN
Fuggerstr. 37
10777 Berlin, D
eku@khm.de

Margaret Watson
EVL/University of Illinois at Chicago
851 S.Morgan St, Room 1120 SEO
Chicago, IL 60607-7053
USA
watson@evl.uic.edu

Olaf Wendt
TRIPMEDIA
7 Leicester Place
London WC2H 7BP, GB
zola@easynet.co.uk

Stephen Wilson
San Francisco State University, Art Dept.
1600 Holloway Avenue
San Francisco, CA 94132
USA
swilson@sfsu.edu

Adrianne Wortzel
19 East 7th ST #5
New York, NY 10003
USA
sphinx@inch.com

Keigo Yamamoto
#202, 2-21-9 Shirasagi-cho
Nakanoku 165, Tokyo, J
keigo@nn.iij4u.or.jp

Greg Zaritski
GZ INTERACTIVE PTY. LTD
Unit 5, 22 Hughenden Road
East St. Kilda,
Victoria 3183, AUS
gregz@gz-interactive.com.au

Gary Zebington
PO Box 394
Surry Hills, NSW 2011
AUS

Stefan Zeyen
Hufelandstr.7
10407 Berlin, D

Helge Zwosta
CYBERSENSORIC ZWOSTA
Asternweg 4a
86399 Bobingen, D
0823441651-1@t-online.de

Participants Computer Animation

Douglas Aberle
ABERLE FILMS
12800 NE 191st. Circle
Battle Ground,
WA 98604, USA
daberle@teleport.com

Mark Baldo
Blue Sky Studios, Inc.
One South Road
Harrison, NY 10528, USA
alison@blueskystudios.com

Beriou
AGAVE
Rue Haute
41800 Trôo, F

Michael Betz
Zweites Deutsches Fernsehen/3Sat
ZDF-Strasse
55100 Mainz, D

Bernd Beyreuter
Hochschule für Film u. Fernsehen 'Konrad Wolf'
Karl-Marx-Str. 33/34
14482 Potsdam, D

John Brigden
5 Nilkerie Street
Brisbane
Queensland 4034, AUS

Buf Compagnie
9, avenue de Villiers
75017 Paris
F
ofely@buf.imaginet.fr

Bilbo Mylène Calvez
Jessnerstr. 19
10247 Berlin, D

Véronique Caraux
ENSAD / AII
31 rue d'Ulm
75240 Paris Cedex 05, F
pierre@ensad.fr

Tom Casey
HOME RUN PICTURES
One Market Street
Pittsburgh, PA 15222, USA
Tom@hrpictures.com

Jaime Castaneda
34 West 13th Street 4th Floor
New York, NY 10011, USA
jaime@blueskystudios.com

Zdenka Cechová
Union of Czech Computer and Multimedia Artist
Litvínovská 520
190 00 Prague, CZ

Anita Cheng
661 Metropolitan Ave #3L
Brooklyn, NY 11211, USA

Raquel Coelho
30 Maple Tree Ave., Apt. 2E
Stamford, CT 06906, USA
raquel@blueskystudios.com

Arno Coenen
Hoendiepskade 23A
9718 BG Groningen, NL

Nikola Cosic
Mirocka 4
11000 Belgrade, YU
ncosic@galeb.etf.bg.ac.yu

Michèle Cournoyer
National Film Board of Canada
3155 Côte de Liesse
St-Laurent,
QC H4N 2N4, CDN

Marc Cuenot
ENSAD / AII
31 rue d'Ulm
75240 Paris Cedex 05, F
pierre@ensad.fr

Martin Dammann
Friedelstr. 51
12047 Berlin, D

Deanan Dasilva
5675A Landregan St.
Emeryville, CA 94608, USA
delusion@delusion.com

Peter De Lorenzo
PO Box 138
Robertson NSW 2577, AUS
pdls@ozemail.com.au

Giacomo De Mas
Alphaville Studio
Via Statuto, 8
20121 Milano
I

Paul E. Debevec
University of California at Berkeley
545 Soda Hall, Berkeley,
CA 94720-1776, USA
debevec@cs.berkeley.edu

Richard Del Rosso
8 Washington St.
Hicksville, NY 11801, USA

Amalia Delicari
422 East 83rd Street,
Apt #2
New York, NY 10028, USA
amalia@panix.com

Sean Donnelly
14 Wabun Ave.
Providence, RI 02908
USAsmdonnel@mailbox.syrdu

Hajo Drott
Platanenstraße 3
82024 Taufkirchen, D

Staceyjoy Elkin
Queen Bee
213 E. 10th St. #14
New York, NY 10003, USA
stacey@tiac.net

Alain Escale
T.E.V.A.
11, rue Torriceli
75017 Paris, F
aescalle@club-internet.frSteve Estes
2230 N. Calvert St.
Baltimore, MD 21218, USA
sestes1@gl.umbc.edu

Stefen Fangmeier
Industrial Light & Magic
PO Box 2459
San Rafael, CA 94912, USA

Anne Farrell
131 Huddleson Street
Santa Fe, NM 87501, USA
afarrell@santa-fe.cc.nm.us

Film And Video Umbrella
2 Rugby St. Off Lambs Conduit
London WCIN 3QD, GB

Markus Fischmann
Wohldorferstraße 39
22081 Hamburg, D

Mark Fontana
5353 N.High St. #1003
Colombus,
OH 43214-1252, USA
fontana@cis.ohio-state.edu

Peter Forgacs
Mese köz 10
1121 Budapest, H

Tsuyoshi Fuyama
IAMAS
3-95 Ryoke-Cho
503 Ogaki, Gifu, J
fuyan@iamas.ac.jp

Alain Galet
ENSAD / AII
31 rue d'Ulm
75240 Paris Cedex 05, F
pierre@ensad.fr

Chriss Ganter
Kunsthochschule für Medien
Peter-Welter-Platz 2
50676 Köln, D

Matt Grenby
MIT Media Lab
20 Ames Street
Cambridge, MA 02139, USA
grenby@media.mit.edu

Peter Gsteu
bad.factory
Rohrbach 40
6850 Dornbirn
A
bad.factory@computer-haus.at

Guinness Plc
Ogilvy & Mather
10 Cabot Square, Canary Wharf
London E14 4QB, GB
info@lostinspace

Ian Haig
PO Box 1049
Cowingwood, VIC 3066, AUS
I.Haig@rmit.edu.au

Josepha Haveman
A/PIX
PO BOX 9063
Berkeley, CA 94709, USA
JosephaH@aol.com

Anna Henckel-Donnersmarck
Filmakademie Baden-Württemberg
Mathildenstr. 20
71634 Ludwigsburg, D

Clay Jaimie
HOTWIRE PRODUCTIONS
327 Ritch Street
San Francisco, CA 94108
USA

Nikolaus Heyduck
Leipzigerstraße 71
60487 Frankfurt, D

Mark Hobbins
Baxter Hobbins Sides
4 Kingly St.
London W1R 5LF, GB
mark@baxter-hobbins-sides.ltd.uk

Karl-Heinz Höhne
Universität Hamburg
Martinistr. 52
20246 Hamburg, D
hoehne@uke.uni-hamburg.de

Vassili Hurmusiadis
National Centre for Computer Animation
Bournemouth University
Fern Barrow,
Poole BH12 5BB, GB
vassili@bmth.ac.uk

Masa Inakage
2-24-7 Shichirigahama-higashi
Kamakura 248 Kanagawa, J
inakage@cyberagenz.com

Seferov Ivailo
15 Elemag
1113 Sofia, BG
declera@aster.net

Sean Jenkins
Washington State University
Pullman,
WA 99164-2220, USA
singhrs@eecs.wsu.edu

Nicolas Kalbache
ENSAD-AII
31 rue d'Ulm
75240 Paris Cedex 05, F

Sei Kazama
3-18-14 Chuo Makano-ku
164 Tokyo, J

Igor Kebel
Faculty of Architecture
Zoisova 12
1000 Ljubljana, SLO
Igor.Kebel@guest.arnes.si

Chee-Hoon Kim
30-44 29th Street,#7R
Astoria, NY 11102
USA
chkim@aaisp.com

Taku Kimura
LINKS Corporation
2-14-1 Higashi-Gotanda
Shinagawa-ku,
Tokyo 141, J
taku@links.imagica.co.jp

Satoshi Kitahara
SEGA Enterprises,Ltd. AM R&D Dept.#3
1-2-12 Haneda, ohta-ku
144 Tokio, J
Kitahara@mars.dtinet.or.jp

John Knoll
Industrial Light & Magic
PO Box 2459
San Rafael, CA 94912, USA

Claudio Knöpli Tamaroglio
Simonstr. 2
3012 Bern, CH

Martin Koch
Lerchenfelderstr. 123/7/12
1070 Wien, A

Reed Kram
MIT Media Lab
20 Ames Street
Cambridge, MA 02139, USA
kram@media.mit.edu

Jerzy Kular
ExMachina
22 Rue Hegesippe Moreau
75018 Paris, F
sophie@exmach.fr

Gerald Laimgruber
Mittersteig 2b/15
1050 Wien, A

Larry Lamb
Lamb & Company
650 3rd Ave. South 17th Floor
Minneapolis,
MN 55402, USA
bob@lamb.com

Olivier Lannaud
13 Rue des Arquebusiers
75003 Paris, F

Jean-Philippe Leclercq
ENSAD / AII
31 rue d'Ulm
75240 Paris Cedex 05, F
pierre@ensad.fr

Kelvin Lee
SONY Pictures Imageworks
9050 West Washington Blvd.
Culver City, CA 90232
USA
kelvin@spimageworks.com

Heidi Leech
Colorado State University
Department of Art
Fort Collins, CO 80523, USA
hleech@manta.colostate.edu

Claudia Liekam
Künstlerhaus Hamburg-Bergedorf
Möörkenweg 18F
21029 Hamburg, D
106110.2166@compuserve.com

Serena Lin
University of Maryland, Baltimore County
1000 Hilltop Circle
Catonsville, MD 21228, USA
slin1@gl.umbc.edu

Robin Lipner
Creation Myth Productions
220 W21st Street 2E
New York City,
NY 10011-3434, USA
binx2@panix.com

Peter Litwinowicz
Apple Computer, Inc.
1 Infinite Loop,
MS 301-3J
Cupertino, CA 94015, USA
litwinow@best.com

Karl Lund
Karl Lund Productions
143 Bedford Ave #2
Brooklyn, NY 11211, USA

Muriel Magenta
PO Box 1854
New York,
NY 10021-1854, USA

Nadia Magnenat-Thalmann
MIRALAB, University of Geneva
24 rue du General Dufour
1211 Geneva, CH
thalmann@cui.unige.ch

Kaarina Mäkinen
Tampere School of Art and Communications
Finlaysoninkuja 5
33210 Tampere, SF
tkunhu@tsac.tampere.fi

Raul Marroquin
FIRST SIGHT FOUNDATION
C/O SALTO
Amstel 82
1017 AC Amsterdam, NL
hksteen@desk.nl

Fuyuki Matsuyama
1-24-13 Nishi-Nippori
Sun Hights 302 Arakawa-ku,
Tokyo 116, J
fuyu@interlink.or.jp

Cecily McKeown
3328 Brittan Ave.#11
San Carlos, CA 94070, USA
amm@midway.uchicago.edu

Me Company
14 Apollo Studios,
London NW5 2SA, GB

Marcello Mercado
Rondeau 548 9°A
5000 Cordoba, RA

Jim Mitchell
Industrial Light & Magic
PO Box 2459
San Rafael, CA 94912, USA

Milla Moilanen
Kroma Productions Ltd.
Magnusborg Studios
06100 Porvoo, SF
milla@magnusborg.fi

Joshua Mosley
1319 W. Estes Apt 16
Chicago, IL 60626, USA
jmosle@artic.edu

Eihachiro Nakamae
2-1-1 Miyake, Saeki-ku
731-51 Hiroshima, J
nakamae@cc.it-hiroshima.ac.jp

Fulvia Nicolini
Art 3000
Château de l'Eglantine
78350 Jouy-en-Josas, F
art3000@calvanet.calvacom.fr

K.Hee Oh
24700 McBean Pkway
Valencia, CA 91355, USA
hee@emsh.calarts.edu

Michael Owens
Industrial Light & Magic
PO Box 2459
San Rafael, CA 94912, USA

Marcus Parker-Rhodes
5 Muswell Hill Road
London, N10 3JB, GB
marcus@prhodes.demon.co.uk

Nancy Paterson
475 The West Mall #1513
Etobicoke,
Ontario M9C 4Z3, CDN
nancy@utcc.utoronto.ca

Fritz Pechmann
Kegelgasse 19/12
1030 Wien, A

Siegmund Petermandl
Hochschule für Künstl. u. Industr. Gestaltung
Reindlstraße
4040 Linz, A

Emanuel Pimenta
Rua Tierno Galvan,
4-11B
1070 Lisbon, P
edmp@mail.telepac.pt

Marine Poirson
ENSAD / AII
31 rue d'Ulm
75240 Paris Cedex 05, F

Konrad Polthier
Technische Universität Berlin
Straße des 17. Juni 136
10623 Berlin, D

Stephen L. Price
Industrial Light & Magic
PO Box 2459
San Rafael, CA 94912, USA

Marti Ribas
A. ClavT 33
08755 Barcelona, E
m.ribas@redestb.es

Roderik Rodenburg
Art And Television Foundation (K.E.T.S)
PO Box 495
7500 AL Enschede, NL
kww@dds.nl

Cynthia-Beth Rubin
64 Keene Street
Providence, RI 02906, USA
cynthia.rubin@uvm.edu

Robert Russett
University of Southwestern Louisiana
University Avenue
Lafayette, LA 70504, USA
rjr6363@usl.edu

Carlos Saldanha
Blue Sky Studios, Inc.
One South Road
Harrison, NY 10528, USA
alison@blueskystudios.com

Laen Sanches
ENSAD / AII
31 rue d'Ulm
75240 Paris Cedex 05, F
pierre@ensad.fr

Gerhard Schedel
Elisabethstraße 39
80796 München, D
100633.2627@compuserve.com

Jan Schönfelder
Silberfundstraße 33
31141 Hildesheim, D

Heidemarie Seblatnig
Krallgasse 6
1220 Wien, A
ferschin@osriris.iemar.tuwien.ac.at

Shouichi Shimizu
3-41-39, Tobitakyu
Chofu, Shi Tokyo 182, J

Tsuji Shinya
Maunac
1-801-108-203 Ogawa-cho
Kodaira, Tokyo 187, J

Andrew Sides
Baxter Hobbins Sides
4 Kingly St.
London W1R 5LF, GB
andrew@baxter-hobbins-sides.ltd.uk

Kim Singhrs
Washington State University Pullman,
WA 99164-2220, USA
singhrs@eecs.wsu.edu

Eduardo Soutullo Garcia
Torrecedeira 13 6 B
36202 Vigo, E

Scott Squires
Industrial Light & Magic
PO Box 2459
San Rafael, CA 94912, USA

Ben Stassen
New Wave International
Oudergemlaan 320
1040 Brussels, B

Eduard Stöckl-Prochazka
Obkirchergasse 33/8
1190 Wien, A

Robert Stodola
Kagraner Anger 22/4/9
1220 Wien, A

Edward Taylor
SVC Television, Filmhouse
142, Wardour St.
London WIV 3AU, GB
post@svctv.co.uk

Kevin Thomason
Texas A&M University Visualization Laboratory
216-A Langford Center
College Station,
TX 77843-3137, USA
kt@viz.tamu.edu

Kawahara Toshifumi
Polygon Pictures
Bond Street T11,2-2-43
Higashi
140 Shinagawa-ku, Tokyo, J
nakaji@ppi.co.jp

Dirk Urbaniak
Stereolize Film + TV Design GmbH
Elisabethstr. 44
80769 München, D

Elene Usdin
ENSAD / AII
31 rue d'Ulm
75240 Paris Cedex 05, F
pierre@ensad.fr

Pablo Valle
School of Visual Arts
209 East 23rd Street
New York, NY 10010, USA
pablo@sva.edu

Johann van der Schiff
SCAN / media-gn
Hoendiepskade 23 A
9718 BG Groningen, NL
schiff@scan.media-gn.nl

Trine Vester
Danish Film Institute Workshop
Vognmagergade 10
1120 Copenhagen, DK

Mark Voelpel
R/Greenberg Associates
350 West 39th Street
New York, NY 10018, USA
mark@rga.com

Francois Vogel
ENSAD / AII
31 rue d'Ulm
75240 Paris Cedex 05, F
pierre@ensad.fr
Frank Voigt
Fritz-Weber-Straße 6
01809 Heidenau, D

Clea Waite
Körnerstr.106
50823 Köln, D
clea@media.mit.edu

Chris Wedge
Blue Sky Studios, Inc.
One South Road
Harrison, NY 10528, USA
alison@blueskystudios.com

Stephen Weston
BERMUDA SHORTS
1 Lower John Street
London W1R 3PD, GB

Stefan Wibbe
Construction-3D-Animation
Hahnenstraße 20
50667 Köln, D

Yin Yin Wong
MIT Media Lab
El5 - 443, 20 Ames Street
Cambridge, MA 02139, USA

Xaos, Inc.
444 Dettaro St.
San Francisco,
CA 94107, USA
Christina@ XAOS.COM

Chunhua Yang
43-06 63 St. Apt. 2A
Woodside, NY 11377, USA

Yellow Fish
Peter Jordanstr. 82
1190 Wien, A
yellow@ikiserve.boku.ac.at

Edward Zajec
516 Kensington Road
Syracuse, NY 13210, USA
ezajec@mailbox.syr.edu

Kerstin Zenner
proDESIGN GmbH, Studio Dresden
Bambergerstr.1
01187 Dresden, D

Participants
Computer Music

Virgile Abela
6, bd Gravitz-St André
13016 Marseille, F

Mathew Adkins
University of East Anglia
Norwich NR4 7TJ, GB
m.adkins@uea.ac.uk

Celso Aguiar
Stanford University / CCRMA
Stanford, CA 94305-8180
USA
aguiar@ccrma.stanford.edu

Roger Alsop
14 Perry Street
Fairfield, Victoria 3078, AUS
ralsop@klang.latrobe.edu.au

Gilberto Alves da Silva
MUNDO NOVO
Lechnerstr. 2-4/7/32
1030 Wien, A

Maryanne Amacher
80 Marius Street
Kingston, NY 12401, USA
citylinks@aol.com

Elizabeth Anderson
11, avenue de Monte Carlo
1190 Bruxelles, B
101712,1444@compuserve.
com

Joseph-L. Anderson
University of Birmingham,
Music Department
Edgbaston,
Birmingham B15 2TT, GB
j.l.anderson@bham.ac.uk

Jorge Antunes
Shin QI 5 -
Conjunto 3-Casa 23
71.505-730 Brasilia, BR
antunes@guarany.cpd.unb.b
r

Dragan Rakic
Art Association Apsolutno
Milana Rakica 15
21000 Novi Sad, YU

Artemiy Artemiev
Ul. Krilatskaya 31-1-321
121614 Moscow, RU

Patrick Ascione
19, rue Cauchy,
75015 Paris, F

Larry Austin
2109 Woodbrook
Denton, Texas 76205, USA
austin@sndart.cemi.unt.edu

Ron Averill
International College of
Music
Lot 126, Sect.85 A
53000 Kuala Lumpur, MAL
ronkari@tm.net.my

Wolf-Harun Bachschneider
C-ART DESIGN STUDIOS
Unertlstr.11
80803 München, D

Christian Banasik
Querstr. 16
40227 Düsseldorf, D
c.banasik@t-online.de

Ros Bandt
14 Collings St., West
Brunswic
Melbourne, Victoria 3055
AUS
rossart@vaxc.cc.monash.edu
.au

Natasha Barrett
15 Bastwick Street, Flat 6
London ECIV 3PE, GB
n.l.barrett@city.ac.uk

Alain Basso
636, les Devants de
Promery
74370 Pringy, F

Peter Batchelor
University of Wales, Music
Department, Bangor
Gwynedd LL57 2DG, GB
mupo14@bangor.ac.uk

Bret Battey
805 Warren Ave North
Apt #5
Seattle, WA 98109, USA
bbattey@u.washington.edu

Mauricio Bejarano
Calle 130 B, No. 8-17
Bogotá, CO

Henning Berg
Försterstraße 25
50825 Köln, D

Oswald Berthold
MEGO / R. Bauer
Stachegasse 16/4
1120 Wien, A

Michele Biasutti
Università di Padova
Piazza Capitaniato 3
35129 Padova, I

Christian Bilger
Hagelbergerstr. 56
10965 Berlin, D

Phillip-Kent Bimstein
PO Box 301
Springdale, UT 84767, USA
songdog@infowest.com

Markus Bless
Lederergasse 9
4861 Schörfling, A
bless@khsa.khs-linz.ac.at

Sergei Bogdanov
Amuratorskaya, 4
220004 Minsk, BY

Isabella Bordoni
Via S. Aquilina 23
47037 Rimini, I

Francesco Boschetto
St. Eriksgatan 66
113 20 Stockholm, S
boschetto@stockholm.mail.t
elia.com

Roman Bossart
Wilerstr. 89
9230 Flawil, CH
100626.1740@compuserve.c
om

Andres Bosshard
Wildbachstr. 62
8008 Zürich, CH
106217.1027@compuserve.c
om

Sergio Claros Brasil
Av. H. Siles #5908
PO Box 7761 La Paz, BOL

Jürgen Bräuninger
University of Natal
King George V Avenue
Durban 4001, ZA

Roland Breitenfeld
Moosgrund 8
79110 Freiburg, D
breitenfeld@t-online.de

Dietmar Bruckmayr
Payergasse 5/2/23
1160 Wien, A

Ludger Brümmer
Hohenzollernstr. 66
45128 Essen, D
ludi@folkwang.uni-essen.de

Warren Burt
c/o ACAT
GPO Box 804
Canberra, ACT 2601, AUS
waburt@melbourne.dialix.u

Christian Calon
Kantstr. 32
10625 Berlin, D
101773.620@compuserve.
com

Lelio Camilleri
Conservatorio di Musica
G.B. Martini
Piazza Rossini 2
40126 Bologna, I
lelioc@mailserver.idg.fi.cnr.it

Antonio Augusto Caminhoto
Neto
Rua Prof. Joao Candido,434,
10 86010-000
Londrina - PR, BR
guto@sercomtel.com.br

Edmund Campion
1750 Arch St.
Berkeley, CA 94709, USA
campion@cnmat.berkeley.
edu

Sebastian Castagna
7 Glebe Road
Norwich, Norfolk NR2 3J6
GB
s.castagna@uea.ac.uk

Damian Catera
Mason Gross School of
Visual Arts
33 Livingston Ave.
New Brunswick, NJ 08903
USA
catera@rci.rutgers.edu

Luigi Ceccarelli
Edison Studio
Via Voghera 7 f
00182 Roma, I
l.ceccarelli@agora.stm.it

Paulo Chagas
WDR Köln - Studio für
elektronische Musik
Annostr. 86
50678 Köln, D
el-studio@wdr.de

Mike Challis
UEA Electroacoustic Music
Studios
University of East Anglia
Norwich NR4 7TJ, GB
harling.Challis@btinternet.co
m

Insook Choi
405 N. Mathews
Urbana, IL 61801, USA
ichoi@ncsa.uiuc.edu

Se-Lien Chuang
Grabenstraße 17
8010 Graz, A

Alessandro Cipriani
Via Voghera 7
00182 Roma, I

Martina Cizek
Rotenlöwengasse 9/22
1090 Wien
A

Michael Clarke
3, Crodingley Farm Court
Huddersfield, HD7 2TY, GB
j.m.clarke@hud.ac.uk

Darren Copeland
1588 Spring Rd.
Mississauga, Ontario L5J 1N3
CDN
d.copeland@bham.ac.uk

Dimitri Coppe
74 rue Bossaerts
1030 Bruxelles, B

Grant Chu Covell
9 Smith Avenue
Somerville,
MA 02143-4309, USA
azalea@world.std.com

Krzystof Czaja
Kasprazaka 5m65
01-211 Warsaw, PL

Igor Czerniawski
Zgoda 13 m.183
00-012 Warsaw, PL

Ricardo Dal Farra
Azcuénaga 2764
1640 Martinez,
Buenos Aires, RA
dalfarra@clacso.edu.ar

Diego Dall'Osto
Corso Palladio 114
36100 Vicenza, I
d.dallosto@vi.nettuno.it

Joachim Damm
Danzigerstr. 19
10435 Berlin, D
hamster@berlin.snafu.de

Xopher Davidson
230 B College Avenue
Palo Alto, CA 94306, USA
antimatter@earthlink.net

Roderik De Man
1e Tuindwarsstraat 3
1015 RT Amsterdam, NL

Anne Deane
5859 Azalea Way
Goleta, CA 93117, USA
annedeane@aol.com

Terence Dermott
PO Box 2025
Fitzroy MDC, VIC 3065, AUS
Terry@klang.latrobe.edu.au

Francis Dhomont
3355, Chemin Queen Mary,
App.3
Montreal,
Quebec H3V 1A5, CDN

Roberto Doati
Via Giorgione 66
35020 Albignasego (PD), I
doati@csc.unipd.it

Rose Dodd
The Flat, Old Barn, Manor
Farm
Drayton, Norwich NR8 6EE, GB
R.Dodd@uea.ac.uk

Kui Dong
34256 Gadwall Common
Fremont, CA 94555, USA
kui@ccrma.stanford.edu

Wolfgang Dorninger
Freistädterstr. 237
4040 Linz, A
fadi@servus.at

Ray Doyle
16 Cedar Ave.
Alfredton 3350, AUS
raydoyle@netconnect.com.a
u

Roger Doyle
Rynville Mews, Killarney
Road
Bray, Co. Wicklow, EI
INFO@CMC.IE

Ingrid Drese
133 Avenue du Val d'Or
1200 Brüssel, B

Shlomo Dubnov
IRCAM
1 Place Igor-Stravinski
75004 Paris, F
dubnov@ircam.fr

László Dubrovay
Gárdonyi G. u. 45/B
1026 Budapest, H

John Duesenberry
514 Harvard St. #3B
Brookline, MA 02146, USA
johndu@world.std.com

Martin Dupras
3 Green Bank, Garth Road
Bangor LL57 2SP, GB
mupo15@bangor.ac.uk

Michael Edwards
Stanford University, CCRMA,
c/o Braun Music Center
Stanford, CA 94305, USA
michael@ccrma.stanford.edu

Christian Eloy
32 Rue Debelleyme
75003 Paris, F
christian.eloy@wanadoo.fr

Rolf Enström
Helgestavägen 127
12541 Älvsjö, S
enstroem@kacor.kth.se

Karlheinz Essl
Am Ölberg 26-30
3400 Klosterneuburg, A
essl@ping.at

Ambrose Field
Department of Music,
University of York
Heslington
York YO1 5DD, GB
ambrose@bliss.york.ac.uk

Orm Finnendahl
Fritschesstr.66
10585 Berlin, D
fdahl@ifnm.hdk-berlin.de

Peter Fleschhut
Clemensstr. 80
80796 München, D

Paul Fretwell
City University
Northampton Square
London EC1V OHB, GB
p.d.fretwell@city.ac.uk

Lawrence Fritts
University of Iowa
1006 VMB
Iowa City, IA 52242, USA
lawrence-fritts@uiowa.edu

Tetsuo Furudate
JABIS Corporation
4F Akasaka Hill-side 2-18-1
Akasaka, Minato-ku,
Tokyo 107, J
jbi@venture-web.or.jp

Aleksander Gabrys
ul. Dworska 17/13
40-584 Katowice, PL

Göran Gamstorp
24 Rue Norvins
75018 Paris, F

Javier Garavaglia
Hendrik-Wittestr. 11
45128 Essen, D
gara@folkwang.uni-essen.de

François Giraudon
GMEB
1 Place André Malraux
18001 Bourges, F
agmebio@calvacom.fr

Gilles Gobeil
5043 A Bordeaux
Montréal,
Québec H2H 2A5, CDN

Keith Goddard
Steinhumergutstr. 1
4050 Traun, A

Annie Gosfield
301 East 12th Street, #3D
New York, NY 10003, USA
agosfield@aol.com

Richard Graf
Nelkenstr.115
2821 Lanzenkirchen, A
a8525223@unet.univie.ac.at

Ragnar Grippe
Öraker
196 93 Kungsängen, S
r.grippe@swipnet.se

Axel Grube
Kronprinzenstr. 108
40217 Düsseldorf, D

Perry Hall
Ocurix Films
309 West 106th Street #5B
New York, NY 10025, USA
ocurix@hypnagogue.com

Simon Hall
166 Coleshill Road
CV9 2AQ Atherstone,
Warwickshire, GB

Jonty Harrison
University of Birmingham
Music Department
Edgbaston,
Birmingham B15 2TT, GB
d.j.t.harrison@bham.ac.uk

Alfred Harth
Oeder Weg 128
60318 Frankfurt/M., D

Christos Hatzis
35 Eaton Avenue,
Unit #2
Toronto,
Ontario M4J 2Z4, CDN
chatzis@chass.utoronto.ca

Martin Hay
26, Kingswood Close,
Lapworth
Solihull B94 6JQ, GB

Matt Heckert
2245 Quesada Ave.
San Francisco, CA 94124
USA
monk@slip.net

Jens Hedman
Creutzgatan 6
112 44 Stockholm, S
jens.hedman@composer.em
s.srk.se

Lynn Hershman-Leeson
HOTWIRE PRODUCTIONS
327 Ritch Street
San Francisco, CA 94108
USA

Saburo Hirano
Akashi-cho 5-13-223,
Chuou-ku
Tokyo 104, J
GAH02163@niftyserve.or.jp

Stephen Horenstein
Jerusalem Institute of
Contemporary Music
Giladi 10
Jerusalem 93385, IL
horen@vms.huji.ac.il

Rupert Huber
Huber Musik
Laaerbergstraße 2/21
1100 Wien, A
werks@thing.or.at

Joseph Hyde
Dartington College of Arts
Dartington Hall
Totnes TQ9 6EJ, GB
j.hyde@dartington.ac.uk

Alejandro Iglesias-Rossi
14 De Julio 1212
Capital Federal 1427, RA

Anna Ikramova
Hustadtring 75
44801 Bochum, D
ikramova@folkwang.uni-
essen.de

Akemi Ishijima
City University
Northampton Square
London EC1 VOHB
GB
a.ishijima@city.ac.uk

Paul Isitt
7 Menai View Terrace
Upper Bangor
Bangor LL57 2HF, GB
mupoo1@bangor.ac.uk

Toru Iwatake
Keio University, Faculty of
Environmemtal Informat
5322 Endo
252 Fujisawa, Kanagawa, JN
iwatake@sfc.keio.ac.jp

David Jaffe
739 Hilldale Ave.
Berkeley, CA 94708, USA
david@jaffe.com

Stanko Juzbasic
HZSU
Ilica 42/2
10000 Zagreb, HR
stanko@woof.music.columbi
a.edu

Frederic Kahn
36, Grande Rue de Vaise
69009 Lyon, F

Ines Kargel
Mannheimstr. 7
4040 Linz, A

Erik-Mikael Karlsson
Sveavägen 121
113 49 Stockholm, S
emk@composer.ems.srk.se

Dieter Kaufmann
Institut für Elektroakustik u.
Musik
Rienösslgasse 12
1040 Wien, A

Lydia Kavina
ul. Astradamskaja 4-232
125422 Moscow, RU
lydia@glas.apc.org

Cyril Kestellikian
11 Avenue des Coccinelles
13012 Marseille, F

Kenneth Kirschner
211 East 11th Street, Apt. #3
New York, NY 10003, USA
jhs6@columbia.edu

Josef Klammer
Neuholdaugasse 51
8010 Graz, A

Katharina Klement
Ramperstorfferg. 8-12/2/28
1050 Wien, A

Knowbotic Research
Schillingstr. 32
50670 Köln, D
kr+cf@khm.de

Paul Koonce
110 N. Stanworth Drive
Princeton, NJ 08540, USA
koonce@music.princeton.ed
u

Patrick Kosk
Ulrikagatan 1A,8
00140 Helsingfors, SF

Tim Kreger
The Australian Centre for the
Arts & Technology
GPO Box 804
Canberra, 2601, AUS
Tim.Kreger@anu.edu.au

Sirius Krenzien
Loop Dundee - Industrial Art
Shows And More
Mittenwalder Str. 61
10961 Berlin, D
skrenzien@aol.com

Charles Kriel
39 Fleming House,
Portland Ris
London, N4 2PX, GB
charles.kriel@virgin.net

Marcin Krzyzanowski
Urzednicza 17 m2
02-511 Cracov, PL

Petri Kuljuntausta
Tunnelitie 9 G 53
00320 Helsinki, SF

Peter Josef Kunz-von Gymnich
Engelsgässchen 24
63075 Offenbach am Main, D

Tobias Kunze
660 Lomita Dr.
Stanford,
CA 94305-8180, USA
t@kunze.stanford.edu

Leo Kupper
Studio de Recherches
23, Avenue Albert-Elisabeth
1200 Bruxelles, B

Jean-François Laporte
1148, Pinard St.
St-Hubert,
Québec J3Y 5K2, CDN
laportej@magellan.umontreal.
ca

Erik Larson
1812 Shubel Ave.
Lansing, MI 48910, USA
elarson@pilot.msu.edu

Simo Lazarov
12 Postojanstro str.
1111 Sofia, BG

Valentin Lazarov
12 Postoyanstro str.
1111 Sofia, BG

Daniel Leduc
15 Rue Waterman
App. 606
Saint-Lambert,
Quebec J4P 1R7, CDN
leduc@ere.umontreal.ca

Junghae Lee
Schönaustr. 54
4058 Basel, CH
LEEJ@uniclu.unibas.ch

Helmut Lenk
Breslauer Straße 8
36151 Burghaun, D
helmutlenk@mgn.maus.de

John Levack Drever
5 Park Road, Eskbank,
Dalkeith
Midlothian EH22 3DF, GB
j.drever@dartington.ac.uk

Andrew Lewis
1 Victoria Park
Bangor,
Gwynedd LL57 2EW, GB
a.p.lewis@bangor.ac.uk

Cort Lippe
University at Buffalo,
Department of Music
222 Baird Hall
Buffalo, NY 14260, USA
lippe@acsu.buffalo.edu

David-Clark Little
Cornelis Springerstr. 14-2
1073 LJ Amsterdam, NL
19521952@xs4all.nl

Yee-On Lo
615 Laurel Ave.
Menlo Park, CA 94025, USA
acoustic@netcom.com

Fernando Lopez-Lezcano
4137 Park Blvd.
Palo Alto, CA 94306, USA
nando@ccrma.stanford.edu

Paul Lorenz
Lorenz Edition
Hauptstr. 137
6233 Kramsach, A
p.lorenz@tirol.com

Robert Mackay
239 Earlsfield Road
Earlsfield,
London SW18 3DE, GB
mupo10@bangor.ac.uk

Heinrich Mader
Buchfeldgasse 13/5
1080 Wien, A

Benoît Maubrey
Die Audio Gruppe
Bahnhofstraße 47
14806 Baitz, D
maubrey@berlin.
snafv.de

Flo Menezes
Studio PANaroma
Al. dos Guainumbis, 1435
04067-003
São Paulo - SP, BR
fmenezes@canalvip.com.br

Marcello Mercado
Rondeau 548 9°A
5000 Cordoba, RA

Frank Michaelis
Düsselstr. 68
40219 Düsseldorf, D
ansgar.kuswik@t-online.de

Jean-François Minjard
Le Meiller
42320 St. Christo en Jarez, F

Raul Minsburg
Bravard 1172 Dto.4
1414 Buenos Aires, RA
Minsburg@pinos.com

Eduardo-Reck Miranda
6 Taylor Gardens, TFR
Leith, EH6 6TG Edinburgh
GB
eduardo@music.gla.ac.uk

Adrian-J. Moore
7, Mossfield Road,
Kings Heath
Birmingham B14 7JE, GB
a.j.moore@bham.ac.uk

Felix Müller
Konradsgraben 22
06493 Schielo, D

Yoichi Nagashima
10-12-301, Sumiyoshi-5
Hamamatsu, Shizuoka 430
J
nagasm@kobe-yamate.ac.jp

Sabine Nagy
Hauzenbergerstr. 20
80687 München, D

Shigenobu Nakamura
Kyoto College of Art
2-116 Uryuyama
Kitashirakawa
Kyoto 606, J
nakmra50@kyoto-art.ac.jp

Jon-Christopher Nelson
University of North Texas,
College of Music
PO Box 13887
Denton,TX 76203-6887, USA
jnelson@sndart.cemi.unt.
edu

Katharine Norman
18 Northcote Road
London E17 7DU, GB
kate@novamara.demon.co.
uk

Shunichiro Okada
Keio University Shonan
Fujisawa Campus
5322 Endou
Fujisawa 252, J
shun@slab.sfc.keio.ac.jp

Jan Oleszkowicz
Jagiellonska 6 m.21
03-721 Warsaw, PL

Ralf-R. Ollertz
Am Friedrichshain 33
10407 Berlin, D

Otakar Olsanik
Digital Audio Studio 32
Kosmakova 30
615 00 Brno, CZ

Daniel Oppenheim
IBM T.J. Watson Research
Center
Route 134
Yorktown Heights,
NY 10598, USA
music@watson.ibm.com

Matthew Ostrowski
Louis Couperusplein 10
2514 HP Den Haag, NL
mattyo@koncon.nl

John Palmer
University of Hertfordshire,
Department of Music
College Lane
Hatfield AL10 9AB, GB
j.palmer@herts.ac.uk

Juan-Carlos Pampin
Barnes 3C Escondido Village
Stanford, CA 94305, USA
juan@ccrma.stanford.
edu

Aquiles Pantaleao
39 Nansen Village, 21
Woodside
London N12 8RW, GB
a.pantaleao@city.ac.uk

Bernard Parmegiani
34 rue de Palaiseau
91140 Villebon s/Yvette, F

Ake Parmerud
Backeskärsg. 27
42159 V. Frönunda, S
ake.parmerud@musik.gu.se

Gianatonio Patella
Via Verdi 3
35012 Camposampiero (PD)
I
toni@nexttis.unipd.it

Leonard Paul
University of East Anglia
Wolfson Close 77
Norwich NR4 7TJ, GB
leonardp@sfu.ca

Maggi Payne
1324 Acton
Berkeley, CA 94706, USA
maggi@mills.edu

Blas Payri
7 rue Ernest Renan
17440 Aytré, F
blas@limsi.fr

Bruce Pennycook
McGill University
555 Sherbrooke St. West
Montréal, QC H3A 1E3, CDN
brp@music.mcgill.ca

Jean Piché
4524 Patricia Ave.
Montréal,
Québec H4B 1Z1, CDN
pichej@ere.umontreal.ca

Emanuel Pimenta
Rua Tierno Galvan, 4-11B
1070 Lisbon, P
edmp@mail.telepac.pt

Gyula Pinter
Szinyei Merse u.16 II/1
1063 Budapest, H
miroy@elender.hu

Sibylle Pomorin
Arndtstr. 18
10965 Berlin, D

Christopher Preissing
406 Peashway
South Bend, IN 46617, USA
71726.2217@compuserve.
com

David Prior
Flat 4, 4 Middleton Grove
Islington
London N7 9LU, GB
d.j.prior@bham.ac.uk

Rostislav Prochovnik
Zednická 953
708 00 Ostrava-Poruba, CZ

Mark Rais
Kentmanni 10-16
EE0001 Tallinn, EV

Miroslaw Rajkowski
ul. St. Dubois 21 m. 17
50 207 Wroclaw, PL
raju@sun20.wcss.wroc.pl

Gerald Raunig
POLYTROPOI
Sechsschimmelgasse 17/17
1090 Wien, A
ptt@tO.or.at

Erwin Redl
16 Clifford Place Apt. 2
Brooklyn, NY 11222, USA
parallel@thing.net

Nathaniel Reichman
FABRICA
Via Ferrarezza
31050 Catena di Villorba
(TV), I
reichman@www.fabrica.it

Joe Reitzer
University of Illinois at
Chicago
851 S. Morgan St.,
Rm. 1120 S
Chicago, IL 60607-7053
USA
reitzer@evl.uic.edu

John Richards
52 Scarcroft Road
York Y02 1NF, GB
jsr103@york.ac.uk

Matt Riedel
999 Hidden Lake Dr, #17D
North Brunswick,
NJ 08902, USA
matthewR41@earthlink.net

Winfried Ritsch
Klangatelier Algorythmics
Leitnergasse 7A
8010 Graz, A
ritsch@iem.mhsg.ac.at

Denise Ritter
Leberstraße 13
66121 Saarbrücken, D

Manuel Rocha Iturbide
Triunto de la Libertad 18
Mexico D.F. 14000, MEX

Harry-Ed Roland
Marchlewski Str. 31
10243 Berlin, D
slashdd@contrib.de

Neil B. Rolnick
Rensselaer Polytechnic
Institute
Dept. of the Arts, DCC 135
Troy, NY 12180, USA
rolnick@rpi.edu

Geraldine Ros
CAMERATA
33 boulevard Rochechouart
75009 Paris, F

Michael Rosas Cobian
72 Redcliffe Gardens
London SW10 9HE, GB

Joseph Rovan
7, Impasse des Trois Soeurs
75011 Paris, F
rovan@ircam.fr

Paul Rudy
1300 S.Pleasant Valley Rd. #15
Austin, TX 78741, USA
rudy@mail.utexas.edu

Bert Rundqvist
Karebacken 2
163 56 Spanga, S
rundqvist@mailbox.swipnet.se

André Ruschkowski
Straußstr. 3
83457 Bayerisch Gmain, D
andre.ruschkowski@moz.ac.at

Rudolf Ruzicka
Serikova 32
637 00 Brno, CZ
ruzicka@ffa.vutbr.cz

Yukio Saegusa
1-5-9-33 Osone, Kohoku-ku
Yokohama 222, J
sa-egusa@baz.so-net.or.jp

Philip Samartzis
62 Clauscen St.
North Fitzroy, VIC 3068, AUS
p.samartzis@rmit.edu.au

Miroslav Savic
Ivana Milutinovica 67
11000 Belgrad, YU
mipi@opennet.classroom.org

Marcus Schmickler
Brüsseler Platz 10a
50674 Köln, D
m.schmickler@proaudio.de

RenT Schneider
8 rue St. Placide
75006 Paris, F
106433.2144@compuserve.
com

Jean Schwarz
13, Rue Duchatel
78220 Viroflay, F

Michal Seta
9091 Godbout #331
LASALLE, P.Q. H8R 2H7, CDN
andrzej@geocities.com

sha
Lederergasse 22/7
1080 Wien, A

Wayne Siegel
Diem, Musikhuset Aarhus
8000 Aarhus, DK
wsiegel@daimi.aau.dk

Randall Smith
15 Spencer Avenue
Toronto, Ontario M6K 2J4
CDN

Laetitia Sonami
36 rue du Gladiateur
27720 Dangu, F
laetitia@cnmat.berkeley.edu

Arthur Stammet
29, rue Léon-Metz
4238 Esch-sur-Alzette, L
arthur.stammet@ci.educ.lu

Karlheinz Stockhausen
Kettenberg 15
51515 Kürten, D

Pete Stollery
Northern College
Hilton Place
Aberdeen AB24 4FA, GB
p.stollery@norcol.ac.uk

Carl Stone
4104 24th Street #410
San Francisco, CA 94114
USA
cstone@sukothai.com

Ulrich Suesse
Staatliche Hochschule für
Musik
Urbanstr. 25
70182 Stuttgart, D

Mark Sullivan
Michigan State University
School of Music - 305 MPB
E. Lansing, MI 48824, USA
sullivan@pilot.msu.edu

Paulina Sundin
Högalidsgatan 38 B IV
117 30 Stockholm, S
Paulina@composer.ems.srk.
se

Takahiko Suzuki
Fukushima University
Electronic Music Studio
2 Sugumichi Asakawa-aza
Matsukawa-chou,
Fukushima 960-12, J
takahiko@shotgun.phys.cst.
nihon-u.ac.jp

Karthik Swaminathan
Edwin Schlossberg Inc.
Design
641 6th Avenue 5th Floor
New York, NY 10010, USA
karthik@echonyc.com

Ryszard Szeremeta
Stachury 1
01-678 Warsaw, PL

Yasuhiro Takenaka
4-9-8 Ichiban-cho
Tachikawa-shi, Tokyo 190, J
vyu00430@niftyserve.or.jp

Jack Tamul
JTM STUDIOS Inc.
2815 Corinthian Ave.
Jacksonville, FL 32210, USA

Suguru Taninaka
M Sound Design Institute
255-6 Hachigasaki,
Matsudo-shi
Chiba, J
nifty-serve@vyu00722

Alain Thibault
ACREQ
1908 Panet Street #302
Montréal,
Québec H2L 3A2, CDN
thibaul@cam.org

Benjamin Thigpen
191, rue du Fbg. St. Martin
75010 Paris, F
101762.3705@compuserve.c
om

Joanne May Thomas
University of Huddersfield
Music Departement
Huddersfield, HD7 2TY, GB
smusjt@hud.ac.uk

Kalev Tiits
Dept. of Musicology
University of Helsinki
Vironkatu 1
000140 Helsinki, SF
kalev@elisir.helsinki.fi

Todor Todoroff
273, rue du Progrès
1030 Brüssel, B
todor@musique.fpms.ac.be

Marco Trevisani
15C Escondido Village
Stanford, CA 94305, USA
marco@ccrma.stanford.edu

Michael Tschögl
Ramsperggasse 9
6900 Bregenz, A
trylok@aon.at

Hans Tutschku
6, rue Meslay
75003 Paris, F
hans.tutschku@ircam.fr

Kazuo Uehara
408-9 Kohnoyama
Abiko, Chiba, J
ueharakz@tky.threewebnet.o
r.jp

René Uijlenhoet
van den Boschstraat 6
3531 GL Utrecht, NL
near@pi.net

Christian Utz
Talgasse 7/12
1150 Wien, A
utz@to.or.at

Sean Varah
Harvard Computer Music
Center
Music Building, North Yard
Cambridge, MA 02138, USA
cello@mario.harvard.edu

Mario Verandi
59 Cruden House, Hillington
St
SE17 3PQ London, GB
mev533@sun1.bham.ac.uk

Nicolas Vérin
41 Bd. de Reuilly
75012 Paris, F
nv@magic.fr

Alejandro Viñao
27 Coolhurst Rd.
London N8 8ET, GB
alej@vinao.u-net.com

John Vincent
VIVID GROUP
317 Adelaide St. W. #302
Toronto,
Ontario M5V 1P9, CDN
vincent@vividgroup.com

Agnieszka Waligórska
Kajavankatu 4 B 46
04230 Kerava, SF

Ming Wang
Fenzlgasse 20/13C
1150 Wien, A

Andreas Weixler
Schaftalbergweg 33
8044 Graz, A
aweixler@iem.mhsg.ac.at

Lennart Westman
Magnus Ladulasgatan 7
11865 Stockholm, S
lennart.westman@composer.
ems.srk.se

Scott Wilson
6380 Constable Dr.
Richmond, BC V7E 3Y5, CDN
sdwilson@sfu.cd

Mark Wingate
1713 East 38th Street
Austin, TX 78722, USA
mwingate@mail.utexas.edu

Gerhard-E. Winkler
Bründlweg 3
5020 Salzburg, A
winklerg@ping.at

Trevor Wishart
83 Heslington Rd., GB

Atsushi Yamaji
International Academy of
Media Arts and Sciences
3-95, Ryouke-cho
Ogaki-shi, Gifu 503, J
sushi@iamas.ac.jp

Dajuin Yao
6429 Hillegass Avenue
Oakland, CA 94618, USA
dajuin@sinologic.com

John Young
Victoria University of
Wellington, School of Music
PO Box 600
Wellington, NZ
john.young@vuw.ac.nz

Christian Zanesi
INA-GRM
116, Ave. du President
Kennedy
75220 Paris, F
grm@worldnet.fr

Maciej Zielinski
Mianowskiego 5/6
02-044 Warsaw, PL
greenski@bigfoot.com

Robert Zwetsloot
Binnenkadyk 197
1018 ZE Amsterdam, NL

12 €

3240/4

2x

Kunst